PHYSIOLOGICAL CHEMISTRY

LONG

BY THE SAME AUTHOR

Elements of General Chemistry

Fourth Edition. 33 Illustrations. x+443 pages. Cloth. $1.50 *net*.

A Text-Book of Elementary Analytical Chemistry

Third Edition. 10 Illustrations. x+297 pages. Cloth. $1.25 *net*.

P. BLAKISTON'S SON & CO.
PHILADELPHIA

A TEXT-BOOK

OF

PHYSIOLOGICAL CHEMISTRY

FOR

STUDENTS OF MEDICINE

BY

JOHN H. LONG, M.S., Sc.D.

PROFESSOR OF CHEMISTRY IN NORTHWESTERN UNIVERSITY MEDICAL SCHOOL, CHICAGO

SECOND EDITION, REVISED

WITH *42* ILLUSTRATIONS

PHILADELPHIA
P. BLAKISTON'S SON & CO.
1012 WALNUT STREET
1909

COPYRIGHT, 1909, BY P. BLAKISTON'S SON & CO.

PRESS OF
THE NEW ERA PRINTING COMPANY
LANCASTER, PA.

PREFACE TO THE SECOND EDITION.

In the preparation of this revision a number of important changes have been made. The new protein classification of the American societies has been added, while a few points based on older notions of protein relations have been dropped. Additions have been made in several other chapters, also, and most frequently in the general text.

A much fuller discussion has been given to the subject of the urine, and a new chapter has been added on the methods of urine analysis. These methods embrace not only the usual clinical tests, but the most important quantitative processes, in certain directions, as well, and are given in sufficient detail for practical metabolism work.

I have endeavored to keep the book within reasonable limits as to size and to make it conform to the needs of the classes of students for whom it is written. I believe it covers the ground which should be required in the chemical courses in medical schools, or in scientific schools where preparation for medicine is given. Some points of minor importance for beginners are printed in smaller type, but in general the details of interest to specialists only are omitted, as there is danger in presenting more to the student than he has time to properly master. This danger is as apparent in the teaching of physiological chemistry as it is in certain other lines of work.

In the preparation of the index and in the reading of the proof I have been greatly aided by my wife, Catherine Stoneman Long, and by my son, Esmond R. Long, to both of whom my thanks are due.

J. H. Long.

Chicago, July, 1909.

FROM THE PREFACE TO THE FIRST EDITION.

"In the following pages I have attempted to present a brief account of the important principles of physiological chemistry in a form suitable for the use of medical students who may be assumed to have completed courses in the elements of general inorganic and organic chemistry. From the very necessities of the case a work of this character, dealing with many topics in an elementary way, must be largely a compilation; in the selection of material, besides consulting the standard hand books, I have made free use of the recent monographs by Cohnheim, Effront and Oppenheimer, as well as of numerous articles in the Zeitschrift für physiologische Chemie, the Beiträge zur chemischen Physiologie und Pathologie and other journals. As the book is intended for beginners I have not thought it necessary to make any special quotations of literature references."

. . . .

"A considerable number of illustrative experiments are given in the text, but distinguished by being printed in smaller type. These experiments are sufficiently numerous and comprehensive to serve the purpose of a laboratory course parallel with the general course."

TABLE OF CONTENTS.

INTRODUCTION

CHAPTER   I. Scope and Methods.......................... 1

SECTION I

THE NUTRIENTS

CHAPTER   II. Inorganic Elements. Water. Air. Salts....... 7
CHAPTER   III. The Carbohydrates and Related Bodies.......... 17
CHAPTER   IV. The Fats and Substances Related to Them...... 40
CHAPTER   V. The Protein Substances...................... 51

SECTION II

FERMENTS AND DIGESTIVE PROCESSES

CHAPTER   VI. Enzymes and Other Ferments. Digestion...... 96
CHAPTER   VII. Saliva and Salivary Digestion.................. 121
CHAPTER   VIII. The Gastric Juice and Changes in the Stomach... 126
CHAPTER   IX. The Products of Pancreatic Digestion.......... 144
CHAPTER   X. Changes in the Intestines. Feces.............. 158

SECTION III

THE CHEMISTRY OF THE BLOOD, THE TISSUES AND SECRETIONS OF THE BODY

CHAPTER   XI. The Blood................................... 175
CHAPTER   XII. The Optical Properties of Blood. The Use of the Spectroscope and Other Instruments.......... 193
CHAPTER   XIII. Further Physical Methods in Blood Examination. Freezing Point and Electrical Conductivity. The Hematocrit............................ 204
CHAPTER   XIV. Some Special Properties of Blood Serum. Bactericidal Action. Precipitins, Agglutinins, Bacteriolysins, Hemolysins...................... 216
CHAPTER   XV. Transudations Related to the Blood............ 230
CHAPTER   XVI. Milk 236

CONTENTS.

CHAPTER XVII. The Chemistry of the Liver. Bile. Cells in General 249
CHAPTER XVIII. Chemistry of the Pancreas and Other Glands. Muscle, Bone, the Hair and Other Tissues...... 272

SECTION IV

THE END PRODUCTS OF METABOLISM. EXCRETIONS. ENERGY BALANCE

CHAPTER XIX. The Excretion of Nitrogen, Sulphur and Phosphorus. The Urine....................... 289
CHAPTER XX. Some Practical Urine Tests.................... 313
CHAPTER XXI. The Gaseous Excretion. Respiration.......... 356
CHAPTER XXII. The Energy Equation....................... 366
Index 380

PHYSIOLOGICAL CHEMISTRY.

INTRODUCTION.

CHAPTER I.

Scope and Methods. In our study of the organized world the most fundamental problems which present themselves are essentially chemical. Beginning with the mysterious transformations wrought through the energy of the sun's rays in such simple substances as the carbon dioxide and aqueous vapor of the atmosphere, when these bodies come in contact within certain vegetable cells, and following the history of the products thus formed through their many changes in the plant organism and later through the highly complex animal structures, for whose formation the plant cell must prepare the raw material, and finally as we note the gradual breaking down of these same elaborate combinations, with liberation of energy and ultimate restoration of carbon dioxide and water and nitrogen to the air and soil which once had held them, we see that, step by step, the various transformations which occur are such as may be represented by the equations of organic chemistry. It may not always be possible to express these equations in simple or exact form, because of the lack of knowledge in details, but the theoretical feasibility of writing such expressions we everywhere recognize.

In following the migrations of atoms of carbon, hydrogen, oxygen and nitrogen through the vegetable and animal worlds, our inquiry naturally widens beyond the field legitimately claimed by chemistry. We find ourselves at the very outset confronted by the question of the final forces inaugurating the changes, the chemical expression of which appears often so extremely simple. It may be the part of wisdom to admit at once that this question is one beyond our power to answer. Then, again, we find ourselves attracted by questions of form, function and general conditions of the existence of organisms upon the earth, in addition to those of composition and mode of formation. In touching these we enter upon the field of General Biology and soon recognize that in this vast and independent science there is much contained which has no possible bearing on our problems of chemistry. But some knowledge of biological science is certainly essential to a proper understanding of the chemistry of living beings.

As proper subjects of inquiry in Physiological Chemistry we recognize mainly the following: (a) The nutrition of plants and animals and the composition and properties of the nutrient substances. (b) The changes which the nutrients undergo before and during the processes of assimilation. (c) The agents of preparation for assimilation and the general conditions of their activity. (d) The fate of the assimilated nutrients and the nature of the products of degradation. (e) The absorption or liberation of energy.

In the broader sense the discussion is extended to the conditions appearing in the life history of plants and the lower animals as well as of man, but ordinarily the narrower field of the life of man and the higher animals alone is considered, and in the present work the latter limits will be observed. But a brief discussion of the general relations of plants to animals will not be out of place.

At one time it was very generally held that the cell activities in the plant are essentially different from those in the animal, the work in the one case being looked upon as wholly synthetic, while in the other it was assumed to be disintegration or analysis. But this is not quite correct. The reactions in the two classes of structures are qualitatively much the same, although the quantitative differences are so great that we almost lose sight of qualitative similarities. The reactions in the plant world are largely endothermal and require for their completion the constant expenditure of external energy. This energy is derived from sunlight and through its agency chlorophyll-bearing plants are able to effect a remarkable condensation, viz.: that of carbon dioxide with water accompanied by liberation of oxygen. In its simplest terms this condensation may be represented as

$$CO_2 + H_2O = H_2CO + O_2;$$

that is, formaldehyde and oxygen result.

Formaldehyde is the first member of the series containing the monosaccharoses and may be the actual starting point in their elaboration by the vegetable cell. Of the mechanism of further transformations by the plant we know but little; it is likely that many of the following changes are brought about by the action of soluble ferments or enzymes, which will be referred to in a subsequent chapter. With the completion of this synthesis a large amount of kinetic energy of the solar rays is transformed into the potential energy of protein, fat or carbohydrate. In the oxidation of the plant as fuel or food the opposite change is accomplished, and the stored up energy in complex organic molecules is liberated as heat, electricity or muscular motion. These reactions are so characteristic for plants and animals that we are apt to lose sight of others which also take place. In plants there is a respiration process as in animals, in which oxygen is absorbed and carbon dioxide liberated, and in the dark this may be readily observed, since then it is not obscured by the much more prominent reduction process. Indeed, this respiration may be followed in the light in the case of those plants which are free from chlorophyll. Further than this there are parasitic plants which, free from chlorophyll, must depend on other plants for their nourishment; they consume organic and not inorganic materials, and in this behavior resemble animals completely.

Then it must be remembered that the activity in the animal is not wholly oxidation or degradation. It is well known that some syntheses are constantly taking place which are commonly overlooked because of the much greater importance of the oxidation reactions. In recent years low forms of animal life have been found which contain chlorophyll grains and which are able to produce oxygen in presence of sunlight. In some of these cases the chlorophyll may be present in a symbiont organism, but in others it appears to be diffuse, and therefore brings the animal structure containing it into close relation with vegetable cells.

In the essential phenomena of life plants and animals have, then, much in common; it is only when we follow them into details that the characteristic differences appear.

From the nature of the materials entering into the structure of plants and animals it follows that the discussions of physiological chemistry are, in the main, but special cases of the general field of organic chemistry. The important nutrients, the carbohydrates, the fats and the protein substances, are all organic and the products appearing as stages in their metabolism are also organic. Therefore much which the student has met with in his study of organic chemistry may be found repeated in his work in physiological chemistry. Inasmuch as many of the relations to be now traced out are quantitative, it is highly important, also, that the student should bring to the work before him a good knowledge of the principles of volumetric analysis, as these will be applied frequently in what is to follow.

Historical. To trace the beginnings of Physiological Chemistry we are not obliged to go far back in the development of science. With the old medical chemistry of the so-called iatro school it has nothing in common, and in fact is in no sense a development of that science of the sixteenth and seventeenth centuries. Before the days of Lavoisier, it is true, some little advance had been made in the study of bodies of animal or vegetable origin, but without a rational theory of chemical combination the isolated facts established led to little of real value. With the nature of respiration explained, however, and the identification of its phenomena with other phenomena of oxidation, the way was opened for true scientific progress. At the same time accurate methods of ultimate organic analysis were suggested and soon developed by the followers of Lavoisier. In the hands of Berzelius, Gmelin and others these soon began to furnish results. This brings us to the end of the first quarter of the nineteenth century, a point which marks the real beginning of our science. A peculiar distinction between inorganic and organic bodies had gradually arisen, and had come to be commonly accepted. This was founded on the notion that while the former might be produced by laboratory processes synthetically, for the latter group nothing similar was possible. By this arbitrary limitation research was naturally greatly curtailed. However, in 1828, Wöhler made the important discovery that urea could be easily formed by warming a solution of ammonium cyanate, and this was in time followed by others of equal value, pointing to the same conclusion, that the production of organic compounds is in no wise dependent on the aid of a so-called vital force. It was soon demonstrated that the chemist's laboratory, no less than nature's laboratory, could take part in the formation of these substances, and in the next ten years, to about 1840, many products of physiological interest were made. The great Wöhler made many of the most fruitful discoveries of this epoch, but it is to Liebig that we owe the most. For many years he was busily engaged in perfecting methods of analysis, and with these developed he turned his attention largely to the chemical phenomena of vegetable and animal life. This led to the publication, in 1840, of his epoch-marking work, "Organic Chemistry in Its Relations to Agriculture and Physiology." This was followed in 1842 by a work giving evidence of his broadened and strengthened views, "Organic Chemistry in Its Relations to Physiology and Pathology." These works passed through many editions and were translated into several languages. In them we find much that is now considered fundamental in physiology and physiological chemistry, and they suggested or called out the active efforts of many succeeding investigators. Not a few of these men are still living, so young is our science, and it will be sufficient to merely call attention to the names of the more prominent workers who followed the active pioneers. In Germany C. G. Lehmann made many important contributions to the chemistry of the blood and published a text-book of Physiological Chemistry which reached a third edition in 1853. In 1858 F. Hoppe-Seyler published the first edition of his Physiological and Pathological Chemical Analysis, and from 1877–81 his Physiological Chemistry in four parts or volumes. This work contributed greatly to our systematic knowledge. C. Voit, since 1863 professor of physiology in Munich, began his valuable studies in nutrition and metabolism about 1856, and continued them nearly forty years. W. Kühne, of Heidelberg, did much to develop the chemistry of the protein substances, his studies dating from 1859. The pupils of these German scholars are to-day among the most active investigators in all fields of physiological chemistry.

In France, Pasteur must be mentioned in this connection on account of his pioneer investigations on fermentation and ferments, a subject of far-reaching importance. Cl. Bernard investigated the chemistry of the digestive secretions and especially the behavior of sugar in the organism. He published valuable works in 1853 and 1855, which went through later editions. Somewhat later P. Schuetzenberger, in Paris, made important additions to our knowledge of the chemistry of the protein bodies and published a work on fermentation which for many years ranked as our only systematic treatise on the subject.

At the present time physiological chemistry has become a recognized department of study in the United States, England and other continental countries as well as in Germany and France, and journals are now published devoted solely to its interests. The rapidly increasing number of investigations published in these journals and elsewhere attests the growing importance of the science from the theoretical standpoint as well as in its practical relations to medicine.

It remains to briefly mention the development of another field of scientific study because of its bearing on certain problems of physiological chemistry. In the last quarter of the eighteenth century Lavoisier clearly showed the nature of combustion and the relation of animal heat to respiration and the oxidation of the tissues. Lavoisier and Laplace carried out the first quantitative experiments in which a calorimeter was employed to measure the evolution of body heat. These were repeated by Despretz in 1824 and later by Dulong. Since then by greatly improved methods many similar investigations have been made.

A little later than the date on which Lavoisier and Laplace announced their important researches on the relation of animal heat to oxidation of foodstuffs, Benjamin Thompson, Count Rumford, announced a discovery of equally far-reaching consequences. He made the observation that the heat of friction between two pieces of metal may be absorbed by water and so measured, and that there is a relation between the mechanical work lost in the friction and the heat generated. He made also the curious observation that the work performed by the horse in one of his experiments in which friction was produced depended in turn on the combustion or oxidation of the food of the horse, from which it followed that indirectly the heating of the water was due to the combustion of a certain amount of food. But all the consequences of his experiments Rumford did not see. He was mainly interested in showing the absurdity of the notion of the material nature of heat, then commonly held, which he did completely. It remained for Joule of England and Mayer of Germany to point out, nearly fifty years later, the true relation between heat and work. In fine, by establishing the work equivalent of heat they made it possible to calculate the food equivalent of work, since the food equivalent of heat had been already proven. These relations are all of the highest value in the study of metabolism, to be considered in the sequel.

The discussions of the earlier part of the eighteenth century placed in clear light finally the full meaning of the doctrine of the Indestructibility of Matter. These later discussions developed a new doctrine, that of the Conservation of Energy, the recognition of which played no small part in the gradual advance of physiological as well as physical science.

It is the intention of the following chapters to present the fundamental facts and theories of physiological chemistry in the simplest possible manner. Much matter found in the larger hand-books must necessarily, therefore, be omitted from a work of this elementary character. But enough will be given to furnish the student, it is hoped, a satisfactory view of that which is most important in the science at the present time. It will be found convenient to make four general divisions of the subject, as follows:

Section I. The Nutrients and Related Substances.
Section II. Ferments and Digestive Processes.
Section III. The Chemistry of the Tissues and Secretions of the Body.
Section IV. The End Products of Metabolism. Excretions. Energy Balance.

SECTION I.

CHAPTER II.

THE NUTRIENTS.

INORGANIC ELEMENTS. WATER. AIR. SALTS.

Composition of the Body. The living animal body is composed in the mean of about 35 to 40 per cent of solids and 60 to 65 per cent of water. In adults the solids are somewhat in excess of this amount, while in infants they are lower, perhaps not over 30 per cent. The elements most abundantly present are carbon, oxygen, hydrogen, nitrogen, phosphorus, sulphur, chlorine, potassium, sodium, calcium, magnesium and iron. In traces only, or in particular tissues, we find iodine, fluorine, bromine, silicon, manganese, copper and lithium. The presence of these in minute amount seems to be necessary for the existence of certain animals. These elements are not present in the free state, but exist combined in more or less complex compounds, the degree of complexity varying between that illustrated in such simple bodies as water or common salt, and that found in the large protein molecules with possibly thousands of atoms present.

In point of abundance these elements are found in the body in about the order given on the following page.

The solids of the body are both organic and inorganic, and approximately the composition of the whole may be thus represented:

	Per Cent.
Water	65
Protein substances	15
Fats	14
Other organic extractives	1
Mineral matters	.5

It must be remembered, however, that the fat may vary widely from the above number and therefore change the ratio, fat: protein. Among the mineral matters calcium phosphate holds the first place, as it makes up the larger part of bone ash; carbonates and chlorides of the alkali metals make up the remainder largely.

In view of this composition of the body, it is important to learn how its waste is replenished, and what substances must be or may be

TABLE OF THE ELEMENTS IN THE BODY.

Name of Element.	Per Cent. Amount.	Occurrence.
Oxygen	66.0	In the water of the body, in the fats, the protein substances and in nearly all the tissues and salts.
Carbon	17.5	In the fats, protein substances and in most of the important compounds produced in the body.
Hydrogen	10.2	In water, the fats, protein substances and in the important products of metabolism.
Nitrogen	2.4	Found mainly in the protein substances of the body. Also in many of the metabolic products derived from these.
Calcium	1.6	This element occurs mainly in the bones, but is found in the blood also and in several secretions in small amount.
Phosphorus	0.9	Is found principally with calcium in the bones, but occurs also in several complex compounds in organic combination.
Potassium	0.4	Found as chloride, carbonate or phosphate in many of the body tissues and secretions. Exists also in organic combination.
Sodium	0.3	Occurs combined, as does potassium; the chloride and carbonate are the most important salts and are found in several body fluids.
Chlorine	0.3	Found in combination with sodium and potassium, also as hydrochloric acid in the gastric juice.
Sulphur	0.2	This element is important as occurring in the protein compounds of the body, and is found also in minute amount in other combinations.
Magnesium	0.05	Is found mainly as phosphate and carbonate in the bones.
Iron	0.004	Iron occurs in the important hemoglobin of the blood, as an integral part of the complex molecule. It is found also in inorganic compounds in traces.
Iodine, Fluorine, Silicon	traces.	Fluorine is found in the teeth, iodine in the thyroid gland, silicon in the hair. Besides these, other elements have been found occasionally, but do not appear to be necessary.

consumed to repair the constant losses and enable the body to do its proper work. This leads to the question of foods or nutrients in the broad sense. Beginning with the inorganic materials used by the body, we have first:

Water. As it appears on the surface of the earth water is classed conveniently as *hard* and *soft*. The descending rain, after the dust is washed from the air, consists of nearly chemically pure water. It holds no mineral matters dissolved, and is contaminated mainly with a small amount of dissolved carbon dioxide. Such water on reaching the earth is soft and can replace distilled water for most purposes. The changes which follow after contact with the soil depend on the composition of the latter. If the strata over which the water flows or through which it percolates consist of sand, quartz or silicate rocks or other insoluble materials the water is left in practically pure condition, and is the water usually spoken of as soft water. But, on the other hand, if the rain water comes in contact with limestone, gypsum, or other slightly soluble substances, something goes into solution and the product is now known as hard water, the degree of "hardness" depending on the amount of dissolved solids. Waters containing the carbonates of calcium and magnesium are described as *temporarily* hard, since the carbonic acid which holds these carbonates in solution may be removed by boiling, which causes precipitation. Calcium sulphate or chloride in water can not be precipitated by boiling and the presence of these and a few other substances produces *permanent* hardness.

Moderate amounts of these mineral matters in water are not objectionable; in fact waters with some lime and magnesia are preferable to absolutely soft water for drinking purposes. But along with the inorganic substances the water may take other things from the soils with which it comes in contact that are not so desirable. These are various partly decomposed organic matters of animal or vegetable origin and, what is more important, minute living vegetable cells, mostly bacteria, which are capable of causing much mischief when taken into the stomach of man. It is very generally believed that several diseases in man have their origin in the consumption of water contaminated in this way.

Natural Purification of Water. But it must not be supposed that these bacteria are always harmful. On the contrary some of them are the common agents which effect the natural purification of waters containing organic matter, in which they incite destructive fermentation or putrefactive changes and finally oxidation. As a result of these changes harmless inert substances such as nitrogen or nitrates, carbon dioxide, methane and water are produced from the relatively complex waste or excreta of the higher organisms. When we speak of the spontaneous or self-purification of water we refer to a series of changes in which these bacteria play a leading part.

Artificial Purification of Water. On a smaller scale water may be rendered safe and suitable for household use by several methods. By *distillation* all objectionable matters may be rejected and a wholesome drinking water obtained. It is possible, also, to separate practically all bacteria and other solid matters by *filtration* through beds of fine sand. In this way the supplies of many cities are obtained at the present time. Frequently the filtration is preceded by *coagulation* or *precipitation* of the organic substances by means of some suitable agent such as alum or salts of iron, or lime.

Tests of Drinking Water. In the sanitary examination of water it is not necessary to make very full analyses to determine its value for household use. A few tests usually suffice to discover the presence or absence of objectionable substances. For example, in uncontaminated waters from ordinary springs, lakes, rivers or wells, chlorine is present in small amount only. Any excess of chlorine suggests contact with sewage or household waste somewhere, and a quantitative test is of prime importance to settle this point. Such a test and a few others will be illustrated below.

Experiment. THE TEST FOR CHLORIDES. A test is often made in this way: Measure out 200 cc. of the water, add to it a few drops of a solution of pure neutral potassium chromate, and then from a burette run in, with constant stirring, solution of tenth normal silver nitrate until a faint reddish precipitate of silver chromate appears. Each cubic centimeter of the silver solution precipitates 3.54 mg. of chlorine from common salt or other chloride, and when the last trace of chlorine is combined, the silver begins to precipitate the chromate with production of red color. The chromate acts here as an " indicator," as it shows just when the chlorine is all combined by beginning to precipitate itself.

In making this test it is well to take two similar beakers, place them side by side on white paper, pour equal amounts of water in each, add to each the same number of drops of the indicator, and then with one make the actual test by adding the silver solution. Note the amount used to give a light shade and then discharge it by adding a drop of salt solution. Now, with this opalescent or turbid liquid for comparison add silver nitrate to the second beaker until the light yellowish red shade just appears. This reading is usually somewhat more accurate than the first.

As a result of the decomposition of various nitrogenous matters ammonia is frequently found in natural waters. Its amount is therefore a measure of contamination to some extent, and tests for its presence are always made in sanitary examinations. In practice the test is usually made on a distillate from the water in question, but the following experiment will illustrate the behavior of the reagent employed.

Experiment. TEST FOR AMMONIA. Solutions of ammonia or ammonium salts possess the peculiar property of giving a yellowish brown color with what is known as Nessler's reagent (a solution of mercuric-potassium iodide, made strongly alkaline with sodium or potassium hydroxide). With more than traces of ammonia a precipitate is formed.

To make the test measure out 50 cc. of the water in a large test-tube, or tall narrow beaker, and add to it 2 cc. of the Nessler solution. By placing the beaker on a sheet of white paper and looking down through it, the depth of color can be observed. A few parts of ammonia in one hundred million parts of water can be readily seen and measured.

THE OXIDATION TESTS. Pure water absorbs free oxygen from the atmosphere but has no tendency to decompose compounds to secure it. On the other hand, waters containing organic matters or certain inorganic contaminations have the power of decomposing oxygen salts to secure the oxygen they desire, and the amount of oxygen so taken up becomes a measure of the impurity of the water. Potassium permanganate is a salt, which, under certain conditions, gives up its oxygen to waters containing organic bodies in solution and is frequently employed in water analysis for this purpose. An experiment will show one way in which it is used.

Experiment. Measure out about 100 cc. of pure, carefully distilled water, pour it into a clean beaker in which water has just been boiled and add 5 cc. of pure dilute sulphuric acid (1 to 3). Place the beaker on wire gauze and heat to boiling. Now add 5 drops of a dilute permanganate solution (300 milligrams to the liter) from a burette or dropping tube and boil five minutes. The pink color persists.

Repeat the experiment, using 100 cc. of common hydrant water to which a trace of egg albumin or urea has been added, and after running in the permanganate boil again. The color fades out and more may be added. Finally, after sufficient has been added the pink color remains. The number of drops or cubic centimeters used is a measure of the contamination of the water, although often, as in this experiment, a very rough one.

THE TESTS FOR NITRITES AND NITRATES. Nitrogenous matters undergoing oxidation in water and soil usually give rise, in time, to nitrites and finally to nitrates. These compounds are therefore looked for in water as evidence of past contamination. In most instances nitrites, as a less advanced stage of oxidation than nitrates, suggest comparatively recent contamination. The tests are especially interesting in the examination of well and spring water.

Chemists are acquainted with a number of methods for the detection of traces of nitrogen in the form of nitrites and nitrates, but at the present time certain color reactions are, because of their simplicity, mainly in favor. These are illustrated by the following tests:

A reagent for nitrites is prepared by dissolving 0.5 gm. of sulphanilic acid in 150 cc. of acetic acid of 25 per cent strength, and mixing this with a solution of 0.1 gm. of pure naphthylamine in 200 cc. of dilute acetic acid. This mixture keeps very well for a time in the dark.

Experiment. To about 50 cc. of water in a clean beaker add 2 cc. of the above solution. If the water is quite free from nitrites the reagent imparts no color to it. One hundredth of a milligram of nitrogen as nitrite in the water gives a faint pink color at the end of five minutes; with large quantities the color may become deep rose red.

Experiment. A nitrate test may be illustrated in this manner: To the residue obtained by evaporating 50 cc. of an ordinary river or lake water to dryness in a porcelain dish add 1 cc. of phenolsulphonic acid. Rub the acid over the bottom of the dish, and add a few drops of dilute sulphuric acid. Warm the dish a few minutes and add 25 cc. of water. This should show now a faint yellow color. By supersaturating with ammonia the color becomes deeper. In this experiment picric acid is at first formed if a nitrate is present and the addition of ammonia yields ammonium picrate, the color of which is more marked.

For the interpretation of all these tests works on sanitary analysis must be consulted.

Physiological Importance of Water. This is suggested by the large proportion in which it is present in the animal body, as shown above. It serves primarily as the general solvent for all the solid foodstuffs taken into the system and assists in the removal of the solid waste products or excreta. To accomplish these ends it must be drunk in sufficient quantity. It is a well recognized fact that most people in the United States drink too little water, from which various ills result. Important chemical changes within the body are dependent on the so-called hydrolytic action of water. These appear mainly in the phenomena of digestion, in which starches, sugars, protein bodies and fats are altered before absorption, and will be discussed in detail in sections to follow.

It must be remembered further that water plays a very important part in the removal of heat from the body. For each gram of water evaporated as perspiration or in the breath nearly 600 units of heat are absorbed, and in this way over 20 per cent of the heat expenditure may be accounted for.

The average amounts of water found in the important tissues is shown in the following table:

	Per Cent.		Per Cent.
Dentine	10.	Pancreas	78
Fatty tissues	20	Blood	79
Bones	50	Kidney	83
Elastic tissue	50	Brain (gray matter)	86
Liver	70	Milk	88
Skin	72	Vitreous humor	98.5
Muscles	75	Cerebro-spinal fluid	99.0
Spleen	76	Saliva	99.5

Air. Besides its content of oxygen, nitrogen and argon the atmosphere contains several other gases in small amount. The most abundant of these is water vapor, with smaller traces of carbon dioxide, helium, neon, etc. As the amount of aqueous vapor present is extremely variable it is customary to give the analysis of the dry air only, which in volume per cent is about this, the rarer gases being included with the nitrogen and the argon:

	Per Cent.
Nitrogen	78.40
Oxygen	20.94
Argon	0.63
Carbon dioxide	.03

The water vapor present varies with the temperature and other physical conditions and may sometimes make up one per cent, or even more, by weight of the whole mass. A cubic meter of fully saturated air contains 30.1 grams of aqueous vapor at 30° C., and 75 per cent of this is frequently present in the hot, "close" weather of our summers. It is this high proportion of moisture which renders further evaporation from the skin so difficult, and which therefore contributes greatly to our bodily discomfort.

The normal carbon dioxide content is given above as 0.03 per cent or three cubic centimeters in ten liters. This amount is greatly exceeded in the air of poorly ventilated houses, but is not in itself the cause of the unpleasant sensations experienced in going into such an atmosphere, although this was long believed. It has been found by experiment that one can breathe, although not comfortably, in a *pure* atmosphere containing as much as 3 per cent of carbon dioxide, while an atmosphere contaminated to the extent of 1 per cent by human respiration would be practically unbearable. This condition is doubtless due to the traces of organic products thrown off in the breath and perspiration, and especially to the decomposition of organic matter on the unclean skin. The carbon dioxide is often made the approximate measure of the contamination of inhabited rooms, because of the practical difficulty of measuring anything else.

In respiration the air is modified about as shown by these figures:

	Inspired Air Per Cent.	Expired Air Per Cent.
Nitrogen, argon, etc	79.0	80.0
Oxygen	21.0	16.0
Carbon dioxide	.03	4.0

The amount of oxygen inhaled each day by a full-grown man is not far from 500 liters, while the volume of carbon dioxide exhaled is somewhat less, about 450 liters in the mean. Later something will be said about the numerical relation existing between the volume of carbon dioxide eliminated and the volume of oxygen absorbed.

The most accurate method of finding the amounts of aqueous vapor and carbon dioxide in the air is to aspirate a measured volume through a series of weighed absorption tubes. The first of these contain dry granular calcium chloride or some other good water absorbent, while the following tubes contain soda-lime or a strong potassium hydroxide solution to absorb the carbon dioxide. The increase in weight of the tubes shows the amount of vapor and gas absorbed from the given volume of air. For quick determinations somewhat less exact methods are often used in practice.

The atmosphere often contains traces of other gases, as ammonia, sulphurous oxide, oxides of nitrogen and ozone, which are of little physiological importance and need not be here considered. Of greater importance are the minute *organized* forms everywhere present to some extent, at least, and which include bacteria and many other agents of putrefaction and fermentation. Most of these are practically harmless in respiration, but the presence of others is an element of the greatest danger, because of the disturbances they occasion when taken into the body.

MINERAL SUBSTANCES REQUIRED.

Salts. The table some pages back gives the percentage amount of the different elements which make up the human body, some being united in organic and the others in inorganic compounds. Aside from water the most abundant and important of the inorganic materials are the phosphates and carbonates of the alkali-earth metals found in the bones, the alkali chlorides and the alkali carbonates. The solid mineral matter or ash of the adult body amounts in the mean to about 5 per cent; not far from four-fifths of this content comes from the skeleton, while about one-tenth of it is derived from the muscles. The proportion of ash in the different tissues, taken in the moist condition, is approximately as follows:

	Per Cent.		Per Cent.
Bones	33	Pancreas, brain	1.0
Cartilage	2	Lung, heart	0.95
Liver and spleen	1.5	Blood	0.93
Muscles	1.3	Skin	0.75
Kidney	1.2	Milk	0.70

Leaving traces out of consideration, it appears that the body contains four metallic elements, calcium, sodium, potassium and magnesium, which exist in combination with four acids, viz., phosphoric, hydrochloric, carbonic and sulphuric. Of all these compounds the calcium phosphate of the bones is the most abundant.

Phosphates. The phosphates of the body are salts of the common or orthophosphoric acid, H_3PO_4. The three kinds of salts possible here are:

Primary phosphates, MH_2PO_4,
Secondary phosphates, M_2HPO_4,
Tertiary phosphates, M_3PO_4.

The alkali salts of the three classes are readily soluble in water. The secondary and tertiary phosphates are mostly insoluble, those of the alkali metals excepted. Secondary phosphates are converted into pyrophosphates by heat and the primary phosphates into metaphosphates. To most indicators the primary phosphates show acid behavior, while the secondary phosphates are feebly alkaline. The soluble tertiary phosphates are strongly alkaline. The action on different indicators must be remembered in attempting to estimate the acidity or alkalinity of urine.

The phosphates furnished us in various animal and vegetable foods are mainly those of calcium and potassium, but it is likely that the larger part of the phosphorus utilized in the body is combined in relatively complex organic compounds, the lecithins and nucleins, for example, which yield phosphoric acid and phosphate in the final oxidation. We find therefore the tertiary calcium and magnesium phosphates, $Ca_3(PO_4)_2$ and $Mg_3(PO_4)_2$ in bones. Acid calcium phosphate of the formula $Ca(H_2PO_4)_2$ occurs in some of the body fluids, and is an important urinary excretion. The phosphate $CaHPO_4$ may sometimes be deposited from the urine. Secondary potassium phosphate, K_2HPO_4, is a constituent of all animal cell structures, possibly in soluble form, but possibly, also, in organic combination. The muscular juice is rich in alkali phosphates.

Chlorides. Chlorine is found in the body as sodium chloride and potassium chloride, also as free hydrochloric acid in the gastric juice. In our foodstuffs it comes to us mainly as sodium chloride, but this by double decomposition may give rise to the potassium chloride later:

$$K_2CO_3 + 2NaCl = 2KCl + Na_2CO_3$$

In the gastric and pancreatic juices and in the blood sodium chloride is more abundant than potassium chloride, but the latter is in excess in the cell structures. Chlorine is utilized in the animal body only as it is found in the metallic compounds or chlorides. The various organic combinations of chlorine can not replace the salts. It has been pointed out by Bunge that sodium chloride is much more necessary in the food of man or animals consuming a vegetable diet than it is when the diet is mainly flesh.

Carbonates. The carbonates found in the human body are produced there from the carbonic acid of oxidation. Hard waters contain the carbonates of calcium and magnesium, but these must suffer decomposition when taken into the stomach, and the traces of acid gas so expelled. Besides the carbon dioxide of tissue oxidation we must consider also that formed by several ferment processes in the intestines. A large amount of the gas is produced in this way and part of this is absorbed into the circulation. Under certain conditions the "weak" carbonic acid is able to decompose sodium chloride and produce sodium carbonate and free hydrochloric acid. The origin of the latter in the gastric juice is now accounted for in this way, while the sodium carbonate formed at the same time is carried into the blood and through this to other parts of the body, where other carbonates may be made by double decompositions. The soluble alkali carbonates are the most abundant, but in the bones and in the teeth calcium carbonate forms an important part. Some carbonates are always excreted by the urine, and the alkali-earth carbonates may occasionally appear in the sediment.

Sulphates. Sulphur may enter the body in a variety of combinations but nearly all of it is finally excreted in the completely oxidized form, that is, as sulphates, by the urine. Sulphur in organic combination is found in all protein substances and in the final oxidation of these compounds in the body sulphuric acid is produced. The manner in which this is combined before excretion will be discussed later. The amount of sulphur normally present in the body is but a small fraction of one per cent of the weight of the latter and practically all of it is found in the protein or protein-like substances. Among these keratin is characterized by its relatively high sulphur content. Of the exact manner in which the sulphur is combined in most of these bodies but little is known.

Bases. The several acid radicals referred to occur in combination with metals and four of these only are present in the body in appreciable quantity. These are calcium, magnesium, sodium and potassium, and they exist in the form of salts with the acid radicals. To these four the iron of the blood must be added, but it is present in organic combination. In a following chapter it will be shown to what extent these bases are present in ordinary foodstuffs. Of some it is important that they enter the body in certain forms only, otherwise their utilization—assimilation—is imperfect or even impossible. This is especially true of the iron, the chief use of which is in the building up of hemoglobin. For this purpose the iron of the mineral salts is probably not available, but it may be taken from certain peculiar organic combinations. Many mineral matters are as essential for the growth of the body as are the organic foods to be described in the next chapters and care must be observed to provide them in sufficient quantity, especially in the feeding of the young. There is some reason for believing that most of the basic material assimilated in the body is in the form of complex salts or organo-metallic combinations of some kind. The sulphur, phosphorus and carbon are so found and possibly the metals also. It has been pointed out that in the oxidation of such organo-metallic compounds carbonates or basic salts must result, and some portion at least of these salts must be available to combine with the sulphuric acid arising from the oxidation of the protein substances. According to Bunge common salt is the only mineral substance, in excess of that furnished by the usual organic foods, which the body actually demands in large amount.

CHAPTER III.

THE CARBOHYDRATES AND RELATED BODIES.

Under the term carbohydrate it has long been customary to include a number of bodies with closely related properties and similar composition, which may be expressed by such simple formulas as $C_6H_{10}O_5$, $C_6H_{12}O_6$, or multiples of these. The term carbohydrate came into use long before the structure of the bodies in question was known. It is now possible to describe these substances in their relations to the fundamental hydrocarbons or alcohols and this classification will be therefore briefly explained.

NATURE OF THE CARBOHYDRATES.

In their chemical behavior these bodies resemble aldehydes or ketones in certain important characteristics. Like the latter they are often strong reducing substances and most of them form combinations with phenyl hydrazine. These and other properties suggest that they may be considered as aldehyde or ketone derivatives of the polyhydric alcohols, which relationship is shown by the table on pages 18 and 19, which contains also some acid derivatives for further illustration.

Some of the bodies in the table are naturally occurring substances and are highly important, but most of them are artificial. The aldohexoses and the ketohexoses are closely related to two groups of more complex bodies, in which cane sugar and starch are the best illustrations, and with them form the important class of carbohydrates in the more restricted sense.

CARBOHYDRATES PROPER.

Following the usual classification we have then:

> Monoses, or monosaccharides,
> Saccharodioses, or disaccharides,
> Saccharotrioses, or trisaccharides,
> Polysaccharides.

These bodies are mostly of vegetable origin, but some, such as sugar of milk, are found in the animal kingdom. The synthetic preparation of some of these sugars has been accomplished, starting from either formaldehyde or the mixture called above glycerose. When formaldehyde, CH_2O, is treated with lime or other weak bases it polymerizes or condenses to a mixture of sugars, one of which has been isolated in pure condition and is known as α-acrose. By the condensation of the mixture of glyceraldehyde and dioxy-acetone

RELATIONS OF THE CARBOHYDRATES.

Polyhydric Alcohols.	Aldehyde Derivatives.	Ketone Derivatives.	Acid Derivatives.	Acid Derivatives.
CH_2OH | CH_2OH Glycol.	CH_2OH | CHO Glycol aldehyde, diose.		CH_2OH | $COOH$ Glycollic acid.	$COOH$ | $COOH$ Oxalic acid.
CH_2OH | $CHOH$ | CH_2OH Glycerol.	CH_2OH | $CHOH$ | CHO Glyceraldehyde.	CH_2OH | CO | CH_2OH Dioxyacetone.	CH_2OH | $CHOH$ | $COOH$ Glyceric acid.	$COOH$ | $CHOH$ | $COOH$ Tartronic acid.
	Glycerose or triose.			
CH_2OH | $CHOH$ | $CHOH$ | CH_2OH Erythrol.	CH_2OH | $CHOH$ | $CHOH$ | CHO Aldotetrose.	CH_2OH | $CHOH$ | CO | CH_2OH Ketotetrose.	CH_2OH | $CHOH$ | $CHOH$ | $COOH$ Erythritic acid.	$COOH$ | $CHOH$ | $CHOH$ | $COOH$ Tartaric acids.
	Erythrose.			
CH_2OH | $CHOH$ | $CHOH$ | $CHOH$ | CH_2OH Pentitols, arabitols, etc.	CH_2OH | $CHOH$ | $CHOH$ | $CHOH$ | CHO Aldopentoses, arabinose, etc.	CH_2OH | $CHOH$ | $CHOH$ | CO | CH_2OH Ketopentoses.	CH_2OH | $CHOH$ | $CHOH$ | $CHOH$ | $COOH$ Tetrahydroxy-monocarboxylic acids, arabonic acid, etc.	$COOH$ | $CHOH$ | $CHOH$ | $CHOH$ | $COOH$ Trihydroxydi-carboxylic acids, trioxyglu-taric acids.
	Pentoses.			
CH_2OH | $CHOH$ | $CHOH$ | $CHOH$ | $CHOH$ | CH_2OH Hexitols, mannitol, etc.	CH_2OH | $CHOH$ | $CHOH$ | $CHOH$ | $CHOH$ | CHO Aldohexoses, glucoses, etc.	CH_2OH | $CHOH$ | $CHOH$ | $CHOH$ | CO | CH_2OH Ketohexoses, fructose.	CH_2OH | $CHOH$ | $CHOH$ | $CHOH$ | $CHOH$ | $COOH$ Pentahydroxy-carboxylic acids, mannonic acids, dextronic acid.	$COOH$ | $CHOH$ | $CHOH$ | $CHOH$ | $CHOH$ | $COOH$ Tetrahydroxy-dicarboxylic acids, saccharic acid, etc.
	Hexoses.			

THE CARBOHYDRATES AND RELATED BODIES.

RELATIONS OF THE CARBOHYDRATES.—*Continued.*

Polyhydric Alcohols.	Aldehyde Derivatives.	Ketone Derivatives.	Acid Derivatives.	Acid Derivatives.
$C_7H_{16}O_7$	$C_7H_{14}O_7$		CH_2OH $(CHOH)_5$ $COOH$	$COOH$ $(CHOH)_5$ $COOH$
$C_8H_{18}O_8$	$C_8H_{16}O_8$		CH_2OH $(CHOH)_6$ $COOH$	$COOH$ $(CHOH)_6$ $COOH$
$C_9H_{20}O_9$	$C_9H_{18}O_9$		CH_2OH $(CHOH)_7$ $COOH$	$COOH$ $(CHOH)_7$ $COOH$

(glycerose or triose) mentioned in the table above the same acrose has been obtained. This acrose is identical with the sugar mixture known as $(d+l)$-fructose.

A number of sugars have also been obtained by a general method of synthesis which depends on the fact that as aldehydes and ketones they have the power to unite with hydrocyanic acid and produce nitriles of acids which may be reduced to new aldehydes with a larger number of carbon atoms than the original substance contained. This may be illustrated by starting with arabinose, $C_5H_{10}O_5$, as figured above. This with hydrocyanic acid yields a cyanide as follows:

$$CH_2OH.(CHOH)_3CHO + HCN = CH_2OH.(CHOH)_3.CHOHCN,$$

and this by the usual reaction gives arabinose carboxylic acid:

$$CH_2OH.(CHOH)_3CHOHCN + 2H_2O = CH_2OH.(CHOH)_3.CHOH.COOH + NH_3.$$

By loss of water from this acid the corresponding lactone is formed:

$$CH_2OH.(CHOH)_4COOH - H_2O =$$
$$CH_2OH.CHOH.CH.CHOH.CHOH.CO = C_6H_{10}O_5.$$
$$\underline{\qquad O \qquad}$$

By reduction with sodium amalgam this lactone becomes a sugar, identical with that obtained by the other condensation:

$$CH_2OH.CHOH.CH.CHOH.CHOH.CO + H_2 =$$
$$\underline{\qquad O \qquad}$$
$$CH_2OH.(CHOH)_4.CHO = C_6H_{12}O_6.$$

By an extension of the principle, sugars with 7, 8 and 9 carbon atoms have been obtained. In what is to follow a brief discussion of the more important natural substances will be given.

THE MONOSES OR MONOSACCHARIDES.

A number of pentose and hexose bodies must be considered here.

Pentoses. Small amounts of these sugar-like compounds exist in nature, but they are mostly derived from simple antecedent substances called pentosans. The pentoses bear the same relation to the pentosans that dextrose bears to starch; by hydration the latter compounds are converted into the former, thus:

$$C_6H_{10}O_5 + H_2O = C_6H_{12}O_6.$$
$$C_{12}H_{22}O_{11} + H_2O = C_6H_{12}O_6.$$

Among the pentoses two, known as *l*-arabinose and *l*-xylose, are the most important.

ARABINOSE, or pectin sugar, is made by warming cherry gum, wheat bran, gum arabic, quince or gedda mucilage, exhausted brewers' grains and various other substances with dilute acids. It has a specific rotation $[\alpha]_D = +104.5°$. When boiled with dilute hydrochloric acid it yields furfuraldehyde.

XYLOSE, or wood sugar, is obtained by boiling wood gum with dilute sulphuric acid. Its specific rotation is $+19.4°$. Like the preceding body it is a reducing sugar, but non-fermentable. The nutritive value of these substances for man is low, but for the herbivora these and the antecedent pentosans are more important, as they appear to be rather easily digested in the alimentary tract of many animals. Traces of pentoses have occasionally been found in human urine.

All these bodies are distinguished from the sugars proper by yielding relatively large quantities of furfuraldehyde when distilled with hydrochloric acid or sulphuric acid, which reaction may be illustrated by the following test:

Experiment. In a small retort or flask fitted with a delivery tube mix 5 gm. of bran and 100 cc. of ten per cent hydrochloric acid. Connect the retort or flask with a condenser, apply heat and distil over about half the liquid. With a few drops of this make the Schiff furfuraldehyde test. Moisten a small strip of paper with aniline acetate obtained by mixing equal volumes of glacial acetic acid and aniline. Touch this test-paper with a glass rod holding a drop of the bran distillate. A bright red color appears, due to the formation of furoaniline, $C_4H_3O.CH.(C_6H_5NH_2)_2$. The reaction is extremely delicate and serves for the detection of traces of the products yielding furfuraldehyde, $C_4H_3O.CHO$.

The Hexoses. These are important substances represented by the general formula $C_6H_{12}O_6$. Several occur widely distributed in nature, being found in ripe fruits and elsewhere. A few are artificial products formed by laboratory operations. Complex combinations of these bodies known as glucosides are also common and are essentially ethereal salts of the hexoses. These hexose bodies are all sweet, soluble in water, nearly insoluble in alcohol and are all reducing substances with oxidizing agents like Fehling's solution. They undergo fermentation readily, and all the important forms yield alcohol and carbon dioxide under the influence of the yeast organism. Some of them yield lactic acid when acted upon by the proper ferment. By partial oxidation they yield monocarboxylic acids, such as mannonic acid or dextronic acid, and by the more pronounced oxidation they are converted into dicarboxylic acids, as saccharic or mucic acid.

A reaction of great importance in connection with the hexoses is that which they exhibit when acted upon by phenyl hydrazine. While this is a general aldehyde or ketone reaction, the behavior of the hexoses is so characteristic as to require mention. With one molecule of phenyl hydrazine, $C_6H_5NH-NH_2$, the hexoses yield *hydrazones*, $C_6H_{12}O_5-N-NH.C_6H_5$, the ketone or aldehyde oxygen being replaced by the hydrazine group. The hydrazones are mostly soluble in water. An excess of phenyl hydrazine, enough to give two molecules of that substance to one of the hexose, yields bodies called *osazones*, which are mostly yellow, insoluble crystalline compounds of great importance for the separation and identification of several of the sugars. They are represented by the formula $C_6H_{10}O_4(N-NH.C_6H_5)_2$, and by warming with strong hydrochloric acid they yield peculiar compounds called *osones*, which are mixed ketone and aldehyde structures. These reactions will all be illustrated below.

GLUCOSE (*d*-glucose), known also as dextrose, grape sugar, or diabetic sugar, is the best known representative of the group. It is found in honey and many fruit juices, often associated with levulose or fruit sugar. It may be produced by the action of weak acids on cellulose or starch and on the large scale is so made from the latter substance. Weak sulphuric acid was at first commonly employed and the hydration or conversion was effected under pressure. Hydrochloric acid is now generally employed in making a commercial glucose, the acid being neutralized with sodium carbonate at the end of the reaction. With other acids the action is much slower or less complete. In the section below on the behavior of starch the nature of the reaction will be explained. The following experiment will illustrate the production of the sugar by the aid of sulphuric acid:

Experiment. Make a paste by boiling about a gram of starch with 100 cc. of water in a glass flask. Add 10 drops of dilute sulphuric acid (1:5) and boil five to ten minutes. Now allow the liquid to cool, remove 5 cc. with a pipette, dilute this to 25 cc. with water and add a few drops of iodine solution; a blue violet color results, showing that starch or a starch-like substance is still present. The remainder of the acid liquid in the flask is next boiled steadily for one hour, a little water being added from time to time to replace that lost by evaporation. At the end of an hour remove 5 cc., dilute and test with iodine solution as before. The characteristic starch reaction is now absent, while the liquid has become thin and transparent.

Neutralize the free sulphuric acid by addition of a slight excess of chalk or fine marble dust, heat gently to complete the reaction. Then filter and evaporate the filtrate nearly to dryness on a water-bath. Allow to cool and notice that the residue has a sweet taste. It is, in fact, glucose and the experiment illustrates the old method of manufacturing glucose on the large scale. Test it by dissolving a little in water and adding a few drops of Fehling's solution, described below. On boiling, a yellowish precipitate appears which becomes bright yellow and finally red.

Under certain conditions it is possible to obtain a pure crystalline product from the syrup made as just illustrated. This is known as crystallized grape sugar or pure anhydrous dextrose. The common commercial product, sold as glucose syrup, often contains much unconverted dextrin from the incomplete hydrolysis of the starch. At the present time large quantities of glucose, both solid and liquid, are made and used in the fermentation industries, by bakers and confectioners and in the household. Glucose is sweet, but not as sweet as cane sugar, and, because of the fact that it readily undergoes fermentation, it can not replace cane sugar for certain purposes, such as the preparation of the syrups of the pharmacopœia or the canning or preserving of fruits.

The typical aldose reactions are well shown with a solution of glucose. For the first of these we require a reagent, referred to above as Fehling's solution, which is made as follows:

Fehling's Solution. Ex. Dissolve 69.28 gm. of pure crystallized copper sulphate in distilled water and dilute to make a liter of solution. In a second portion of distilled water dissolve 100 gm. of pure solid sodium hydroxide and 350 gm. of pure recrystallized Rochelle salt by aid of heat. Cool and dilute to make one liter. These solutions when mixed yield the Fehling's solution proper. It is best to mix equal volumes, *quite accurately measured,* just before the reagent is needed for use.

This Fehling's solution is commonly employed as a qualitative test for reducing sugars in general, but the reaction may be first illustrated by a simpler method:

Experiment. TROMMER'S TEST. Add to a dilute solution of glucose a considerable excess of strong potassium hydroxide solution and then a very few drops of a dilute copper sulphate solution. This produces no precipitation but imparts a deep blue color. On warming the solution a yellowish precipitate forms, which grows bright red by boiling. This is cuprous oxide, and the test is known as "Trommer's" test. It is frequently employed to detect the presence of sugar in liquids, especially in urine, but on the whole is not as satisfactory as the next one.

Experiment. FEHLING'S TEST. To a very weak glucose solution add an equal volume of diluted Fehling's solution. The mixture remains deep blue in the cold, but on heating a yellowish precipitate turning to red is soon produced. This precipitate of cuprous oxide comes from the reduction of the cupric compound held in solution in the test reagent. In the first or Trommer test the first indication of the presence of a sugar is the formation of a deep blue clear solution rather than a greenish blue precipitate of cupric hydroxide which would result from the action of the copper sulphate and alkali alone. But cupric hydroxide dissolves in solutions of sugars and other polyhydric alcohols, the solution being deep blue, and stable in the cold. In the case of glucose and other *reducing* (aldehyde or ketone) sugars this stability is only temporary, since reduction follows on boiling. If the polyhydric alcohol employed to produce the deep blue solution is not a reducing substance the liquid remains clear and stable, even on boiling. This is the case with the Fehling's solution, in which the cupric hydroxide is held dissolved through the alcoholic behavior of the Rochelle salt. Similar solutions are made by the aid of glycerol (trihydric) and mannitol (hexahydric). The Fehling test has this advantage over the Trommer test, that in the latter if too much copper sulphate is used, and little sugar is present the precipitate on boiling may be mainly black cupric oxide instead of the red cuprous oxide. With the Fehling liquid no black precipitate can form.

Experiment. BISMUTH REDUCTION TEST. Add to a glucose solution some strong potassium hydroxide solution and then a very small amount of bismuth subnitrate. For an ordinary test a few milligrams will be enough. On boiling, a black precipitate appears, which frequently forms a bright mirror on the walls of the test-tube. This precipitate seems to be a mixture of metallic bismuth with some oxide, and shows the strong reducing power of the sugar. Similar reductions may be obtained from alkaline solutions of several heavy metals, but these tests illustrate the general principle.

Another test which serves for the recognition of even minute traces of glucose and other sugars, is the following, proposed by Molisch:

Experiment. To a small amount of a dilute sugar solution add two drops of a solution of α-naphthol, containing about 20 gm. in 100 cc. On shaking the liquid becomes turbid. Now add to it an equal, or slightly greater, volume of pure strong sulphuric acid and shake. A deep violet color appears, which gives place to a violet precipitate on addition of water. This reaction has been shown to be due to the combination of the α-naphthol with furfuraldehyde produced by the action of sulphuric acid on the sugar present.

Experiment. PHENYL HYDRAZINE TEST. A characteristic reaction of great practical value, referred to above, is given on the addition of phenyl hydrazine to a solution of glucose under certain definite conditions.

To 20 cc. of a dilute glucose solution add about a gram of phenyl hydrazine hydrochloride, and two grams of sodium acetate. Heat on the water-bath half an hour, and then allow the liquid to cool. There will now be found a beautiful yellow crystalline precipitate of phenyl glucosazone, the nature of which is best seen under the microscope. This test is one of great delicacy, and has been applied to the detection of traces of sugar in urine. But care must be taken to keep the reagent in considerable excess, since otherwise the soluble hydrazone may be formed. The melting point of the pure osazone is 205° C. The following reactions illustrate the combination of the phenyl hydrazine:

$$C_6H_{12}O_6 + C_6H_5.NH.NH_2 = C_6H_{12}O_5.N.NH.C_6H_5 + H_2O,$$

$$C_6H_{12}O_6 + 2C_6H_5.NH.NH_2 = C_6H_{10}O_4.(N.NH.C_6H_5)_2 + 2H_2O + H_2$$
$$\text{Phenyl glucosazone}$$

$$C_6H_5.NH.NH_2 + H_2 = C_6H_5.NH_2 + NH_3.$$

By treatment with strong hydrochloric acid the osazone decomposes to yield an osone and phenyl hydrazine hydrochloride:

$$C_6H_{10}O_4.(N.NH.C_6H_5)_2 + 2HCl + 2H_2O =$$
$$2C_6H_5.NH.NH_2.HCl + CH_2OH.(CHOH)_3.CO.CHO.$$
$$\text{Glucosone}$$

This osone on reduction with nascent hydrogen yields a sugar, not glucose, but d-fructose, or levulose:

$$CH_2OH.(CHOH)_3.CO.CHO + H_2 = CH_2OH(CHOH)_3.CO.CH_2OH.$$

Glucose and levulose (d-fructose) yield the same glucosazone; we are therefore able by this reaction to pass from one sugar to the other.

The ready fermentation of glucose will be shown later in the discussion of fermentation reactions in general. The production of glucose from cane sugar will also be explained. The specific rotation of glucose in 20 per cent solution is given by the formula $[\alpha]_D = 53°$, and increases slightly with the concentration.

d-FRUCTOSE, fruit sugar, levulose, is a ketohexose similar to d-glucose in some respects but very different in others. It occurs in honey and sweet fruits, but is not easily separated in a pure state because it is very soluble and does not readily crystallize. The preparation of glucose from ordinary starch has been referred to above. In like manner fructose may be obtained from certain less common starches, especially from inulin by hydrolysis with very weak acid. In pure condition this sugar has no technical importance.

The various reduction and fermentation reactions are shown by fructose as well as by glucose, but the quantitative relation between copper hydroxide and fructose is not quite the same as with glucose. As they yield the same osazone the phenyl hydrazine test can not be employed to distinguish between them. The most characteristic property of fructose is found in its optical behavior. While the specific rotation of d-glucose is about 53° to the right, that of d-fructose is, at 20° C., and for a strength of 20 per cent, about 93° to the left. Because of this behavior the sugar is commonly called *levulose*. Another reaction which may be applied is this:

Experiment. Dissolve resorcin in 20 per cent hydrochloric acid and heat a little of this solution with the levulose solution to be tested. A red color results. At the same time a precipitate forms which may be dissolved in alcohol with a red color. One-tenth gram of resorcin to 5 cc. of the acid is sufficient.

d-GALACTOSE. This is the third hexose of importance, but it is not a natural substance. In the inversion of milk sugar by weak acids galactose is formed along with glucose, and it results also from the action of acids on several gums. The sugar is readily soluble in water, fermentable and dextro-rotatory like glucose. It forms a characteristic osazone. It reduces Fehling's solution but not in the same proportion as glucose. On reduction it yields dulcitol, which shows its chemical relations most characteristically. On oxidation it yields galactonic and mucic acids.

d-TALOSE is an unimportant aldose of artificial origin.

SORBINOSE is a ketone sugar obtained from the juice of the mountain ash berry. It is levorotatory and non-fermentable with yeast.

INVERT SUGAR. This name is given technically to the mixture of glucose and fructose, equal molecules, produced by the action of weak acids on cane sugar, as described below.

THE CANE SUGAR GROUP.

The Saccharobioses or Disaccharides. These are sugars of the formula $C_{12}H_{22}O_{11}$ and are important substances. The best known representatives of the class are cane sugar, milk sugar and malt sugar, all of which are natural products, with cane sugar the most abundant. These bodies are closely related to the hexoses, two molecules of the latter being in some manner condensed or united to produce one of the former. The disaccharides, on the other hand, by treatment with weak acids or certain ferments break up easily into two molecules of a hexose, water being added in the reaction:

$$C_{12}H_{22}O_{11} + H_2O = C_6H_{12}O_6 + C_6H_{12}O_6.$$

The hexose molecules formed may be alike or different, and the process of converting the disaccharides into monosaccharides in this manner is called "inversion." By this inversion the following changes should be noted:

Saccharose or cane sugar yields glucose and fructose.

Lactose or milk sugar yields glucose and galactose.

Maltose or malt sugar yields glucose and glucose.

SACCHAROSE. This sugar has been known from earliest times to some peoples, but did not become an article of commerce until after the discovery of the Americas. It is found in the juices of various canes, several kinds of beets, the saps of many trees and in many seeds and nuts. On the commercial scale it is produced from the beet and canes, and in smaller amount from maple sap.

Cane sugar does not undergo fermentation directly with pure yeast, but by prolonged action of common yeast on a dilute solution of the sugar fermentation appears. This is due to the fact that the crude yeast contains an inverting enzyme known as *invertase* which produces glucose and fructose which then yield to the true fermentation. Cane sugar gives no combination with phenyl hydrazine, and is not a reducing sugar. These facts point to the absence of aldehyde or ketone groups in the large molecule. An experiment illustrates this:

Experiment. Prepare a dilute solution of pure cane sugar and boil it with Fehling solution in the usual manner. Observe that no reduction of the copper compound takes place. Next boil a similar cane sugar solution with a few drops of dilute hydrochloric or sulphuric acid several minutes, neutralize with sodium carbonate, and then apply the Fehling test. The characteristic red precipitate now appears. In this reaction the cane sugar is broken up by the acid into a molecule of glucose, and a molecule of fructose, both reducing sugars as explained above.

The behavior of cane sugar solutions with polarized light is characteristic and affords the simplest and most accurate means for quantitative determination. The specific rotation is practically independent of the concentration and is represented by the formula $[\alpha]_D = +66.5°$.

Strong solutions of cane sugar, "syrups," are used in the household and in pharmacy to prevent fermentation. Hence the use of this sugar in the canning or preserving of fruit.

LACTOSE. This is the characteristic sugar of all kinds of milk, with possibly one or two exceptions. It may be separated from the "whey" which is the product remaining after skimming and precipitating the casein. It is made commercially in large quantities as a by-product in the cheese industry, and in pure crystallized form has the formula $C_{12}H_{22}O_{11}.H_2O$.

Milk sugar resembles cane sugar in respect to the conditions under which it may be fermented, but it is a reducing sugar directly, acting strongly on copper or bismuth solutions. In its behavior with polarized light it resembles glucose closely, having a specific rotation, $[\alpha]_D = +52.5°$. Inverted milk sugar ferments readily, and products known as *kumyss,* from mare's milk, and *kephir,* from cow's milk, are made in this way. In digestion lactose splits up into glucose and galactose readily, while cane sugar yields glucose and fructose, but less readily. With phenyl hydrazine a yellow lactosazone is formed.

Milk sugar is much less soluble in water than is cane sugar and has but a slightly sweet taste. It is used mainly in the production of infant and invalid foods and in manufacturing pharmacy in tablets, pills, etc.

MALTOSE, or the sugar of malt, is produced by the action of malt diastase on starch. It therefore occurs in germinating seeds and grains, and is present wherever a diastase acts on starch. In the action of weak acids on starch paste malt sugar is produced as a transition stage, glucose finally resulting by inversion. In this country malt sugar is not a common article of commerce, but in several European countries it has been produced in considerable quantities to be used as an article of food in the place of glucose or cane sugar. The manufacture of a relatively pure sugar by the use of malt diastase and a starchy material, such as corn, seems to be attended, however, with great practical difficulties.

Maltose is readily soluble in water, sweet but not to the same degree as cane sugar, and is not directly fermentable. But an inverting enzyme in common yeast changes it so quickly that it was long classed among the true fermenting sugars. The view is now generally held that the disaccharides must first be converted into monosaccharides before real fermentation can take place. In the industries malt sugar is thus fermented on the large scale. Toward oxidizing solutions its behavior is like that of glucose, although its reducing power is not quite as great. With phenyl hydrazine it forms a maltosazone, and on polarized light its rotating power is very great, the specific rotation being at 20° C. $[\alpha]_D = +137°$. In the body maltose is changed into glucose by action of an inverting enzyme occurring both in the pancreas and in the true intestinal juice. This inversion seems to take place much more readily than in the case of cane sugar, which is a fact of considerable physiological importance.

ISOMALTOSE. This is a sugar which has been made by the action of fuming hydrochloric acid on glucose. It also accompanies the true maltose in the products formed by the action of diastase on starch. It differs from maltose in rotating power, which is much less, in the character of its phenyl osazone, and in water solubility. It reduces copper and bismuth solutions but undergoes fermentation with yeast very slowly.

Other disaccharides known have but little importance. *Mycose* or *trehalose* is found in certain fungi, *agarose* is obtained from the juice of the agave plant. *Melibiose* and *turanose* are formed in the hydrolysis of certain polysaccharides.

The Saccharotrioses or Trisaccharides. This group contains a few sugars and but one of these is important at the present time.

MELITOSE or raffinose. This sugar, having the formula $C_{18}H_{32}O_{16} + 5H_2O$, is found in certain kinds of manna and also in sugar beets in small amount. It is characterized by having a strong rotation, $[\alpha]_D = 104.5°$. Being more soluble than saccharose it is found in the last crystallizations from beet juice, and thus sometimes contaminates the beet sugar. Its high rotation may cause an error in the estimation of sugar by the polarimeter. When inverted with acids it yields fructose, and the disaccharide melibiose.

THE DETERMINATION OF SUGARS.

This is carried out in several ways. In one method the reactions depend on the reducing power of sugars on alkaline copper or other metallic solutions. The Fehling reagent is usually employed. Methods with the polariscope will be described later.

Method with Fehling's Solution. Fehling's solution, as described above, is made arbitrarily of such a strength that one cubic centimeter is reduced by 5 milligrams of glucose, on the supposition that the sugar and copper salt react on each other in the proportion of one molecule of glucose to five molecules of crystallized copper sulphate.

It was formerly held that the reaction was a perfectly definite and simple one, and could be expressed in this manner, but it is now known that the dilution of the solutions is a very important factor in determining the amount of copper reduced. The best conditions to be employed in practice have been determined by Soxhlet, who found the reducing power of several sugars to vary as follows, when they were tested in solutions of 1 per cent strength:

0.5 gm. of invert sugar in 1 per cent solution reduces 101.2 cc. of Fehling's solution, undiluted.

0.5 gm. of invert sugar in 1 per cent solution reduces 97.0 cc. of Fehling's solution, diluted with 4 volumes of water.

0.5 gm. of glucose in 1 per cent solution reduces 105.2 cc. of Fehling's solution, undiluted.

0.5 gm. of glucose in 1 per cent solution reduces 101.1 cc. of Fehling's solution, diluted with 4 volumes of water.

0.5 gm. of milk sugar in 1 per cent solution reduces 74 cc. of Fehling's solution, undiluted. The reducing power in diluted solution is the same.

0.5 gm. of maltose in 1 per cent solution reduces 64.2 cc. of Fehling's solution, undiluted.

0.5 gm. of maltose in 1 per cent solution reduces 67.5 cc. of Fehling's solution, diluted with 4 volumes of water.

The oxidizing power of 1 cc. of Fehling's solution with each kind of sugar may be tabulated as follows, assuming the sugars to be in solutions of approximately 1 per cent strength when acted upon.

One cubic centimeter of Fehling's solution oxidizes:

	When Undiluted.	When Diluted with 4 Vols. of Water.
Glucose	4.75 mg.	4.94 mg.
Invert Sugar	4.94 "	5.15 "
Milk Sugar	6.76 "	6.76 "
Maltose	7.78 "	7.40 "

The practical application of the test is best shown by an experiment.

Experiment. Measure out accurately into a flask holding about 250 cc., 25 cc. of the copper solution and the same volume of the alkaline tartrate. Heat the mixture, or Fehling's solution, on a wire gauze and note that it remains clear. Fill a 50 cc. burette with a dilute glucose solution and run 10 cc. into the hot liquid. Boil one minute, shaking the flask continuously, and allow the mixture to settle. If the supernatant liquid appears yellow this indicates that the sugar solution is much too strong and must be diluted with at least an equal volume of water before beginning another test. If, on the other hand, the liquid is still blue, add 2 cc. more of the sugar solution, boil again for a minute and allow to settle. If the color is now yellow an approximate value for the amount of sugar in the solution becomes known, but if still blue, the operation of adding solution and boiling must be continued until, after settling, a yellow color appears. Approximately 250 mg. of glucose is required to reduce the Fehling solution taken, and this must be contained in the sugar solution added. From this preliminary experiment calculate the amount of sugar present in each cubic centimeter.

Experiment. With the data obtained in the above experiment as a basis, make now a new sugar solution, having a strength of about 1 per cent. Measure out 50 cc. of the Fehling's solution, heat to boiling and run the new sugar solution from the burette as before, the first addition being about 20 cc. Boil and note the color after settling and then cautiously continue the addition of sugar solution, a few tenths of a cc. at a time, boiling after each addition, until the blue color gives place to a yellowish green and then, by the addition of a drop or two, to a pale yellow.

Sometimes the final disappearance of the copper from the solution is determined by filtering a few drops through a very small filter and adding a drop of acetic acid and a drop of ferrocyanide solution to the filtrate, when the characteristic reddish color is given if a trace of copper is present.

To determine cane sugar by the Fehling's solution it must first be converted or "inverted" into a mixture of glucose and fructose. If the sugar is in the dry condition the inversion can be accomplished as follows: Weigh out 9.5 gm., dissolve in 700 cc. of water, add 20 cc. of normal hydrochloric acid and heat for 30 minutes on the water-bath. Then neutralize with 20 cc. of normal sodium hydroxide solution and make up to 1000 cc. on cooling. This gives now a 1 per cent solution, which is employed as given for glucose, using the factor 4.94 instead of 4.74 as the amount of sugar oxidized by each cc. of the copper solution. On completion of the experiment calculate 95 parts of cane sugar for each 100 parts of invert sugar found.

The attention of the student is directed to the fact that malt sugar and milk sugar may be determined by the aid of Fehling's solution without previous inversion. This should be verified by experiment.

Method by Use of Ammoniacal Copper Solutions. The determination of glucose in pure aqueous solution by the above method is simple and accurate, but in liquids containing other organic matters the precipitate sometimes fails to settle readily, so that the recognition of the end point is difficult. This is often the case with urine and other physiologically important liquids. Advantage may be taken of the fact that cuprous oxide dissolves in ammonia without color to prepare a quantitative solution with which this difficulty may be largely overcome. Pavy was the first to employ such a reagent practically and his solution was made by diluting the ordinary Fehling's solution with ammonia in certain proportions. His suggestion has received several modifications. In all of these the weak sugar solution is added to the boiling ammoniacal copper solution until the color of the latter is just discharged, at which point the reduction of the copper hydroxide by the sugar is complete. In place of using the Fehling's solution it is well to make the Loewe solution with glycerol the basis of the dilution. A solution of this kind may be made by the formula below, in which the proportions have been found by the present writer to give the best result in practical work. One cubic centimeter of the solution oxidizes one milligram of glucose in 0.2 per cent solution.

It is made with the following amounts per liter:

Copper sulphate, cryst	8.166 gm.
Sodium hydroxide (100 per cent)	15.000 "
Glycerol	25.000 cc.
Ammonia water, 0.9 sp. gr	350.000 "
Water to make	1,000.000 "

Experiment. Of this solution, measure 50 cc. into a flask and dilute with water to 100 cc. To prevent too rapid an escape of ammonia and avoid reoxidation to some extent, add to the mixture, while warming, enough pure white solid paraffin to make a layer 3 or 4 millimeters in thickness when melted. The burette tip for discharging the sugar solution is made long enough to pass down the neck of the flask and below this paraffin. By boiling gently and adding the weak saccharine liquid slowly, very close and constant results may be obtained. At the end of the titration the paraffin is solidified by inclining the flask and immersing it in cold water, or by flowing cold water over it. The reduced liquid is then poured out and the cake of paraffin is thoroughly washed for the next test. A flask so prepared may be used for a hundred titrations. To prevent bumping and facilitate easy and uniform boiling, it is well to add a few very small fragments of pumicestone.

A solution made as above is not too strong in copper for accurate work, but the volume of ammonia necessary to hold a much larger amount of the reduced oxide in solution would render the process very inconvenient. Some practice is necessary to show just how fast the saccharine solution may be safely added. If added too rapidly the end point may be overlooked and the sugar content made to appear too low.

Polarization Tests and the Use of the Polariscope. This is the proper place to show the applications of the polariscope in the examination of sugars and other substances. For a description of the various forms of polariscopes and discussion of the optical principles involved in their construction the reader is referred to the

Fig. 1. A common form of Laurent polariscope. The polarizing prism is situated in the tube below H, the analyzer at E, B is the reading microscope and C a vernier.

author's translation of Landolt's work, "The Optical Rotation of Organic Substances and its Practical Applications," but a few words of elementary explanation may be in order here. A simple form of polarimeter in common use is shown in the illustrations.

In the polariscopes in common use for general scientific studies homogeneous yellow light is employed and this is first polarized by passing through a specially designed prism in the front part of the instrument. This prism is usually some form of a Nicol prism and is so constructed that only one of the polarized rays produced at the start is allowed to emerge and pass through the instrument. The plane in which this ray vibrates is called the plane of polarization. Such plane polarized light passes through air, water, alcohol, ether, glass and many other transparent substances without change; that is the direction in which the light vibrates remains unaltered. But many organic substances, liquids or solids dissolved, have the remarkable property of causing this plane of polarization to change direction; in other words the plane of vibration of the light suffers a twist or rotation in passing through a column of the liquids. Substances which have the power of changing the direction of the plane of vibration of polarized light passing through them are called "active" substances and the extent of the rotation is dependent on

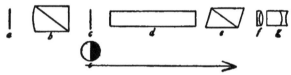

FIG. 2. This represents the course of the light through the Laurent polariscope, the direction being reversed, however, from that of the last figure. *a* is a bichromate plate to purify the light, *b* the polarizing Nicol, *c* a thin quartz plate covering half the field and essential in producing a second polarized plane, *d* the tube to contain the liquid under examination, *e* the analyzing Nicol and *f* and *g* the ocular lenses.

the number of molecules which the light passes. In the case of homogeneous liquids like oil of turpentine the rotation varies with the length of the column through which the light must pass, while in the case of dissolved solids, sugar solutions for example, the amount of the twist or rotation varies with the length of the column of solution, and also with its concentration or number of molecules in a given volume. An instrument which has some device which enables the observer to read off this rotation in degrees is called a *polarimeter*, and the number of degrees read constitutes a measure of the strength or concentration of the substance.

In order to compare the rotation of substances the term "specific rotation" has been introduced. This, as applied to liquids, may be defined as the rotation which a substance would exhibit if examined in a column 100 millimeters in length having a concentration of 1 gram of active substance to each cubic centimeter. This rotation must therefore be a calculated one, and is found as illustrated by this concrete case. Consider a solution made by dissolving 25 gm. of pure cane sugar in distilled water and diluted to make exactly 100 cc. This is then examined in a polarization tube, which is a long tube of glass or metal having ends of plane polished glass perfectly parallel to each other. The sugar solution forms then a clear transparent column of definite length, which, assume in this case, is 200 millimeters. By examination in the polarimeter it is found now that this solution rotates the plane of polarized sodium light through 33.25°. For a solution with 100 grams to 100 cc. the rotation by calculation should be four times this, or 133°, in the 200 mm. tube or 66.5° in the 100 mm. or standard tube. This is then the specific rotation, and we express it by the formula:

$$[a]_D = 66.5°,$$

in which [a] is the usual symbol for the specific rotation, and the *D* the indication that the observation is made with sodium light. a without the brackets is the angle of rotation as read off. Remembering the definition of specific rotation we have this general formula as applied to solutions:

$$[a] = \frac{100 \times 100a}{l \times c} = \frac{10^4 a}{l \cdot c}$$

in which l expresses the length of the observation tube in millimeters and c the concentration or strength of the solution in grams per 100 cubic centimeters.

For many substances this rotation is so characteristic and so easily observed that it constitutes a good test of purity or identity. With the specific rotation known the following relation enables us to find the amount of active substance in solution:

$$c = \frac{10^4 a}{[a] \cdot l}$$

The following are some specific rotations which have importance from the standpoint of physiological chemistry, the temperature being 20° C. in each case:

Cane sugar,	$[a]_D = +\ 66.5°$	$c = 10$ to 30
Milk sugar ($+ H_2O$),	$[a]_D = +\ 52.5°$	$c = 3$ to 40
Malt sugar ($+ H_2O$),	$[a]_D = +\ 137.0°$	$c = 2$ to 20
Glucose,	$[a]_D = +\ 53.0°$	$c = 20$
Levulose,	$[a]_D = -\ 93.0°$	$c = 10$ to 20
Invert sugar,	$[a]_D = -\ 20.2°$	$c = 15$

The protein substances, dextrin, glycogen and a number of other compounds to be referred to later have also a high rotating power, which finds application in investigations.

THE POLYSACCHARIDES.

We have here a very important group of bodies, some of which appear to have an extremely complex structure. Formerly these compounds were assumed to be simpler than the sugars and were represented by the general formula $C_6H_{10}O_5$. The action of water in producing glucose was assumed to consist merely in the addition of one molecule as shown by the formula:

$$C_6H_{10}O_5 + H_2O = C_6H_{12}O_6.$$

But this view is no longer held; the starches, cellulose bodies and certain gums belonging to the group have been shown to exist in the form of large and probably very complex molecular aggregations, and the formula $(C_6H_{10}O_5)_n$ is now usually employed to indicate this fact.

These polysaccharides are related to the real sugars by several reactions. By certain treatment most of them may be converted more or less readily into maltose, glucose or fructose, and besides this they yield the ester derivatives characteristic of polyhydric alcohols. In their natural condition they are mostly insoluble in water and other solvents. It is customary to make three classes of these compounds, of which the starches or amyloses, as the most important, will be treated first.

The Amyloses. In the vegetable kingdom starch is a common and widely distributed reserve material, a sugar, probably saccharose, being first formed as a synthetic product. Starch is found in many seeds, grains and tubers in the form of minute granules which are often characteristic in shape or size. They may be extremely small, pepper starch for example, or relatively large, as in the case of arrowroot starch. Under the microscope these granules often appear to be built up of concentric layers, and furthermore they are not homo-

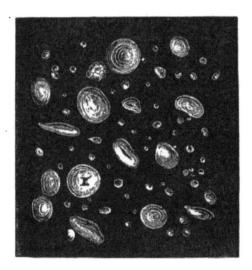

FIG. 3. Wheat starch magnified about 350 diameters.

geneous in composition. The outer part of the granule consists of a covering or sheath of *starch cellulose* within which is the large mass of *starch granulose*. The cellulose sheath is insoluble in water at the ordinary temperature, but with elevation of temperature in presence of an excess of water the protective layer breaks and allows the granulose to form a more or less perfect solution of so-called soluble starch.

On the technical scale starch may be obtained from a variety of substances. The common sources are potatoes, corn, rice and arrowroot. The manufacture is largely a mechanical operation, which may be illustrated as follows:

Experiment. Grate a potato to a pulp by means of an ordinary tin grater, mix the pulp with water and squeeze through a piece of coarse unbleached muslin, collecting the strained liquids in a large beaker. Allow the mixture to settle a half hour or longer and pour off the water, which contains some soluble albuminous substances, some cellular floating matter, but very little starch. Most of this will be found in the bottom of the beaker. Add some fresh water, stir up and allow to settle. Now pour the water off again and repeat these operations until the starch appears perfectly clean and white. Transfer this starch to a clean shallow dish and allow what is not intended for immediate use to dry spontaneously in an atmosphere free from dust. The dried product will consist of minute glistening particles resembling small crystals. Save this starch for tests given below.

Experiment. Examine starch from several sources under the microscope, employing a power of about 300 diameters. Clean a glass slide thoroughly, place in the center of it a small drop of water, and stir into this by means of a needle, or glass rod, a minute quantity of starch. Now drop on a clean cover glass in such a manner as to exclude air bubbles, and place under the microscope for observation.

Experiment. Repeat the last experiment, using an aqueous solution of iodine instead of water. The starch granules will now appear blue. For the detection of starch in mixtures the use of iodine is often indispensable.

Some idea of the size of the starch cells may be obtained from this table which gives the mean diameter in fractions of a millimeter. For starch granules which are oval instead of circular, the averages of the longer and shorter diameters is given:

Potato	0.06 –0.10
Common arrowroot	0.01 –0.07
Corn	0.007–0.02
Wheat	0.002–0.05
Rice	0.005–0.008
Pea	0.016–0.028
Bean	0.035
Barley	0.013–0.040
Rye	0.002–0.038

Starch may be recognized by a number of chemical tests, the best of which are the following:

Experiment. Boil a small amount of starch with water so as to make a thin paste. Allow this to cool, and add a few drops of an aqueous, or alcoholic solution, of iodine. A deep blue color is formed, which disappears on boiling the mixture. This test is exceedingly delicate and characteristic, and serves for the detection of minute traces of iodine as well as starch. The blue color is destroyed by alkalies or much alcohol as well as by heat.

Experiment. That starch is insoluble in cold water may be shown by stirring some with water in a beaker, allowing to settle, and pouring the liquid through a paper filter. The filtrate tested with the iodine solution does not give a blue color. Use the potato starch of the experiment for this test.

When boiled with dilute acids starch is converted into soluble compounds. The nature of these compounds depends on the acid used and on the duration of the heating. By prolonged heating glucose is the main product of the reaction, as already illustrated, but various intermediate steps may be recognized, maltose and forms of dextrin being readily demonstrated. With strong acids the results are quite different. With sulphuric acid the reaction is completely destructive, water, carbon dioxide and sulphurous acid from reduction being formed. Strong nitric acid acts as an oxidizing agent and by proper manipulation oxalic acid may be obtained in quantity as a product of the oxidation.

Experiment. Add 15 cc. of strong sulphuric acid to a gram of starch in a flask holding about 200 cc. Heat to the boiling point and observe that a black mass is soon produced. By prolonged heating this is further decomposed, while fumes of sulphurous oxide escape, leaving finally a colorless liquid.

Experiment. Add 15 cc. of strong nitric acid to one gram of starch in a flask holding 200 to 300 cc., place this on a sand-bath in a fume chamber and apply heat. After a time copious red fumes are given off. Remove the lamp and allow the reaction to continue until the fumes cease to be evolved. Finally, transfer the liquid to a porcelain dish and evaporate to a small volume. On cooling, a crystalline residue remains which consists mainly of oxalic acid.

When carefully heated starch may be largely converted into a form of dextrin, which, as will be fully explained later, is one of the important stages in the common transformations of starch. The reaction is employed on the large scale in the manufacture of British gum which is used in the preparation of size and paste for various purposes.

Experiment. Heat about 10 gm. of starch in a porcelain dish on a sand-bath to a temperature short of the point where it begins to scorch. It is necessary to stir well all the time, and continue the heat ten minutes after the starch has become uniformly yellowish brown. Then allow the dish to cool, add water and boil thoroughly, which brings part of the product into solution. When sufficiently diluted this solution can be filtered. The filtrate is precipitated by alcohol. The addition of a few drops of iodine solution to the aqueous liquid gives rise to a reddish color characteristic of dextrin.

The chief uses of starch have been referred to in other connections. Much is directly employed as food and large quantities are converted into glucose as shown above. The production of various kinds of dextrin and British gum is also extremely important and consumes enormous quantities of starch. In the form in which it occurs in nature, that is mixed with other substances in small amount, starch is the most abundant of our foodstuffs, and the one consumed in largest amount. Great interest therefore attaches to the reactions by which this starch is made soluble or digested as a step in its assimilation. The discussion of this fundamental point will be left, however, for a following chapter, when the theory of digestive operations can be explained as a whole.

In certain plants a variety of starch called inulin occurs. It is best obtained from tubers of the dahlia, and is interesting from the fact that by hydrolysis it yields fructose instead of glucose. It differs from the ordinary starch in yielding a true solution with hot water, and in giving a yellow instead of a blue color with iodine.

GLYCOGEN, or animal starch. This product, which is formed in the liver, is related in many ways, both chemically and physiologically, to common starch. In some respects it resembles also the simple sugars, from which it is indeed derived, and may be said to stand between them and vegetable starch. It is readily soluble in water, giving, however, an opalescent solution. This is especially characterized by a strong action on polarized light, $[a]_D = +196°$ to $213°$, according to different authorities. Like common starch glycogen is a reserve material, being formed from the absorbed sugar of the digestive process, and, in turn, being reconverted into sugar from the liver as this is required for oxidation in the body. After death the store of glycogen in the liver rapidly diminishes, glucose being produced. The amount of glycogen present in the liver varies greatly with the diet and time after eating. It may make up 12 to 16 per cent of the total weight of the organ, but is usually much below this, perhaps in the mean 2 to 3 per cent. In addition to its occurrence in the liver glycogen is found in variable amount in the muscles, and in traces in other body tissues. It occurs also in the vegetable kingdom, and has been recognized in certain fungi. The laboratory production of glycogen and some of the simple reactions are illustrated by the following experiments, while the physiological relations will be reserved for further discussion in a later chapter.

Experiment. Kill a rat or a rabbit which has been well fed; remove the liver as quickly as possible, and without delay cut it into small bits, which throw into a vessel of boiling water. The weight of the water should be about five times that of the minced liver. Boil five minutes, then remove from the water and rub up in a mortar with fine clean quartz sand. In this way the fragments of liver become thoroughly disintegrated. The contents of the mortar, sand as well as liver, are thrown into boiling water again and kept at 100° 15 minutes. At the end of this time enough dilute acetic acid must be added to impart a faint acid reaction. This coagulates and precipitates some albuminous matters, which are separated when the hot mixture is filtered. In the opalescent filtrate, which must be collected in a cold beaker, a further precipitation of albuminous matter is effected by adding a few drops of hydrochloric acid and some potassium mercuric iodide as long as a precipitate forms.

Filter again and use the dilute aqueous solution of glycogen resulting for tests below.

Experiment. Evaporate about half of the liquid above to a small volume and precipitate impure glycogen as an amorphous white powder by addition of strong alcohol.

Experiment. Add a little tincture of iodine to a small portion of the solution, and note the red color produced. This color is discharged by heat. Boil some of the solution with dilute hydrochloric acid ten minutes; neutralize the acid nearly, cool and again test with iodine. No color is now produced, as glycogen has disappeared under the treatment, having been converted into sugar.

It has been remarked above that after death the store of glycogen in the liver rapidly disappears, so that tests applied at the end of a day or two fail to show its presence. This may be shown as follows:

Experiment. Cut some beef liver from the market into small bits and extract with boiling water. Boil longer to coagulate protein bodies, after adding some sodium sulphate. Apply the iodine test for glycogen, which is found absent, and the Fehling test for sugar, which is found present in quantity. It is an excellent exercise also to determine the amount of sugar which may be obtained from a given weight of liver. In this test the extraction must be repeated with several portions of water.

The Gums. Some of these occur in nature as products related to the pentose group of sugars referred to some pages back. Others are related to starch, and on transformation yield finally hexoses. The group of gums includes further the dextrin bodies formed from starch by several reactions, one of which has been already illustrated. The transformation of starch by the action of weak acids or enzymes is far from being a simple process and much uncertainty still exists as to the number of intermediate products between the parent substance and the final maltose or glucose. Some writers have attempted to distinguish several well defined stages in the reaction and describe as definite bodies *erythrodextrin, achroödextrin, amylodextrin* and *maltodextrin*. The first gives a violet-red color with iodine and is easily precipitated by alcohol; the second gives no color with iodine, but is still precipitated by alcohol. It shows reducing power with Fehling's solution, and may be looked upon as one of the end products of the action of diastase on starch, maltose being the other. The name amylodextrin is given to a product of diastatic action and also to a dextrin formed by the treatment of starch with very dilute acids. It is said to show a purple color with iodine, and to exhibit very strong dextro rotation. The existence of maltodextrin is affirmed by several writers, but the properties of the substance are not well established. Some authors have gone so far as to recognize several modifications of achroödextrin, which are described as α, β and γ forms, and which differ in optical properties and reducing power. The more recent extended investigations seem to disprove this notion, however, and the most that can be safely said is that along with maltose an end product is produced by diastatic conversion of starch which is probably a single substance. What is called erythrodextrin is more likely to be a mixture, possibly, of soluble starch and achroödextrin. Under the name *erythrogranulose* a similar complex has been described. In a later chapter on the action of ferments more will be said about the theory of the transformation of starch into these products.

The true dextrins are not directly fermentable with yeast, but they appear to be aldehyde bodies and as such have reducing power. They react also with phenyl hydrazine and yield osazones, which, however, are not easily purified, because of their solubility. The dextrins have a slightly sweetish taste and all show a specific rotation about $[a]_D = +196°$. Beyond the empirical formula $C_6H_{10}O_5$ it is not possible to go in describing the constitution of these bodies.

The natural vegetable gums are often mixtures of several substances, and but few of them have been studied. Gum arabic and gum senegal are the potassium and calcium salts of arabic acid to which the formula $(C_6H_{10}O_5)_2 + H_2O$ is given. On treatment with weak sulphuric acid both arabinose and galactose appear to be formed. Agar-agar is said to yield lactose and then galactose, while cherry gum yields arabinose.

Cellulose. The cell walls of vegetable substances consist of cellulose mixed always with related compounds of which a body called lignin is the most important. The cellulose resists the action of strong oxidizing or other agents much more perfectly than do the accompanying bodies, and may therefore be freed from them by various treatments. In washed Swedish filter paper we have an illustration of nearly pure cellulose, as all the other bodies in the original fibers have been removed by the bleaching and washing processes to which the raw material was subjected in the manufacture of the paper. A pure cellulose paper may be made from wood also, but only by more complicated operations.

The pure cellulose is characterized by insolubility in water, weak acids, alkalies, alcohol or ether. It may be dissolved rather readily in a solution known as Schweitzer's reagent and by prolonged digestion with acids is converted into hexoses and pentoses.

The natural celluloses may be divided roughly into three groups: (a) those which resist hydrolytic action very perfectly and are not capable of serving as foodstuffs for any animals; in this group we have linen and cotton fibers, hemp, China grass, etc. (b) Those which are less resistant to hydrolytic action and which contain active CO groups. These bodies may be called oxycelluloses; they yield also furfuraldehyde by distillation with hydrochloric acid. In this group we have the mass of the material found in the fundamental tissues of flowering plants, and a large part of ordinary woody tissue. This lignified tissue is made up of compound celluloses or lignocelluloses from which the cellulose proper may be isolated in a variety of ways. Some of the bodies in this group are partly digestible and have some value as foods for the herbivora. (c) In this third group we have substances described as pseudocelluloses or hemicelluloses and which offer little resistance to hydrolysis. They are easily attacked by weak acids and alkalies and also suffer digestion by enzymes, so that they may be classed among the foodstuffs of limited value for the herbivora. These bodies resemble starch much more than they resemble the lignocelluloses. By action of weak acids fermentable sugars are formed almost quantitatively from pseudocelluloses, while from the lignocelluloses not over about 20 per cent of fermentable sugars may be obtained.

Experiment. Prepare Schweitzer's reagent by first precipitating copper hydroxide from copper sulphate solution, in the presence of a little ammonium chloride, by addition of sodium hydroxide in excess. Wash the precipitate thoroughly and then dissolve it in the smallest possible quantity of strong ammonia water. This yields a deep blue solution, the reagent in question. It dissolves cotton rather easily. This solution may be filtered after dilution, and from the filtrate a pure cellulose is thrown down by addition of acids.

By action of strong nitric acid, aided by sulphuric acid, cellulose is converted into a series of nitrates or "nitro-celluloses." The number of NO_2 groups added depends on the strength of the acid mixture and time of its action. The more highly nitrated products constitute the well-known explosives. Products not so highly nitrated are used in the preparation of collodion and celluloid. This latter is essentially a mixture of camphor and nitrated cellulose from cotton or paper. These bodies are esters and therefore may be decomposed by alkalies.

CHAPTER IV.

THE FATS AND SUBSTANCES RELATED TO THEM.

In nature we find a large number of esters composed of the fatty acids united to glyceryl. These are the ordinary fats and as foodstuffs they are nearly as important as the carbohydrates. In structure they are practically all of the type $C_3H_5(C_nH_{2n-1}O_2)_3$, but include bodies of widely different physical properties. Some are liquids, while others at the ordinary temperature are hard solids. Nearly all vegetable products contain fats of some kind; often the amount is very small, but frequently it constitutes fully 50 per cent by weight of the seed, nut or fruit in question. In the animal kingdom fats are always present, in some amount, in all organisms. The animal fats are often derived from the vegetable fats consumed as food.

THE NATURAL FATS.

The important fatty acids combined with the radical of glycerol, $C_3H_5(OH)_3$, are given in the following table. The combinations are essentially like that illustrated by this structural formula of stearin:

$$CH_2-O-C_{18}H_{35}O$$
$$CH\ -O-C_{18}H_{35}O$$
$$CH_2-O-C_{18}H_{35}O$$

SATURATED ACIDS, $C_nH_{2n}O_2$.

Formic acid, $HCHO_2$ ⎫
Acetic acid, $HC_2H_3O_2$ ⎬ glycerides not natural substances.
Propionic acid, $HC_3H_5O_2$ ⎭
Butyric acid, $HC_4H_7O_2$, occurs in butter fat as glyceride.
Pentoic acid, $HC_5H_9O_2$, valeric acid occurs as a natural compound.
Caproic acid, $HC_6H_{11}O_2$, in butter fat as glyceride.
Œnanthylic acid, $HC_7H_{13}O_2$, does not occur as glyceride.
Caprylic acid, $HC_8H_{15}O_2$, as glyceride in butter and other fats.
Pelargonic acid, $HC_9H_{17}O_2$, in vegetable kingdom, but not as glyceride.
Capric acid, $HC_{10}H_{19}O_2$, in butter and other fats as glyceride.
Undecylic acid, $HC_{11}H_{21}O_2$, not found as natural glyceride.
Lauric acid, $HC_{12}H_{23}O_2$, as glycerol ester in several fats.
Myristic acid, $HC_{14}H_{27}O_2$, in nutmeg butter and other fats as glyceride.
Palmitic acid, $HC_{16}H_{31}O_2$, as glyceride in many fats.
Margaric acid, $HC_{17}H_{33}O_2$ obtained as artificial glyceride.
Stearic acid, $HC_{18}H_{35}O_2$, as glyceride in many fats.
Arachidic acid, $HC_{20}H_{39}O_2$, as glyceride in peanut oil.
Behenic acid, $HC_{22}H_{43}O_2$, as glyceride in certain oils.

A few other acids of this series are known in glycerol combinations but they are unimportant.

NON-SATURATED ACIDS, $C_nH_{2n-2}O_2$ AND $C_nH_{2n-4}O_2$.

Not many of these acids occur as natural glycerides.

    Hypogæic acid, $C_{16}H_{30}O_2$, as glyceride in peanut oil.
    Oleic acid, $C_{18}H_{34}O_2$, as glyceride in many oils.
    Linoleic acid, $C_{18}H_{32}O_2$, as glyceride in drying oils.
    Ricinoleic acid, $C_{18}H_{34}O_3$, as glyceride in castor oil.

A large number of the acids in the first list and two in the second occur in the edible fats, while some are found in products of little importance, or in such as have technical uses only. The fats which are most commonly used as foods are those consisting largely or wholly of stearin, palmitin and olein. Butter fat, however, contains relatively large amounts of other glycerides.

Reactions of Fats. Certain general reactions are common to practically all fats and will be explained here in detail.

SAPONIFICATION. This term is applied to the change which follows when fats are treated with alkali solutions, usually with application of heat. The fats decompose more or less readily, and as products we have soaps and glycerol, according to these equations:

$$C_3H_5(C_{18}H_{35}O_2)_3 + 3KOH = C_3H_5(OH)_3 + 3KC_{18}H_{35}O_2$$
$$2C_3H_5(C_{18}H_{33}O_2)_3 + 3PbO + 3H_2O = 2C_3H_5(OH)_3 + 3Pb(C_{18}H_{33}O_2)_2$$

In the first case potassium stearate is formed and in the second lead oleate, which is the important constituent of lead plaster. Here lead oxide and water are equivalent to lead hydroxide. Much of the glycerol of commerce is produced by such decompositions. When sodium hydroxide is used with the common fats ordinary hard soap results.

An analogous change occurs when fats are subjected to the action of water at a high temperature or superheated steam. We have here hydrolysis purely, although the term *saponification by steam* is sometimes applied. The same reaction is brought about by certain enzymes at the ordinary temperature, for example by the enzyme known as lipase or steapsin in the pancreatic secretion. This will be discussed fully later as it is important in the digestion of fats. Sometimes the reaction is complete as shown by this equation:

$$C_3H_5(C_{18}H_{35}O_2)_3 + 3H_2O = C_3H_5(OH)_3 + 3C_{18}H_{36}O_2$$

But products of partial hydrolysis, *monostearin* and *distearin*, for example, may be left, as:

$$C_3H_5 \begin{cases} C_{18}H_{35}O_2 \\ C_{18}H_{35}O_2 \\ C_{18}H_{35}O_2 \end{cases} + H_2O = C_3H_5 \begin{cases} OH \\ C_{18}H_{35}O_2 \\ C_{18}H_{35}O_2 \end{cases} + HC_{18}H_{35}O_2$$

ILLUSTRATIVE TESTS.

Some of the saponifications are illustrated by these experiments:

Experiment. Boil about 25 gm. of tallow with a solution of 10 gm. of potassium hydroxide in 100 cc. of water. Stir the mixture thoroughly until it becomes homogeneous; that is, until no oil globules are seen floating on top of the aqueous liquid, which may require half an hour. Add water from time to time, to make up for that lost by evaporation. The resulting mass is a mixture of glycerol, excess of alkali and soft soap. To this now add a solution of 15 gm. of common salt in 75 cc. of water and heat again, which brings about a conversion of the soft soap into the hard or sodium soap. On cooling, this separates and floats on the excess of spent lye and salt solution.

Experiment. The presence of fatty acids in the above soap may be shown by adding to a portion of it enough hydrochloric acid to decompose the soap. Use about half the product of the experiment, dilute with water, and add the acid in slight excess, about 10 cc. of the strong commercial acid. Warm on the water-bath, which will cause the liberated fatty acids to collect on the surface as a liquid layer as soon as the temperature becomes high enough. Add more water and allow the whole to cool. A semi-solid layer of fatty acids can now be lifted from the surface of the liquid. The hardness of the mixed acids depends on the fat taken for experiment. Mutton and beef tallows yield very solid acids; with lard the mass is softer, while with some oils the acids do not solidify at all at the ordinary temperature.

Experiment. Dissolve a small portion of the fatty acids in warm alcohol, nearly to saturation. On cooling, the acids separate in crystalline scales.

Experiment. The presence of glycerol as one of the products formed by the saponification of fats is best shown as follows:

Mix 50 cc. of cottonseed oil with 25 gm. of litharge and 100 cc. of water in a porcelain dish. Place over a Bunsen burner on gauze and stir until all oil globules have disappeared, adding a little water from time to time. The litharge with water acts as lead hydroxide and saponifies the fat, forming an insoluble lead soap, or plaster, and glycerol. When the saponification is complete add more water, heat and stir well to dissolve glycerol. Allow to settle a short time and pour the aqueous solution through a filter. To the residue add water again, heat, allow to settle and pour through the same filter. Concentrate the mixed filtrates to a small volume and after cooling observe the sweet taste of the thickish residue.

Experiment. Dissolve a portion of the sodium soap in water with aid of a little alcohol. Then add some solution of calcium chloride or lead acetate. A white precipitate is formed, as the calcium and lead salts of the fatty acids are not soluble in water. Hard waters, which contain salts of calcium and magnesium, decompose soap in the same manner.

Other Reactions. The common fats are insoluble in water and when mixed with the latter tend to separate immediately. However, it is possible to convert the fats and water into a peculiar mixture called an emulsion which does not separate into two layers on standing. In this condition the fat consists of extremely minute globules which remain in suspension and which may be passed through the pores of coarse filter paper. It does not seem possible to secure an emulsion with perfectly neutral fats, and in most cases the phenomenon depends on the presence of a trace of soap formed. In the processes of

THE FATS AND SUBSTANCES RELATED TO THEM.   43

digestion of fats in the animal body emulsification plays a very important part as will be shown later. It follows, probably, the partial hydrolysis of the fats by lipase, referred to above.

As they exist in the animal or vegetable organism the fats are doubtless all amorphous substances, but in the separated condition the solid fats always become more or less crystalline. This may be

FIG. 4. Mutton tallow crystallized from chloroform.   300 diameters.

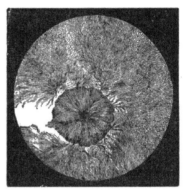

FIG. 5. Cat fat crystallized from chloroform.   300 diameters.

FIG. 6. Beef tallow crystallized from chloroform.   300 diameters.

FIG. 7. Beef tallow crystallized from chloroform.   300 diameters.

readily shown by dissolving some fat in a proper solvent which is then allowed to evaporate slowly.

The common fats can not be distilled under the ordinary pressure without decomposition, and the distillation of the fatty acids is also difficult. When the common fats are strongly heated they emit a peculiar odor, due to the acrolein formed by the partial destruction of glycerol. The following experiments illustrate some of the points referred to:

Experiment. Note the solubility of small bits of tallow in ether, chloroform, benzine and alcohol, using in each case the same volume of liquid, with equal weights of fat. The solubility in alcohol will be found much less than in the other menstrua.

Experiment. Dissolve some mutton or beef tallow in chloroform and with a glass rod put two or three drops of the nearly saturated solution on the center of a glass slide. As the chloroform evaporates a film begins to form on the top of the drop. Now put on a perfectly clean cover glass and allow to stand until crystallization is complete, which may require only a few minutes or some hours, the time necessary depending on the temperature and on the concentration of the solution. Examine the crystals with a microscope. Use a power of 200 to 300 diameters. By repeating the experiment with different fats considerable variation in the form of the crystals may be noticed, which is shown in the annexed cuts.

Experiment. Add to 5 cc. of cottonseed oil half its volume of strong white of egg solution and shake thoroughly. The liquids mix and form a white mass or emulsion which, however, is not usually stable.

Experiment. To 5 cc. of cottonseed oil containing a little free fatty acid add 10 drops of strong sodium carbonate solution and shake. A good stable emulsion is made in this way, as the sodium of the alkali solution forms a soap with the free acid and this appears to form a film around the little fat globules which prevents their flowing together again.

Origin of Fats in the Body. The question of the formation of fats in the animal organism has been much discussed. It was once assumed that, like protein substances, the fats are products of the vegetable world only, and that the animal has not the power of building them up from compounds which are not fats. But this view is not correct, as we have abundant proof that fats may be made in other ways. Much has been learned from the results of cattle feeding experiments carried out in agricultural experiment stations, where the gain in fat is often much in excess of what could be accounted for by the amount of fats in the food consumed. This gain must in some way be due to the effect of the carbohydrate and protein substances in the rations fed. The fattening power of sugar has long been recognized, but this has been in part accounted for on the theory that the sugar acts to protect the fats of the body from oxidation, by being readily oxidized itself to keep up the body energy. But much evidence has been accumulated to show that carbohydrates take part directly in the production of fats. How this is accomplished is not known, but in the processes syntheses and oxidations both must be concerned, since the fat molecules are more complex than those of the carbohydrates.

It is further likely that protein compounds are important factors in fat production. Many writers have indeed assumed that we have in the breaking down of protein molecules the chief sources of fats, but this view has been strongly combated. Indirectly the transformation may follow in this way: It is known that sugars are formed as cleavage products of certain albumins in the ordinary katabolic processes of the body and possibly a portion of the sugars thus formed may be then built up to produce fats.

The existence of the substance known as *adipocere* has long been held to furnish a pretty strong proof of the production of fatty acids from protein. This adipocere or cadaver wax is often found in large masses in old cemeteries and consists of fatty acids, calcium and ammonium soaps in the main. It is held that this substance could not possibly have come from the small amounts of fat ordinarily present in cadavers but must have been produced from the muscular portions of the body. This view has met with objections, however, and attempts are made to account for the development of the adipocere in other ways.

In the body fats constitute a reserve material in which potential energy is conveniently stored up. In sickness or in wasting diseases where there is a partial or complete failure in nutrition this fat is called upon to supply the needs of the body. The fats are oxidized while the muscular tissue is preserved.

PHYSIOLOGICALLY IMPORTANT FATS.

STEARIN or TRISTEARIN, $C_3H_5(C_{18}H_{35}O_2)_3$. This is a simple fat which does not occur in nature unmixed with other fats. When separated in purest possible condition it shows a melting point of 55° to 58°. It is the hardest of the common simple fats and apparently the least soluble in alcohol or ether. It may be separated in the form of rectangular plates by crystallizing from hot alcohol. Stearic acid may be obtained in the form of pearly crystalline plates or scales. It melts at 71°.

PALMITIN or TRIPALMITIN, $C_3H_5(C_{16}H_{31}O_2)_3$. The perfect separation of this fat from stearin, with which it is usually associated, is a matter of considerable difficulty. The fats are much alike. The melting point is variously stated by different observers, but appears to be about 51°. A mixture of stearin and palmitin was formerly supposed to be a distinct fat and was called *margarin*, $C_3H_5(C_{17}H_{33}O_2)_3$. This fat has been produced artificially but it does not appear to be a natural product. Palmitic acid resembles stearic acid in appearance and solubility; both acids are slowly soluble in strong hot alcohol and yield crystalline plates on cooling. The melting point of palmitic acid is about 62°.

OLEIN or TRIOLEIN, $C_3H_5(C_{18}H_{33}O_2)_3$. This is a liquid fat at the ordinary temperature and is a constituent of most of the natural fats and oils. Some fatty oils are nearly pure olein and become solid at a low temperature. The soft consistence of lard, human fat and several other natural mixtures is due to the olein present. Olein is a nonsaturated fat and will therefore show an absorption coefficient as explained below. By the action of reagents yielding nitrous acid it is converted into an isomeric substance known as *elaidin.*

Oleic acid in pure condition is not very stable, because of its unsaturated structure. It is an oily liquid at the ordinary temperature, but below 14° is converted into a crystalline solid. By treatment with nitrous acid it yields the isomeric *elaidic acid.* Oleic acid is characterized by forming a lead salt which is soluble in ether while lead palmitate and stearate are practically insoluble.

LARD and TALLOW are essentially mixtures of the three fats, palmitin, stearin and olein. By heating lard to its melting point, cooling slowly and subjecting the warm mass to pressure in a filter press the softer portion, consisting mainly of olein, may be separated. This is known as lard oil, while the harder residue is sometimes called lard stearin. It has about the consistence of butter. By subjecting beef suet to the same treatment a soft portion known as *oleo* oil is separated from a solid residue called beef stearin. The oleo oil is the material most often employed under the name of *oleomargarin* as a substitute for butter. A mixture of somewhat similar consistence is made in other ways; for example by combining cottonseed oil with beef stearin.

OLEOMARGARIN is the name given by law to these butter substitutes in the United States. Sometimes the fats are churned with milk or mixed with a certain amount of real butter to furnish a product with flavor suggesting butter. The name butterine is usually given to such mixtures and when properly made they are wholesome and in every way as good as butter from the standpoint of nutritive value.

BUTTER. The fat of milk is a very complex mixture and an exact separation has not yet been made by the methods of chemical analysis. According to the older notion butter fat contains essentially olein, stearin and palmitin, with a little butyrin, to which the flavor and odor are largely due, but it has been shown that other glycerol esters are certainly present. The results of some recent examinations may be approximately expressed as follows:

```
Glyceryl butyrate ......................................... 7.0
Glyceryl caproate, caprylate and caprate ................. 2.0
Glyceryl oleate ........................................... 36.0
Glyceryl myristate, palmitate and stearate ............... 55.0
                                                          100.0
```

From various investigations it appears that these fats are not present as simple esters, but may possibly exist in combinations represented by formulas like this:

$$C_3H_4 \begin{cases} C_{12}H_{23}O_2 \\ C_{16}H_{31}O_2 \\ C_4H_7O_2 \end{cases}$$

The melting point of butter fat is between 38° and 45°. On melting 100 parts by weight of pure butter fat, saponifying, separating the fatty acids and washing out everything soluble in hot water (butyric acid, etc.) it is found that the insoluble residue left amounts in the mean to 88 per cent, but may be more or less with different grades of butter. Commercial butter contains in the mean about 85 per cent of fat, 10 per cent of water and 5 per cent of salt.

HUMAN FAT. This consists essentially of olein, palmitin and stearin. In the fat of children the solid glycerides apparently are in excess, while in later life the proportion of olein increases, so that the separated fat may appear quite soft. In the human adult the olein may amount to 75 per cent of the whole fat, but the proportion varies with different parts of the body.

GLYCEROL. Since it is a constituent of all the true fats a few words about this alcohol will be in order here. The substance was first recognized in the aqueous liquid left in the preparation of lead plaster and for many years all used in pharmacy and in the manufacture of cosmetics was made by the same reactions. Since its importance in technology was recognized it has been produced on the large scale by other kinds of saponification or hydrolysis. In pure condition it is a thick, sweetish liquid with a specific gravity of 1.266 at 15°, referred to water at the same temperature. It mixes with water and alcohol in all proportions, but is not soluble in pure chloroform, benzene, carbon disulphide or petroleum ether. It is very slightly soluble in ether. Like other alcohols it may be combined with acids to form esters. With stearic acid mono-, di- and tri-stearin are produced, the last being identical with the natural fat. When fed to animals glycerol may be oxidized to a limited extent only. Any excess of it fails to be assimilated and soon produces disorders in digestion and absorption. Its food value, in free form, is therefore very slight.

Recognition and Determination of Fats. In physiological chemical investigations fats are separated from accompanying substances through their solubility in warm ether, chloroform or petroleum ether. The carbohydrates, protein bodies and salts are not soluble in these liquids. . The saponification test is also of value in identification. Many of the fats contain unsaturated acid groups and are therefore able to absorb certain amounts of halogens from specially prepared solutions.

Oleic acid, $C_{18}H_{34}O_2$, absorbs iodine or bromine to form $C_{18}H_{34}I_2O_2$ or $C_{18}H_{34}Br_2O_2$. The fat to be examined is dissolved in chloroform and treated with the standard solution. After a time the excess of iodine (or bromine) is determined, and that absorbed by the fat is a measure of the non-saturated acid present. Linoleic acid absorbs twice as much iodine or bromine as oleic acid, as the formula $C_{18}H_{32}O_2$ becomes $C_{18}H_{32}I_4O_2$.

The determination of the amount of insoluble fatty acids furnished by a given weight of fat is also a valuable factor in the study of these bodies. In another method the fat is saponified, the soap formed decomposed by dilute sulphuric acid and the resultant product subjected to distillation. Fatty acids which are volatile pass over and collect in the distillate, where their amount may be determined in terms of KOH or NaOH, by titration. In this treatment stearin and palmitin yield acids not volatile with steam. This is a valuable test and is applied in the examination of butter supposed to be adulterated with other fats.

In certain lines of investigation and especially in the examination of tissues by aid of the microscope, fat is recognized by its coloration through the coaltar product known as Sudan III. This is one of the few colors which are soluble in fats and fatty oils. A yellowish-red color is imparted.

LECITHIN. This is a peculiar complex body which contains phosphoric acid and an organic basic group in place of one of the fatty acid radicals in the common fats. It is found in the vegetable kingdom, but commonly and most characteristically in many animal tissues, in the brain and nerve tissue, blood, lymph, milk, pus, yolk of egg, etc. It is most readily prepared from the last named substance. The following formula represents the supposed structure of the body,

$$C_3H_5 \begin{cases} -O-C_{18}H_{35}O \\ -O-C_{18}H_{35}O \\ -O-PO \cdot HO \cdot O \cdot C_2H_4 -N \begin{cases} (CH_3)_3 \\ OH \end{cases} \end{cases}$$

in which distearylglycero-phosphoric acid is combined with the base choline:

$$(CH_3)_3 \equiv N \begin{cases} OH \\ C_2H_4OH \end{cases}$$

It appears that several forms of lecithin exist, containing oleic and palmitic acid as well as stearic. They undergo saponification, yielding fatty acids, glycero-phosphoric acid and choline. They are soluble in ether and alcohol and in other respects resemble the true fats. With water lecithin swells to a gelatinous mass which under the microscope presents a characteristic appearance. The function of lecithin in the body is not understood, but from the fact of its wide distribution and its occurrence in milk it is reasonable to assume that it performs some important part.

The above formula represents the simple or typical lecithin. As recent investigations have shown that a number of related bodies exist containing other proportions of nitrogen and phosphorus it has been proposed to give the name *phosphatides* to the whole group. The group name *lecithan* is also used. The chemistry of these bodies is far from simple. Some of them appear to be associated with sugars, and others with proteins in the animal and vegetable tissues, but as they suffer decomposition very easily their separation in pure form for study is an extremely difficult problem.

The Waxes. These bodies bear some resemblance to the fats and will be briefly mentioned here. They consist largely of esters of the higher monohydric alcohols of the saturated series, and in most cases are complex mixtures of which the composition is not exactly known. Spermaceti seems to consist largely of cetyl palmitate, $C_{16}H_{33}OC_{16}H_{31}O$. Beeswax contains some free acid, cerotic acid, in addition to the esters. The most important constituent is apparently myricin or myricyl palmitate, $C_{30}H_{61}OC_{16}H_{31}O$. The waxes are not easily saponified and as a rule clear soap solutions are not obtained.

Cholesterol. This substance is an alcohol, but in appearance it resembles some of the solid fats and is associated with them in several natural products. Hence it is in place to describe it here. The formula $C_{27}H_{45}OH$ probably represents the composition of the body which is found in the brain, yolk of egg, the liver, blood and other tissues and is especially abundant in the fat of wool. It constitutes also the main substance in certain gall-stones from which it may be separated in nearly pure condition. In wool fat it exists partly in the free state and partly in combination with fatty acids in the form of esters.

It is readily soluble in hot alcohol, ether, benzene and chloroform, but not in water or alkali solutions. It is therefore left as an insoluble residue in the saponification of fatty mixtures containing it. Under the microscope pure cholesterol appears as a mass of white plates with sharp angles. The cholesterol esters combine with water to form stable emulsions, and it is probably on account of the presence of these esters that the substance known as lanolin is practically valuable. This lanolin is made from purified wool fat and is largely used in the preparation of salves and ointments.

An isomeric substance known as isocholesterol is often found associated with the true cholesterol, especially in wool fat. In the vegetable kingdom other forms of cholesterol are widely distributed in small quantities, being found in most oils and seeds. All forms of cholesterol have a marked action on polarized light.

Experiment. If gall-stones are obtainable the following reactions may be carried out in illustration of properties of cholesterol. Crush the stones to a powder and boil with water to remove anything soluble. Extract the residue several times with hot alcohol, filter, unite the solutions and allow the cholesterol to crystallize on cooling. As some fat may be present this must be removed by saponifying with a little alcoholic potassa solution. After saponification boil off the alcohol and extract the dry residue with ether, in which soaps are insoluble. On evaporation of the ether a nearly pure cholesterol is obtained. It may be further purified by recrystallization from hot alcohol. With the substance these tests may be made:

SALKOWSKI'S TEST. Dissolve about 10 milligrams of cholesterol in 2 cubic centimeters of chloroform and shake with an equal volume of strong sulphuric acid. The chloroform becomes colored blood red, then cherry red and finally purple. The acid shows a dark green fluorescence. If the chloroform is poured into a dish the color changes to blue, then green and finally yellow.

BURCHARD-LIEBERMANN TEST. Dissolve about 10 milligrams of cholesterol in 2 cubic centimeters of chloroform, add 20 drops of acetic anhydride and 1 drop of strong sulphuric acid. A violet pink color results.

The appearance of cholesterol plates should be studied under the microscope.

The presence of cholesterol in the ester form in lanolin and similar preparations of wool fat may be shown by the above tests.

CHAPTER V.

THE PROTEIN SUBSTANCES.

These extremely important bodies, usually called albuminous bodies, are found in vegetable and animal organisms of all kinds, and in some form are essential elements in cell growth. Unlike the fats and carbohydrates, they seem to be elaborated in the vegetable kingdom only; or, at any rate, the fundamental structures in them appear to be formed in vegetable growth only. The animal is able to modify and transform to some extent, but apparently can not build them up from simple materials. In composition the protein bodies are extremely complex; qualitatively they contain carbon, hydrogen, oxygen, nitrogen and sulphur. In an important group of these bodies phosphorus is also present, and a few contain iron. The quantitative composition of some of the best known protein compounds is expressed approximately as follows:

	Per Cent.
C	50.0–55.0
H	6.5– 7.3
O	19.0–23.0
N	15.0–17.0
S	0.3– 2.4

Attempts have been made to calculate formulas for certain protein bodies from the results of analyses, but no great importance attaches to the empirical formulas so reached. The best analyses made of the compounds differ among themselves to an extent that makes a definite result quite impossible. This is largely due to the fact that there are great practical difficulties in the way of properly purifying the substances as a preliminary to analysis; they all occur mixed with other compounds, such as fats, carbohydrates and mineral matters, and to remove these without in any way altering the composition of the complex albuminous molecule is extremely difficult, if not impossible. The formulas which have been published are interesting chiefly in showing roughly how complex the structures certainly are. For serum albumin Hofmeister has given this minimum formula:

$$C_{450}H_{720}N_{116}S_6O_{140},$$

while for egg albumin he has given this:

$$C_{239}H_{386}N_{58}S_2O_{78}.$$

Even more complex formulas are given, for example this:

$$C_{720}H_{1218}N_{196}S_{10}O_{248}.$$

These formulas are in a measure based on an assumption as to the number of sulphur atoms present, about which something will be said later.

The usual methods of fixing organic formulas by aid of a molecular weight determination can not be successfully applied in these cases. In the cryoscopic method, for example, the traces of mineral impurities present have possibly as much influence on the result as the whole weight of dry protein. Because of changes in composition at a high temperature the boiling point method, even if otherwise reliable, can not be applied, and methods based on osmotic pressure observations lead to results of no value.

CLASSIFICATION OF THE PROTEIN BODIES.

The substances thus far studied have been divided into groups or classes dependent on chemical composition or structure. With the protein compounds this is only partially possible because of our lack of full knowledge in this direction. Of the structural relations of the protein molecules nothing whatever is known, while of composition only a few general facts are clearly enough established to be available in a scheme of classification. The first efforts at classification, which we owe largely to the work of Hoppe-Seyler, were therefore essentially empirical. Other schemes were later proposed as more facts were brought to light, so that finally a grouping like the following came to be gradually accepted by physiological chemists, with slight differences in details only. The arrangement below is that of Hammarsten, in the form given by Cohnheim. He makes four principal divisions as follows:

PROTEIN BODIES
{
 TRUE OR NATIVE ALBUMINS.
 DERIVED ALBUMINS OR TRANSFORMATION PRODUCTS.
 PROTEIDS.
 ALBUMOIDS.
}

These four great divisions may be further subdivided:

TRUE OR NATIVE ALBUMINS.
- **Albumins proper.**
  Serum albumin, egg albumin, lactalbumin.
- **Globulins.**
  Serum globulin, egg globulin, lactoglobulin, cell globulin, vegetable globulin.
- **Coagulating albumins.**
  Fibrinogen, myosin, gluten protein.
- **Nucleoalbumins.**
  Casein, vitellins, mucin-like nucleoalbumins.
- **Histones.**
  Scomber-histone, salmo-histone.
- **Protamines.**
  Salmin, clupein, sturin, scombrin.

DERIVED OR TRANSFORMATION PRODUCTS.
- **Coagulated or Modified Albumin.**
- **Acid and Alkali Albumins, Albuminates.**
- **Albumoses, Peptones.**

PROTEIDS.
- **Nucleoproteids.**
  Nucleic acid with histone, protamine, etc.
- **Hemoglobins.**
  Hematin with histone.
- **Glucoproteids.**
  Combination of a protein and carbohydrate group, mucin, mucoids, phosphoglucoproteids.
- **Lecithoproteids.**
  Combination of a protein with a lecithin body.

ALBUMOIDS.
- **Collagen,** forming gelatin, glue.
- **Keratin,** in horn, hair, nails, etc.
- **Elastin,** elastic tissue.
- **Amyloid,** in pathological formations.
- **Spongin,** in sponge.

In recent years enormous additions have been made to the literature of the proteins, and many new substances have been described. Our knowledge of certain groups has been advanced largely through the labors of the Yale school of chemists, and especially by Chittenden and Osborne. Much of our systematic knowledge of vegetable proteins must be credited to these investigators. In view of these important extensions of our acquaintance with the details of protein chemistry a committee representing the American Society of Biological Chemists and the American Physiological Society has recommended the following classification of the bodies in question. The known substances are thrown into three main groups in place of four, as above. The term *proteid* is abandoned.

SIMPLE PROTEINS
- Albumins.
- Globulins.
- Glutelins.
- Alcohol-soluble Proteins.
- Albumoids.
- Histones.
- Protamines.

CONJUGATED PROTEINS.
- Nucleoproteins.
- Glycoproteins.
- Phosphoproteins.
- Hemoglobins.
- Lecithoproteins.

DERIVED PROTEINS.
- Primary Derivatives.
  - Proteans.
  - Metaproteins.
  - Coagulated proteins.
- Secondary Derivatives.
  - Proteoses.
  - Peptones.
  - Peptides.

The differences in the two classifications are not great. The albumoids are here considered as simple proteins, which is probably an advantage. The term *phosphoprotein* in the new classification covers substances like the *nucleoalbumins* of the old.

GENERAL REACTIONS OF THE PROTEINS.

The various substances in the protein group respond to a number of reactions which, taken together, are sufficient to characterize and identify the bodies in question. They all contain nitrogen in a form to be liberated as ammonia when the dry substance is heated with soda-lime. A positive result with this test does not, of course, prove the presence of a protein compound, since all ammonium salts and amino compounds in general respond to it; but with a negative result proteins as well as these other compounds are certainly excluded. The reaction therefore serves as a preliminary test in the examination of unknown substances for proteins. The test may be easily carried out and is delicate.

Experiment. Mix some dried albumin or some wheat flour with an equal bulk of soda-lime in a dry test-tube. Apply heat and note the escape of ammoniacal vapors as shown by the odor, or reaction on moist litmus paper. The fixed alkali decomposes the protein matter very quickly, and ammonia always results.

COAGULATION TESTS.

Many of the protein substances undergo a peculiar change known as coagulation when heated, or treated with certain reagents. The test is characteristic of most of the bodies except some of the products of transformation. This coagulation is usually accompanied by precipitation, that is, the body is thrown out of solution and as a rule can not be restored to its original condition. But there are cases of precipitation without coagulation; the terms must not, therefore, be used as synonymous. In coagulation proper the protein body becomes permanently altered, so that it can not be brought into its original condition again by addition of reagents or by other means. On the other hand, it is in many cases possible to throw a protein body out of solution by simply adding an excess of some inorganic salt, without at the same time producing any decided alteration in the character of the protein precipitate. By largely diluting with water the precipitate may be brought into the soluble condition again. This will be illustrated later by the use of ammonium sulphate which behaves in a characteristic manner with different proteins. Many of these have definite *precipitation limits* with the sulphate. That is, they begin to separate when the amount of the salt reaches a certain value, and precipitate completely with a greater concentration.

Experiment. The simplest coagulation may be shown by boiling a dilute white of egg solution. As long as it is perfectly neutral coagulation follows at once, but in presence of alkali acid must be added to the point of neutrality. This behavior finds practical application in the ordinary test for serum albumin as it occurs pathologically in urine. The precipitate may be redissolved only by some digestion or chemical process which produces a new substance.

Coagulation by Reagents. By the addition of certain chemicals many of the protein compounds are easily thrown into the coagulated condition. Some of the most characteristic reactions in this direction are shown by simple experiments with mineral acids, salts of heavy metals and the so-called alkaloid reagents.

Experiment. Among the acids which bring about coagulation, nitric acid is the most certain in its action and is commonly used in practical cases where it is desired to recognize a small amount of serum or egg albumin in solution. The test may be made by adding about a cubic centimeter of strong nitric acid to four or five cubic centimeters of white of egg solution and warming. Coagulation follows at once. With a very dilute albumin solution the substance separates in flakes, while with a strong solution a stiff, jelly-like mass may result. The test is a common one in urine analysis, but must be conducted with certain precautions.

Experiment. Solutions of most protein substances are precipitated by addition of alcohol in excess. This may be shown by mixing white of egg solution with strong alcohol, the latter being added gradually until the maximum of precipitate is obtained. With dilute alcohol precipitates are usually not secured, and besides the alcohol precipitation is usually not a permanent coagulation as in the above case with the acid.

Experiment. Precipitation by Salts. Some of the salts of heavy metals give characteristic precipitates with protein solutions. The behavior may be illustrated by adding to dilute egg albumin solution small amounts of solution of *mercuric* *chloride, lead acetate, copper sulphate* or *ferric chloride*. The reagents must not be added in excess, as in some cases this causes a resolution of the precipitate. Similar reactions may be obtained with solutions of most of the heavy metals, but the salts mentioned are often used in practice. The behavior of mercuric chloride or corrosive sublimate as an active disinfecting agent depends on its property of coagulating the protein matter of the pathogenic bacteria, to destroy which it is used. The precipitates may be formed in neutral, acid or alkaline solution as a rule, and chemically may be regarded as salts of the metals used as precipitants.

Experiment. Precipitation by Alkaloid Reagents. In acid solution the protein bodies are very generally precipitated by addition of solutions of *picric acid, tannic acid, potassium-mercuric iodide, phosphomolybdic acid* and other reagents employed in the detection of alkaloids. The precipitates are voluminous, and in most cases complete in presence of sufficient acid.

White of egg, much diluted, may be used in illustration.

The above reactions may be explained on the assumption that the proteins here act as pseudo-acids or pseudo-bases. In perfectly pure solution they are neutral and indifferent to some indicators, but the addition of a mineral acid imparts to them the character of pseudo-ammonium bases which yield precipitates as the alkaloids do under like circumstances. On the other hand, with salt solutions they become pseudo-acids and form now insoluble precipitates of complex salts. But it has been shown that while some of the proteins may be neutral to litmus they may at the same time be quite distinctly acid to phenolphthalein, and require a decided amount of decinormal alkali solution to give a reaction by displacing the acid in combination with them. This is taken to indicate that they should be looked upon as true bases, rather than as pseudo-bases. The amount of acid which may unite with certain proteins has recently been found with considerable accuracy, and becomes in some degree a measure of the basicity. In other cases it may be shown that they have a true rather than a pseudo-acid character and unite with alkali to form real salts.

Behavior with Millon's Reagent. In this we have one of the most characteristic reactions of the protein bodies. Millon's reagent is made by dissolving mercury in twice its weight of strong nitric acid, completing the reaction by heat. The solution obtained is diluted with twice its volume of water. It contains a little nitrous acid. When warmed with white of egg and other proteins it imparts a deep red color to the coagulum produced and often to the containing solution. The reaction is common to benzene derivatives which have a hydroxyl group attached to the nucleus, and is given by phenol for example. The reaction in the protein substances is due to the presence of the tyrosine group in the complex molecule. This group seems to be present in all protein bodies with the exception of gelatin, so that the reaction is a nearly universal one. The protein derivatives which still contain the tyrosine complex likewise show the reaction. Tyrosine is represented by the formula $C_6H_4OH.CH_2.CHNH_2.COOH$, and will be referred to later, as it is an important decomposition product of proteins.

Experiment. Test the behavior of Millon's reagent by adding some to white of egg solution, milk or flour, and applying heat. The characteristic color appears almost immediately. Its depth depends somewhat on the concentration of the protein substance used. Presence of much salt interferes with the test or may even prevent the reaction.

Experiment. Apply the same test to weak solutions of phenol, salicylic acid and thymol. Note the color and character of the reaction. These bodies all have a benzene nucleus with hydroxyl combination. If pure tyrosine is available a very dilute solution may be employed for tests. It is said that 1 part in 1000 of water gives a distinct reaction. Hydroquinol, resorcinol and α- and β-naphthol give likewise decided reactions, but the colors are orange yellow.

The Biuret Reaction. This, like the above, is a protein test depending on the presence of certain groups in the complex molecule. When biuret and several substances of related composition are mixed with an excess of alkali solution, either sodium or potassium hydroxide, and then a few drops of a weak copper sulphate solution are added, a blue-violet to reddish-violet color is produced. The shade depends on the concentration of the solution, and on the composition of the reacting group. It has been shown that the reaction seems to follow with compounds which contain two groups—$CONH_2$ directly united or joined by a carbon or nitrogen atom, as for example:

$$\begin{array}{cccc} CONH_2 & HN\!\!<\!\!\begin{array}{c}CONH_2\\CONH_2\end{array} & H_2C\!\!<\!\!\begin{array}{c}CONH_2\\CONH_2\end{array} & \begin{array}{c}CO\cdot CONH_2\\NH\cdot CONH_2\end{array}\\ CONH_2 & & & \\ \text{Oxamide} & \text{Biuret} & \text{Malondiamide} & \text{Oxaluramide} \end{array}$$

The combination of copper and alkali with these bodies has been recently studied and formulas determined. If, in place of using a solution of one of these compounds, some white of egg, or other protein solution, is employed the same color appears. The absorption spectra from pure biuret and egg albumin solution, treated in this manner, are the same, which shows that the albumin must split off this group under the influence of the alkali used. The reaction is one of extreme delicacy and may be employed for the recognition of traces of protein compounds. It is used especially in the detection of peptone, one of the derived protein substances. Derivatives of simpler nature, that is, the products of the decomposition of proteins, do not give the reaction. It is therefore of value in following the course of experiments on the digestion or hydrolysis of proteins, as the reaction disappears with the breaking down of the last protein complex.

Nickel salts exhibit an analogous behavior, but show orange yellow colors. Cobalt solutions give reddish colors, but not very strong.

Experiment. Prepare a dilute white of egg solution and add to 5 cc. of it some solution of potassium or sodium hydroxide. Then add one or two drops of weak copper sulphate solution, or enough to impart the characteristic color. An excess of the copper yields a precipitate and must be avoided. The reaction is much sharper with albumose and peptone derivatives than with the original native protein. Repeat the test with solution of nickel sulphate. The test is obscured by the presence of ammonium salts, which is sometimes a matter of importance in practical work.

The α-Naphthol Test of Molisch. In the chapter on the sugars it was shown that a very marked color reaction is given by mixing a few drops of a weak alcoholic solution of α-naphthol with the sugar solution and then adding some strong sulphuric acid. The same behavior is shown by solutions of some protein substances, which indicates that they must contain a carbohydrate group of some kind. The reaction depends on the formation of furfuraldehyde by the decomposition of the sugar by the strong acid. This furfuraldehyde combines then with the α-naphthol to produce a deep violet color, the reaction being similar to that between furfuraldehyde and aniline acetate described in the pentose test in a former chapter.

Experiment. To a few cubic centimeters of white of egg solution add five drops of 10 per cent. solution of α-naphthol in alcohol. Then carefully add three or four cubic centimeters of strong sulphuric acid, which sinks below the lighter solution. Note the color at the zone of contact and throughout the liquid on shaking. Alkalies change the color to yellow. Thymol solution is sometimes used instead of α-naphthol. This gives a deep red color. These furfuraldehyde reactions are extremely delicate, and appear in a great variety of tests. Their general character and importance should therefore be recognized.

The Xanthoproteic Reaction. This is a delicate test, depending on the formation of yellow nitro derivatives of the phenol groups in the protein complex. Similar reactions are given by many simpler organic substances where nitric acid is mixed with them and heat applied. The color produced by nitric acid in contact with the skin is due to the same general reaction.

Experiment. In illustration, add some strong nitric acid to white of egg solution. On application of heat the yellow color appears. By neutralizing with ammonia the color changes to orange yellow.

Make a similar test by warming some phenol solution with nitric acid. In this case a nitro-phenol is formed. Pure phenol and strong nitric acid, it will be recalled, yield trinitrophenol or picric acid, $C_6H_2 \cdot OH \cdot (NO_2)_3$. Add ammonia to neutralize, as before.

The Lead Hydroxide Test. The protein bodies contain sulphur which may be removed by action of an alkaline lead solution, or

THE PROTEIN SUBSTANCES. 59

alkaline bismuth solution or mixture. The second reaction has some importance, as it is the source of a fallacy in the so-called bismuth test for sugar in urine. In presence of albumin, sulphide of bismuth is formed in place of the reduction product indicative of sugar. As all protein bodies contain sulphur the test is a general one. It may be made as follows:

Experiment. Produce first a soluble alkaline compound of lead by adding to a few cubic centimeters of lead acetate solution enough strong alkali, sodium or potassium hydroxide, to form a precipitate and redissolve it. Then add the protein substance, white of egg for example, and boil. A brown or black color appears and sometimes even a precipitate of black lead sulphide. Only a part of the sulphur, however, may be separated in this simple manner. Another portion seems to be much more firmly combined in the protein molecule.

The reactions which have just been explained are the most important and characteristic of all which have been suggested for the recognition and identification of the proteins. Numerous other reactions are known, however, which are easily observed. Several of these are color tests, depending on the formation and combination of furfuraldehyde, but they need not be described as in principle they do not differ much from the Molisch test.

QUANTITATIVE DETERMINATION OF PROTEINS.

The above tests serve for the recognition of proteins but not for their determination, and for the latter purpose it may be said further that no one method is perfectly suited to all cases. Many of the simpler protein bodies are determined by complete coagulation, followed by weighing of the precipitate formed. This involves several operations, all of which must be very carefully performed. For example, a pure native albumin in solution may be coagulated by adding a few drops of acetic acid and boiling thoroughly. The coagulum is collected on a weighed paper filter or in a Gooch funnel, thoroughly washed, dried and weighed. Instead of drying and weighing the precipitate it may be decomposed according to the Kjeldahl process, in which the nitrogen is converted into ammonia by digestion with sulphuric acid. The ammonia is easily separated and measured. The nitrogen is $14/17$ of it. By multiplying the nitrogen found by the factor 6.25 we obtain the original protein content on the assumption that these bodies contain 16 per cent of nitrogen. This method is now commonly followed in the determination of crude protein for many technical and scientific purposes. But in many cases an error is naturally introduced because of the uncertainty of the factor; 6.25 is the mean value for the native proteins and the closely related bodies.

COMPONENT GROUPS IN THE PROTEIN COMPLEX.

In the study of the protein molecule as a whole, a limit is soon reached in any attempt to fix its composition, but much may be learned by observing the various products formed in reactions by which the molecule is broken down under the influence of different agents. Some of these reactions are apparently largely hydrolytic in character, and in a degree may be compared to the decomposition of a fat by superheated steam. In this very simple case glycerol and fatty acids are obtained and we conclude that they were not actually formed in the process, but that they were present in combination in the original fat. In treating protein bodies in a similar manner or in subjecting them to the decomposing influences of acids or alkalies, a number of products are formed. These must be either results of peculiar disintegration and subsequent synthesis, or they must represent groups in some way existent in the original complex. The latter view is strengthened by the fact long observed that certain products result, whatever the method of decomposition. Leucine, for example, is found abundantly among the products liberated by subjecting protein to the action of superheated steam, hot hydrochloric, nitric or dilute sulphuric acid, concentrated alkali solutions, bromine water under pressure, or to prolonged pancreatic digestion. The almost necessary conclusion must be that in these varied reactions the leucine could not have formed from smaller disintegration groups, but must have been set free from something holding it in the protein complex.

The decomposition reactions are therefore considered very important as suggesting probably the component groups in the large molecule. In the following pages a few of the most important of these reactions and their products will be described.

Decomposition by Steam under Pressure. By prolonged heating of protein substances with water certain changes take place, even below a temperature of 100° C. Following the coagulation, which appears in most cases, a gradual hydration and solution begins, and a small portion of the substance is brought into the form of albumose or possibly peptone. At a higher temperature, that is, by heating with water or steam under pressure, more profound changes take place. Ammonia and hydrogen sulphide are split off from the molecule and relatively large amounts of albumose and peptone are formed. If the temperature is high enough the reaction extends to the complete destruction of the molecule and such bodies as leucine and tyrosine are produced in quantity.

Effect of Alkalies. Much more decided changes are noticed when the protein body is heated with alkali solutions. Experiments of this kind were long ago carried out by Schützenberger and have since been frequently repeated. Numerous compounds have been identified among the decomposition products, and of these the most important are leucine in quantity, tyrosine, ammonia, carbonic acid, butyric acid, formic acid, acetic acid, oxalic acid, aspartic acid, amino-valeric and amino-butyric acid. Barium, potassium and sodium hydroxide solutions have been used for the purpose. By melting the dry proteins with alkali some of the same products are formed, especially leucine and tyrosine.

Effect of Acids. Many experiments have been made on the decomposition of protein bodies by boiling with acids, and particularly with strong hydrochloric acid, the hydrolyzing power of which is very great. The most important of the products isolated in this way are the following:

THE HEXONE BASES. The term hexone bases has been given by Kossel to a group of bodies which occur commonly in the decomposition products of practically all the protein substances. We have here *arginine*, $C_6H_{14}N_4O_2$, *lysine*, $C_6H_{14}N_2O_2$, and *histidine*, $C_6H_9N_3O_2$. The first appears to be a guanidine derivative of amino-valeric acid, the second is diamino-caproic acid, while the third appears to be a diamino acid of composition as yet unknown, or is, possibly, an iminoazol derivative of amino-propionic acid. The isolation of these compounds was a very important step in the direction of clearing up the constitution of the proteins, inasmuch as some of the simplest of these bodies, the protamines, seem to consist almost wholly of the hexones. More will be said about this relation later. The hexones are soluble, crystalline, optically active, compounds, and because of their wide occurrence have been very thoroughly studied. They contain the amino group in the α position, and in this respect resemble the other common disintegration products. All the α amino acids appear to have a sweetish taste, which is illustrated by the first of the products to follow.

GLYCOCOLL OR GLYCINE, $C_2H_5NO_2$, amino-acetic acid. Obtained abundantly from gelatin and also from a few other proteins. It is very soluble in water to which it imparts a sweetish taste, and is insoluble in alcohol or ether. From a theoretical standpoint the importance of glycine is very great, as it is the starting point in various syntheses, to be explained later. It also combines with benzoic acid to form hippuric acid in the body metabolism. Hippuric acid is benzoyl glycine.

AMINO PROPIONIC ACID, OR ALANINE, $C_3H_7NO_2$. This is in a sense the nucleus substance corresponding to tyrosine and phenylalanine, to be referred to. It is a soluble product rather widely distributed in protein bodies, and because of this marked solubility is hard to isolate.

AMINO VALERIC ACID, $C_5H_{11}NO_2$. This substance is apparently the α product. It is usually mixed with leucine and is separated only with difficulty from this body. Combined with guanidine, $NHC(NH_2)_2$, it yields the important arginine, referred to above. Diamino valeric acid is known as ornithine.

LEUCINE, $C_6H_{13}NO_2$, α-amino-caproic acid, or α-amino-isobutyl-acetic acid. This has been already mentioned as found abundantly among the products of protein decomposition by other agencies. In some cases it appears to constitute 30 per cent of the reaction products, and must therefore play a very important part in the original complex molecule. Leucine is found in several different forms; the common product obtained by acid hydrolysis is right rotating and shows in hydrochloric acid solution $[\alpha]_D = +18.9°$. In water solution it rotates in the opposite direction.

SERINE, $C_3H_7NO_3$, amino-hydroxy-propionic acid, was first discovered in silk, and hence the name. But it is present in many of the protein bodies as well, although in not very large amount.

ASPARTIC ACID, $C_4H_7NO_4$, amino-succinic acid. Slightly soluble in water, the solutions being apparently left rotating at the ordinary temperature. But in hydrochloric acid it is strongly right rotating. While the acid is found commonly in proteins the amounts are not large.

GLUTAMINIC ACID, $C_5H_9NO_4$, α-aminoglutaric acid.. The acid is found in several optical modifications. The common form is but slightly soluble in water and is right rotating. In hydrochloric acid the right rotation is marked. This acid is obtainable in considerable quantities from many proteins, and is one of the extremely important constituent groups. It is probably more abundant than leucine.

PROLINE, $C_5H_9NO_2$, ($C_4H_7.NHCOOH$), α-pyrrolidine carboxylic acid. From the conditions under which it has been found this interesting body is supposed to be a primary product. It is an *imino* not an *amino* derivative. It was first obtained from casein and later from other proteins. It has a very sweet taste and a high left rotation. The closely related *hydroxy-α-pyrrolidine carboxylic acid* has been obtained from gelatin.

TYROSINE, $C_9H_{11}NO_3$, p-oxyphenyl amino-propionic acid. This appears to be a component part of all the common proteins, with the exception of gelatin, and from some of them has been obtained in quantity. As already mentioned this is probably the substance which reacts commonly with Millon's reagent. Tyrosine is but slightly soluble in water, from which it crystallizes in bundles of fine needles. Its solutions are optically active. In presence of hydrochloric acid $[\alpha]_D = -8°$ to $-15°$, the rotation varying with the amount of acid.

PHENYLALANINE, $C_9H_{11}NO_2$, phenylamino propionic acid. This substance resembles tyrosine closely in structure and behavior and is a common product of protein decomposition. It is slightly soluble in water and has a sweetish taste. It appears to be present in many cases when tyrosine is lacking, and must be considered as a very important decomposition product.

TRYPTOPHANE, $C_{11}H_{12}N_2O_2$, indol-amino-propionic acid. This complex product has been obtained from several of the proteins, and probably occurs in most of them. Many of the peculiar color reactions of proteins are due to the small amount of tryptophane present. The most characteristic of these color reactions is given by the addition of bromine water to a liquid containing the substance. A marked violet color results. This is shown well in advanced stages of the tryptic digestion of proteins. More will be said about the body later.

From the above list of decomposition products it will be seen that the comparatively simple amino acids predominate; most of them, possibly all, are the α compounds, and as will be shown below, they make up a very large portion of the whole protein molecule.

GLUCOSAMINE, $C_6H_{11}O_5(NH_2)$. This appears to be an important constituent in some groups of protein bodies. It has been obtained in quantity from the glucoproteids and is possibly present in small amount in all. It is usually obtained as a salt, hydrochloride or hydrobromide, which is readily soluble in water and optically active. It is regarded, usually, as a secondary product of dissociation.

AMMONIA, NH_3. This is always found in relatively large amount, but in the main may be a secondary product.

SULPHUR COMPOUNDS. Hydrogen sulphide, ethyl sulphide, thiolactic acid, $C_3H_6SO_2$, cystin, $C_6H_{12}N_2S_2O_4$, and traces of other bodies which contain sulphur have been identified in small amount among the decomposition products. Of these sulphur compounds cystin is the most important. It exists in two isomeric forms, one of which is found in certain calculi.

CARBONIC ACID is apparently a constant derivative, but may appear as a result of some secondary reaction. Its significance, therefore, is obscure.

With acids other than hydrochloric very similar reaction products are secured. It will be shown below that the effects of prolonged tryptic digestion are very nearly the same as observed with hydrochloric acid. This is a point of the highest practical importance, as it gives us some insight into the complex physiological process. And in peptic digestion also, where very weak hydrochloric acid and pepsin are employed, essentially the same products result provided the time of the action be made sufficiently long.

RESULTANT CHARACTER OF THE PROTEIN MOLECULE.

While the various decompositions detailed above give us some insight into the number and kind of groups combined in the protein complex, they do not, unfortunately, show us much as to the manner in which these groups are combined. We are not, as yet, able to picture to ourselves a large molecule in which the leucine, tyrosine, aspartic acid, glycocoll, and so on, are united to form a molecule with the general properties and molecular weight as large as we assign to even the simplest proteins, but a step has been made in that direction through the synthesis of various *polypeptides* carried out by Fischer and Curtius, who have succeeded in condensing several amino acids into one molecule with certain properties suggesting those of the peptones. These bodies will be referred to in a later chapter. Hofmeister has suggested the possibility of the combination of amino acids in large groups by the following general scheme:

$$-NHCHCO-NHCHCO-NHCHCO-NHCHCO-$$

CH_2	CH_2	CH_2	$(CH_2)_4$
$CH_3 \cdot CH \cdot CH_3$	C_6H_4OH	$COOH$	CH_2NH_2
Leucine	Tyrosine	Aspartic Acid	Lysine

The recognition of the various component groups suggests some reasons why the proteins may exhibit acid and basic behavior at the same time. Of most of these protein compounds the basic character is the more pronounced and more readily observed; that is, their acid combining power. Some writers consider these bodies as so-called *pseudo bases* and *pseudo acids,* because of the very peculiar manner in which they unite with acids and bases. But several investigations of the last few years indicate that they are more properly true bases and acids, but so weak in their combinations that hydrolysis follows very readily. This hydrolysis obscures the reactions which must take place in the formation of salts. In aqueous solutions the pure proteins and the component amino acids are practically non-electrolytes, which has been explained on the assumption that the basic part of one group is linked to the acid part of another, with little or no dissociation. Possibly, also, a ring-like structure is formed by a kind of internal saturation. By various methods it may be shown that the simple proteins and the amino acids combine in rather definite proportions with the inorganic acids and bases, as will be pointed out in later chapters. But the salts so formed suffer marked hydrolytic dissociation and conduct the electric current essentially as the acid or base used. Glycocoll hydrochloride, for example, $CH_2NH_2.HCl—COOH$, hydrolyzes so as to leave free hydrochloric acid, while sodium glycocollate, $CH_2NH_2—COONa$, hydrolyzes to yield sodium hydroxide, and the conductivity observed is due to this latter essentially. With casein, which is a protein easily obtainable in pure condition from milk, the phenomena of salt formation both with acids and bases may be very easily observed. Something will be said about this later. One molecule of casein combines, apparently, with four or five molecules of sodium hydroxide to form a salt. In their basic capacity serum and egg albumins combine with a large number of molecules of hydrochloric acid.

The presence of sulphur in the proteins was shown by a test referred to some pages back. Investigations have shown that sulphur is present in at least two kinds of combinations in the protein complex; there must be at least two sulphur atoms in the molecule. Some of the sulphur is easily separated by hot alkali solutions, while the rest of it is not. No part of this element appears to be combined in oxidized form, that is, in the condition of a sulphite or sulphate. The sulphur compounds which have been obtained in protein decomposition are such as may be derived from a breaking down of the cystin group. It has been shown that cystin gives up its sulphur very slowly to boiling alkali, and only in part as sulphide.

The general reactions and characteristics of the protein bodies having been discussed, a brief description of the more important individual substances will now follow.

TRUE OR NATIVE ALBUMINS.

In the scheme of classification given some pages back the true or native albumins have the first place. The best known representatives of the protein group are included here.

ALBUMINS PROPER.

These bodies are characterized by solubility in water and in weak cold acid or alkali solutions. They are readily coagulated by heat and by shaking with strong alcohol. Although usually considered as amorphous, the albumins have been obtained in well crystallized form.

The characteristic color reactions previously referred to are all given by the true albumins and they are precipitated by ammonium sulphate or zinc sulphate added to saturation. With strong sodium chloride precipitation follows only after addition of acid.

SERUM ALBUMIN. This is the important protein body of blood serum, of which it constitutes three to four per cent by weight, the related substance, *serum globulin,* making up nearly as much. While closely resembling each other, it is not definitely known that the serum albumins of different animals are identical. In fact, certain reactions to be referred to later suggest peculiar points of difference. In blood, the albumins are associated with globulins, fibrinogen, mucoids, salts and other bodies, the *perfect* separation of which is practically impossible. The purification of serum albumin by crystallization is not easily carried out with all blood serums; in some cases the formation of crystals is slow and incomplete.

Serum albumin contains a relatively large amount of sulphur, about two per cent in the mean, and is characterized further by a high specific rotation. The values which have been given for this are not constant, but in the mean are about $[\alpha]_D = -60°$.

Crude serum albumin is now an article of commerce, being made in large quantities from blood collected at the slaughtering houses. It is usually mixed with globulin, and besides is partly insoluble because of the high temperature employed in drying it. For the following experiments fresh blood must be used.

Experiment. Collect blood in a clean vessel and stir it thoroughly to separate the fibrin and part of the corpuscles, as a clot. Some of the corpuscles, however, remain with the serum and may be separated by allowing the latter to stand in a tall, narrow jar, or better, by rotating the serum in a centrifugal machine. Most of the corpuscles may be deposited in this way, leaving a *yellowish* liquid. A pure white serum can not be obtained because a little of the hemoglobin dissolves from the corpuscles and remains in solution. With this prepared serum make the following tests:

Experiment. To a little of the serum add finely powdered magnesium sulphate to saturation; this produces a precipitation of serum globulin, which separates on standing. Pour off the clear liquid and add to it powdered ammonium sulphate, which gives now a precipitate of albumin.

Experiment. Mix a little of the serum with two or three volumes of water in a test-tube, and test the temperature of coagulation. It will be found near 70° C.

LACTALBUMIN. Milk contains two protein substances, the most important of which is casein. The other is a true albumin which is present to the extent of about one-half per cent in cow's milk. It resembles serum albumin very closely but appears to have a much lower specific rotation, $[\alpha]_D = -38°$.

EGG ALBUMIN. White of egg contains this body as its characteristic constituent along with some globulin and mucoid, and traces of salts. Common albumin reactions are usually made with white of egg solution. Although this substance is always described as a true albumin, some of its reactions seem to suggest that it may belong to the group of glucoproteids, or, at any rate, may contain such a compound in relatively large amount. On heating egg albumin with weak acid glucosamine is split off and in quantity sufficient to indicate a rather large sugar content in the original substance. The specific rotation is much lower than that of serum albumin and may be taken at $[\alpha]_D = -38°$, as for milk albumin. Besides this difference, egg albumin has a much lower coagulating temperature than has been given for serum albumin, viz., 56°. Egg albumin is much more easily coagulated by ether than is serum albumin. Egg albumin becomes very quickly insoluble when mixed with strong alcohol. From serum albumin it differs, further, by this interesting property. When its solution is injected into the blood circulation it passes unchanged through the kidneys into the urine; the same thing happens when large quantities of white of egg are eaten. It seems to escape digestion in this latter case and be absorbed in pure condition, to be later discarded by the kidneys. These various points of behavior indicate, then, a rather marked difference between the two kinds of albumin.

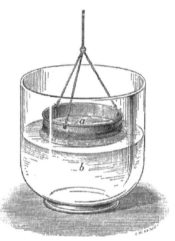

FIG. 8. Typical form of Graham dialyzer frequently used in purification of proteins. The substance to be purified is placed in the cell a, which has a parchment bottom and floats on water in the large vessel b. The simple parchment tube dialyzers now obtainable are more efficient.

Experiment. So-called pure egg albumin may be obtained in this way: The white of egg is shaken in a bottle with some broken glass to thoroughly break up the membranes. The foamy mass is filtered through fine, unsized muslin, and to the filtrate an equal volume of saturated ammonium sulphate solution is added. This produces a precipitate of globulin which after 24 hours is filtered off. Ammonium sulphate in this strength does not precipitate the true albumin. To this filtrate a little more saturated ammonium sulphate is added and until a precipitate or turbidity just begins to show. This is caused to disappear by the cautious addition of water, a few drops at a time. Finally, acetic acid of ten per cent strength saturated with ammonium sulphate is added until a turbidity again appears, and then the mixture is allowed to stand 24 hours in a cool place. A part of the albumin separates in the crystalline form. This is collected, redissolved in a very little cold water and reprecipitated with ammonium sulphate and acetic acid as before. The crystals are collected on a filter, then transferred to a dialyzer with water for the separation of the sulphate by dialysis. In this way a nearly pure albumin may be obtained in solution, but the crystallized substance has not been secured free from salts.

White of egg contains in the mean about 86 per cent of water, 13 per cent of proteins, 0.6 per cent of mineral matters and a little fat. The yellow of egg is a substance of very different composition. The water present amounts to about 50 per cent, the proteins to 16 per cent, the fat to 30 per cent, or more, while the ash is about 1 per cent. The fat contains a notable quantity of lecithin.

GLOBULINS.

The proteins of this group differ from the albumins mainly with respect to solubility in water. In pure water they are practically insoluble, but they dissolve in moderately dilute salt solutions. On diluting a globulin solution of this kind precipitation follows. Globulin solutions coagulate by heat in much the same manner as observed with albumins, but in general they become permanently insoluble even more readily than do the albumins.

The preparation of pure globulin is even more difficult than the preparation of pure albumin. The globulin must first be separated by precipitation with some salt; as the salt is later removed by dialysis the globulin remaining becomes insoluble, which makes further treatment difficult. Globulins are not well known in crystalline condition.

SERUM GLOBULIN. This substance makes up a large fraction of the protein in blood serum, amounting to nearly as much as the serum albumin. For a long time it was confounded with the latter, and it was only after a lengthy series of investigations by different chemists that its true nature was recognized. This globulin may be discovered easily in the serum when the latter body is diluted with water, but the separation is never quite complete by the water treatment alone, as a portion always remains in solution. By salting out with ammonium sulphate to half saturation, or with magnesium sulphate completely, the desired end is reached.

The coagulation temperature of serum globulin is given as 75° and the specific rotation as $[\alpha]_D = -48°$, but these numbers are somewhat uncertain, especially the latter.

BENCE-JONES PROTEID. This is the name given to a substance occasionally found in pathological urines, and which has usually been considered an albumose. When purified, however, it has been found to have the properties of a globulin. It may be held in solution in the urine by the salts present.

Other Globulins. Several other bodies are described as globulins. The most important of these is the so-called *cell* globulin, which is possibly identical with serum globulin. This substance has been obtained from different organs, from the liver, from the pancreas, from muscle plasma, etc. Some of the globulins described as cell globulins have a lower coagulating temperature than the true serum globulin.

In the crystalline lens of the eye a body has been long known which is called *crystallin*. In coagulation temperature and specific rotation this crystallin appears distinct from serum globulin, and further, it seems to be made up of two related substances, α and β crystallins.

Globulins have been described in milk and egg and also in the vegetable kingdom under the name of *phytoglobulins* or *phytovitellins*. This last designation indicates that they may be classed under the head of the nucleo-albumins, with which bodies they have much in common. Among the best known of these bodies we have the abundant protein called edestin.

EDESTIN. According to Osborne this is a true globulin, and of the vegetable products of this class has been among the most thoroughly studied. It has been obtained from many seeds and nuts, but most readily from hemp seed. On analysis it shows, in the mean, about 18.7 per cent of nitrogen, and 0.9 per cent of sulphur. Its specific rotation is about $-44°$. Edestin can be secured in the crystalline condition, which has facilitated greatly its study. When hemp seed meal is extracted with sodium chloride solution and this is followed by dialysis or sharp cooling a portion of the edestin separates in the crystalline form. The name *edestan* is given to a slightly hydrolyzed form of the original substance. The ending *an* is employed in describing primary protein derivatives, formed by the action of water or weak acids. These *proteans* are insoluble in salt solution, as well as in water.

COAGULATING PROTEINS.

Several extremely important substances belong in this group, which, like fibrinogen, have the property of spontaneous coagulation. In nature they exist normally in the soluble and dissolved form, from which, under certain influences, not always well understood, they pass to the solidified condition. This coagulation is a different thing from that produced by heating to a high temperature or by the addition of reagents; the changes in the latter case seem to be more profound. We use in English the term coagulation to describe both classes of alterations, which are really of a very different character, as will appear from what follows.

FIBRINOGEN. Blood contains a peculiar protein body in small amount, to which it owes its property of spontaneous coagulation. This body is called *fibrinogen* and the product of coagulation is known as *fibrin*. The nature of and important factors in this change have been long subjects of investigation and discussion; it can not be said that the matter has been fully explained in all its bearings. The essential points of what is known will be given in the chapter on the blood.

As a chemical substance fibrinogen is not known in perfectly pure condition, since to hold it in soluble form various agents must be added to the blood. But the *fibrin* formed, doubtless through ferment action, is easily obtained and its properties are well established. As usually prepared it is a white, elastic, stringy mass, insoluble in water, but somewhat soluble in salt solutions. Like other proteins it undergoes true coagulation through elevation of temperature or action of various reagents. Fibrinogen, as prepared by salting out from plasma at a low temperature, coagulates when warmed to 56°. Its specific rotation has been found only in presence of salt or alkali and varies from $[\alpha]_D = -36°$ to $-53°$ according to the nature of the admixture or method of preparation. It undergoes digestion with the body ferments very readily and has therefore often been used as a starting point in digestion experiments.

MYOSIN AND MYOGEN. The living muscle plasma contains a number of protein substances, one of which, at least, possesses the property of spontaneous coagulation as observed in the solidification of the muscle after death. At one time the term *myosin* was applied to this body and it was supposed to be very simple in nature. Numerous investigations, however, have shown that the chemistry of the muscle proteins is comparatively complex and that the results of experiments do not well agree. In the older sense this myosin was assumed to be derived from a preëxisting body, *myosinogen*, in the living muscle, much as fibrin is considered as derived from fibrinogen. The solidified myosin behaves as a globulin, which may be illustrated by the following experiment:

Experiment. Free muscle (round steak) as far as possible from traces of fat and sinews, and then thoroughly disintegrate it by passing through a sausage mill. Then wash it repeatedly with cold water until the latter is no longer reddened, and the residue appears white. This is placed in a ten per cent solution of ammonium chloride and allowed to remain about a day, with occasional shaking. Myosin dissolves in the ammonium chloride and is found in the filtrate when the mixture is filtered. Pour the filtrate into twenty times its volume of distilled water, which causes a precipitation of the insoluble myosin. Allow to settle and wash three times by decantation. Collect the precipitate and observe that portions of it dissolve readily in ten per cent solutions of sodium chloride and ammonium chloride, or in a 0.1 per cent solution of hydrochloric acid. The solution in salt is precipitated by the addition of more to saturation.

By this treatment with the dilute ammonium chloride solution nearly all of the protein of the muscle plasma may be removed, leaving the stroma. It is now pretty generally recognized that this solution contains two substances instead of one. The first of these is still called myosin, and is said to make up about 20 per cent of the plasma protein, while the name myogen is given to the other, constituting 80 per cent of the soluble protein. Myosin is the part of the plasma which coagulates or solidifies the most readily and may be separated from the plasma by adding ammonium sulphate to make 28 per cent of the solution. On filtering, the myogen may be separated by adding ammonium sulphate nearly to complete saturation. The coagulation temperature of myosin is given as $47°$, while that of myogen is $56°$. The former becomes quickly insoluble on addition of alcohol, while myogen seems to be partly soluble in alcohol. Myosin-fibrin and myogen-fibrin are the names given to the coagulated forms of these bodies. More will be said of these relations when we come to consider the muscular substance as a whole.

NUCLEO-ALBUMINS.

This group contains bodies which in the pure state are rather markedly acid in character. They are called nucleo-albumins because of the earlier fancied resemblance to the nucleo-proteids. The characteristics of the latter group, such as the presence of nucleic acid and the xanthine bases among the decomposition products, are wholly wanting in the nucleo-albumins. Both groups contain phosphorus, and in both cases the phosphorus is separated in complex combinations on digestion with pepsin and hydrochloric acid; the character of the phosphorus compound separated is very different in the one case, however, from what it is in the other.

The free acids are but slightly soluble in water, but in the salt form they are very soluble and these solutions do not coagulate on boiling, as shown by the behavior of casein in milk. The addition of weak acids to these salt solutions forms precipitates of the free nucleo-albumin acids. From very weak solutions the precipitate may not separate until after heating. A large number of bodies have been described as nucleo-albumins, but only those will be mentioned here which are well known. In the newer classification referred to above all these compounds are described simply as *phospho-proteins*. No assumption is made regarding the exact form in which the phosphorus is held, but the combination may be in a general way that of an ester of phosphoric acid.

CASEIN. Of all the nucleo-albumins this is the best known and most important. It occurs in milk as a neutral calcium salt, and in the case of cow's milk makes up nearly 4 per cent by weight. It may be readily separated from milk by the addition of a little acetic acid. In precipitating, the fat is usually carried down too, but may be removed after drying by treatment with ether or petroleum spirit. Rennin, a peculiar enzyme of the stomach, to be described later, causes a kind of coagulation in casein solutions; if lime salts are present, which is practically the case in milk, the coagulation extends to the formation of a curd or cheesy mass which is very characteristic. The first product formed by the rennin is known as *paracasein* and the curd, or cheese, is the calcium combination of this.

Casein was formerly considered as an alkali albuminate because of its behavior with acids and alkali solutions. Many of its alkali combinations are now produced in a technical way as by-products in the butter and cream industries. Plasmon and nutrose are apparently sodium-casein compounds. These are used as foods, but some of the others find application in other directions. Casein forms two series of salts with calcium hydroxide and other bases and the amount of metal in several of these has been found with considerable accuracy. Most of these salts form opalescent rather than perfectly clear solutions. The addition of sodium chloride or magnesium sulphate to these solutions in sufficient amount completely precipitates the casein. Like the other nucleo-albumins, casein leaves a pseudo-nuclein residue on digestion with pepsin and hydrochloric acid.

In combining casein with alkali 1 gram of the former may be dissolved in 4.5 cc. of N/10 sodium hydroxide or equivalent solution. But this is still acid toward phenol-phthalein. To obtain a solution neutral with phenol-phthalein just twice as much alkali must be used. The second reaction corresponds to an equivalent weight of 1111 for the casein. Casein shows also a basic behavior and unites readily with many acids. 1 gram combines with 7 cc. almost exactly, of N/10 hydrochloric or equivalent strong acid, the reactions being completed without the aid of heat. These reactions illustrate very beautifully the chemical behavior of complex groups of amino acids. Something will be said later about the method of preparing pure casein used in such tests.

VITELLIN. While white of egg contains essentially albumin proper and globulin, the yellow part is extremely complex, containing many substances. At least two of these compounds hold phosphorus in combination; one of these is lecithin, referred to earlier, and the other is the nucleo-albumin called *vitellin*. The separation of these substances from each other is extremely difficult. Vitellin is not soluble in water, but dissolves in weak alkali solutions; on digestion with pepsin and hydrochloric acid it yields a pseudo-nuclein residue which contains iron as well as phosphorus. The name *hematogen* has been given to this, and it is considered as of great physiological importance because of its iron content. It is possibly one of the parent substances of hemoglobin.

Other Nucleo-albumins. In the eggs of fishes there is found a peculiar vitellin called *ichthulin*, which has been obtained in crystalline form. It is not soluble in water, but yields a clear solution with weak alkalies.

In *cell protoplasm* several different nucleo-albumins are found. These bodies contain iron, are insoluble in water in pure condition, but with alkalies form salts which are readily soluble.

Vegetable Proteins. Most of the protein bodies thus far referred to have belonged to the animal kingdom, but as plant constituents fully as great a number occur. The exact nature of some of these is obscure, but many valuable observations have been made by Osborne and other chemists in the last few years which have cleared up some of the points in dispute. Only brief mention can be made here.

In wheat flour, for example, four or five protein bodies appear to be present. The most abundant of these is called by Osborne *glutenin* and makes up over 4 per cent of the weight of the grain. Next in abundance is another important compound known as *gliadin*, amounting to about 4 per cent of the grain weight. These two proteins unite in the formation of *gluten* which is essential in the production of an elastic dough, which on leavening yields a porous and light bread. Gliadin is soluble in dilute alcohol and forms an opalescent solution with water. In some respects it resembles a globulin. In its behavior with weak alkalies glutenin bears some resemblance to casein. Wheat flour contains also a true globulin in small amount.

A peculiar protein body known as zein, or maize fibrin, is found in corn meal. It is soluble in alcohol but not in water, and is not soluble in dilute alkali solutions. Corn contains also three globulin-like bodies and one or more substances to be classed with the albumins proper.

Legumin is found in peas, beans and related seeds; it was formerly placed in the group of nucleo-albumins, but in its solubility conditions resembles the typical globulins and is now so included. The legumin obtained from vetches does not coagulate on boiling. On boiling a solution of pea legumin a jelly-like substance is formed.

Recently Osborne has proposed the name *prolamins* for the seed proteins soluble in alcohol. As the best representatives of this class we have the gliadin of wheat and the zein of corn, just mentioned, and the hordein from barley. These proteins, which are soluble in all proportions in alcohol of 70 to 80 per cent, are found in the seeds of all cereals, apparently, and constitute a relatively large proportion of their reserve material. They do not appear to occur in other parts of the plant. On decomposition these prolamins yield relatively large amounts of glutaminic acid.

The *glutelins*, according to the same author, make up a large part of the protein matter of cereals. They are said to be insoluble in all neutral solvents, but dissolve in weak acid or alkalies. The glutenin, mentioned above from wheat flour, is the best known member of the group, because of its ready accessibility and ease of preparation. It is difficult to separate the glutelins from other seeds in a form pure enough for study, because they yield no coherent *gluten*, to begin with.

Seeds contain, also, compounds which appear to be true nucleo-proteins, that is combinations of nucleic acid with a protein group. But the separation and identification of these bodies has not been, thus far, satisfactorily carried out.

THE HISTONES.

These are relatively simple proteins which, apparently, always occur in combination with certain groups to form the nucleo proteids, or conjugated proteins. They behave as rather strong bases and yield basic groups on cleavage. In consequence of their basic character they are precipitated from solution by addition of alkalies, especially by ammonia. In presence of salts they are coagulated by boiling, and are also precipitated in cold solution by nitric acid; this precipitate disappears on warming, to return on cooling. They yield precipitates with the alkaloid reagents in neutral as well as in acid solutions. The nitrogen content of the histones is relatively high and the sulphur content low. They contain no phosphorus. Histones are obtained from several sources, and the best known are the following:

GLOBIN. This makes up about 96 per cent of the hemoglobin of the red blood corpuscle, existing in combination with the iron-containing constituent, hematin. It is precipitated by a relatively small amount of ammonia, and redissolved by a slight excess. On cleavage it yields much histidine and leucine. Of all the histones this is the one most readily obtained for experiment.

SALMO-HISTONE, SCOMBER-HISTONE, AND GADUS-HISTONE. These bodies are obtained from the immature testicles of the salmon, the mackerel and the codfish, and were first classed as albumoses. But their precipitation reactions throw them into the group of histones. Similar products have been obtained from the testicles of other animals.

NUCLEO-HISTONE. This name was given to a product separated from the thymus glands of the calf and was one of the first studied. On cleavage it yields much arginine and tyrosine, and is characterized by easy digestibility.

As strongly basic bodies the histones show the interesting property of forming precipitates with many of the other simple albumins, especially with casein, egg albumin and serum albumin. Their precipitates contain the component proteins in definite proportions.

PROTAMINES.

We come here to the simplest of all the naturally occurring proteins. They do not exist free in nature but, like the histones, in combination with nucleic acids, hematin or other simple "prosthetic group." The protamines contain no sulphur but are very rich in nitrogen and low in carbon as compared with the ordinary proteins. They are not coagulated by heat and do not give the Millon's reagent reaction or that of Adamkiewicz. The biuret reaction is marked and the alkaloid reagents produce precipitates. Some of the groups in the common proteins are therefore wanting in the protamines. Several of these bodies have been isolated, particularly from the nucleo-proteids of fish spermatozoa and the names given to them suggest their origin. Thus, we have *salmin, sturin, scombrin* and *clupein*. In recent analyses the following formulas have been found for the more important protamines:

Salmin .... $C_{30}H_{57}N_{17}O_6$
Clupein .... $C_{30}H_{62}H_{14}O_2$
Scombrin .... $C_{32}H_{72}N_{16}O_6$
Sturin .... $C_{24}H_{72}N_{17}O_6$

When warmed with weak acid, or when subjected to pancreatic digestion, they yield at first *protones*, corresponding to the peptones of ordinary digestion and finally simpler products, among which the hexone bases, arginine, lysine and histidine predominate. From salmin, for example, over 80 per cent of arginine has been obtained.

In some cases of decomposition the cleavage into the hexone bases has been nearly quantitative, which is an important step toward establishing the empirical formula of the parent protamine. The protamines appear to have rather marked toxic properties.

The histones are more complex bodies than the protamines, and possibly contain the latter as a component part. It is also possible that the histones represent a stage in the development of the protamines, since while the former are found in immature spermatozoa, the latter are commonly obtained from the mature organisms. In basic properties the protamines are more marked than are the histones, and are precipitated easily by alkalies. They do not seem to be altered by peptic digestion, but by trypsin and erepsin they may be reduced to crystalline products. The *protones,* referred to above, are stages in this cleavage.

From a purely scientific standpoint these bodies possess great interest and importance, since they represent, apparently, the beginnings in the formation of protein molecules. On cleavage they yield groups of amino acids which are quantitatively more readily measured than are the products from the more complex proteins.

TRANSFORMATION PRODUCTS.

The protein bodies which have been described in the foregoing pages are natural unmodified substances or primary products. We have now to consider briefly a class of important protein compounds which includes secondary or modified substances which in the main are derived from the native albumins just discussed. These modified forms may be obtained in various ways, but for convenience three groups of transformation products may be made, as shown below.

COAGULATED OR MODIFIED ALBUMINS.

It has been shown already that white of egg dissolves easily in water. The solution so made undergoes a change when heated or when treated with certain reagents. This change is called coagulation and the resultant product is so essentially altered that it may no longer be brought into the original form, or a similar form, by any known means. Some of the conditions of coagulation have been explained above and illustrated by experiments. While the simple or native egg albumin is soluble in water the modified product is insoluble. It is, however, soluble in weak acids or alkalies, but is insoluble in solutions of neutral salts. It follows, therefore, that while coagulation or modification of a native albumin always follows on heating, *precipi-*

tation may not result. This depends on the reaction of the mixture. Coagulation or modification on the one hand and precipitation on the other are perfectly distinct phenomena. In the case of egg albumin in solution, for example, a precipitate forms on heating as long as the solution is nearly neutral. In presence of salts the precipitation is more complete. But if the original solution is alkaline modification of the albumin takes place but without precipitation, as soluble alkali albuminate is now formed. In presence of acid in proper amount soluble acid albumin is formed. Although often used synonymously the terms coagulation and precipitation have here distinct meanings.

The exact nature of the change which takes place when native albumins are heated is not known. Hence the terms used in describing the phenomenon are somewhat indefinite. They are "modified," or, to freely render a German expression, "denatured." To bring them again into the original condition is not possible. White of egg may sometimes be modified or altered without becoming opaque, and the same is true of clear blood serum. In both cases we have coagulation without precipitation.

Some of these changes in condition of the protein are termed *reversible*, and others *irreversible*. Many of the precipitation reactions are reversible; that is, the protein may afterwards be returned to its former condition. But the change produced in a protein by coagulation, for example, is irreversible.

ACID AND ALKALI ALBUMINS.

These products represent the most important forms of the coagulated modified albumin, and may be looked upon as forming salts of the albumin nucleus acting as an acid or basic ion. They are most readily secured by the action of acid or alkali in excess on some native albumin, usually white of egg. These actions of alkali or acid are but the beginning of the profound changes in which the protein molecule finally breaks down into small groups. They may not be looked upon, therefore, as absolutely sharp and definitely limited conditions, which may always be exactly duplicated.

Alkali Albuminates. Strong alkali solutions act very energetically on white of egg and the reaction is always accompanied by some decomposition of the latter. There is a loss of nitrogen in the form of ammonia, and of sulphur as hydrogen sulphide. The reaction with lead solution, production of lead sulphide, disappears after the alkali treatment. The most characteristic product of alkali action on native albumin is a thick jelly-like mass and is known as "Lieberkuehn's jelly." It may be obtained as follows:

Experiment. Add strong sodium hydroxide solution to white of egg, with constant stirring, until a thick jelly is formed. Too much alkali must not be added here, but just enough to make the maximum of jelly. This is next cut into small pieces and washed in distilled water several times until the lumps are white throughout. They are then heated with fresh pure water, but very gently, until they go into solution. This is then filtered and the filtrate precipitated by acetic acid, *avoiding any excess*. The precipitate is washed with pure water, and used for experiments below.

This precipitate is the modified alkali albumin or alkali protein proper. It is likewise insoluble in salt solutions. In the treatment with the alkali a salt of the modified protein is formed, and this is called an alkali albuminate. The salt is readily soluble, while the alkali-protein itself is not.

Experiment. Use some of the alkali albumin of the last experiment to test other properties. Dissolve a portion in weak hydrochloric or sulphuric acid and observe that the solution does not coagulate on boiling. An acid solution is precipitated by addition of sodium chloride to saturation, and it is also precipitated by adding weak alkali to the point of neutrality. When this neutral point is reached more alkali brings about solution again.

The formation of Lieberkuehn's jelly illustrates the production of the alkali albumin at once in the cold. A similar result is obtained by heating some white of egg solution for a time with very weak alkali. A clear solution is finally obtained.

Experiment. Dilute white of egg with water and add a small amount of N/10 alkali solution. A few cubic centimeters will suffice. Keep the mixture at a temperature of about $40°$ to $45°$ on the water-bath through an hour, and then test some of it by boiling in a test-tube. It should not coagulate. To a portion of the clear solution add a few drops of phenol-phthalein indicator and then run in dilute sulphuric acid to neutralization. A precipitate forms as shown above.

Acid Albumin. According to the view held at one time the solution of the alkali albuminate in water yields an acid albumin on acid treatment. But the weight of evidence now indicates that the group in the albuminate having an acid function is different from the group in the so-called acid albumin which certainly plays the part of a basic radical. Although the albuminate and the acid albumin have certain points in common, as will be shown, they are not identical. It appears, however, that while the albuminate may not be converted into acid albumin by action of weak acid, the opposite conversion is possible; that is, weak alkali will change acid albumin into albuminate. Some simple experiments may be made here:

Experiment. Dilute white of egg with four volumes of water, take 25 cc. of the mixture, add 5 cc. of 0.2 per cent hydrochloric acid and warm it on the water-bath for about two hours to a temperature of $45°$ C. Then carefully neutralize the solution with dilute sodium hydroxide, using phenol-phthalein as indicator.

This precipitates insoluble acid albumin, which can be washed with water by decantation. It is essential that just the right amount of alkali be added here; an excess would redissolve the precipitated acid albumin with formation of alkali albuminate. The washed acid albumin can be used for a number of tests.

Experiment. Dissolve a little of the washed acid albumin in water by the aid of weak hydrochloric acid, and note that the solution does not coagulate on boiling. Observe, however, that the addition of common salt to the acid solution brings about precipitation. The same thing was found to be true with the solution of alkali albuminate in acid.

In forming acid albumin from a native albumin the action of the weak hydrochloric acid employed is much less destructive than is the action of the alkali in producing albuminate. The actual modification of the protein molecule is much less profound. Nothing is split off as is the ammonia or hydrogen sulphide in the other case, and this may account for the observed fact that the acid albumin may be changed into albuminate by use of weak alkali. It must of course be remembered that a stronger acid may not be used in making the acid albumin, since here too the reaction may become destructive.

SYNTONIN. This appears to be an acid albumin, resulting from the action of dilute acids on muscle, and is very readily formed in presence of the ferment pepsin. The name is often applied to all acid albumins, but it is perhaps preferable to restrict its use to describe the product from muscle.

Experiment. Free the muscle part of meat from fat as for as possible and run it through a sausage mill several times to bring it to a fine state of subdivision. Wash this chopped mass with distilled water until the washings remain clear. Now, to about 5 gm. of the moist residue in a small flask, add 50 cc. of dilute hydrochloric acid, containing 0.1 per cent of the true acid. Warm the mixture slightly (to 35° or 40° C.), and keep at this temperature about three hours. Then filter and test the filtrate. It contains the soluble syntonin, held by the excess of weak acid used.

Experiment. To a small portion of the filtrate add weak caustic soda, which produces a precipitate soluble in excess of the alkali. This latter solution contains albuminate.

Boil another portion of the filtrate. It does not coagulate directly, but after the addition of common salt precipitation follows.

It must be remembered that the action of both acids and alkalies on the native albumins may easily extend beyond the formation of the simple products here mentioned. These are merely limiting cases. It has been already shown that by more prolonged action various products of disintegration are obtained and the substances just described represent the first stages. With slightly stronger acids or alkalies or by elevation of temperature the more easily separated of the amino complexes begin to split off. The condition of stability is only relative. With molecules as large as these it may even be possible to separate some of the outlying groups without greatly impairing the integrity of the whole.

It will be recalled that in the second classification of the protein substances, given at the outset, a group of so-called *metaproteins* was mentioned. This group includes the alkali albumins and the acid albumins, but not the salts. That is, it does not include the so-called albuminates or the opposite class of bodies, which consists of combinations of acid proteins with acids. This distinction should be kept in mind. It will be recalled, further, that the less highly modified protein, formed by the action of water alone, is called in this classification a *protean*. The proteans are, like the acid and alkali albumins, insoluble in water.

ALBUMOSES AND PEPTONES.

By the simple treatment with weak acids or alkalies alone, the changes in the native protein bodies are of the character described in the last paragraph. But in presence of certain enzymes further modifications are reached and these have received the names of *albumoses* and *peptones* when they are produced by the ferments of the digestive tract. It is indeed true that these substances may be produced in fairly large amount by the simple chemical treatment or by heating the protein substances with water under pressure. But the names, in practice, are usually restricted to the products of enzymic formation.

Of the exact nature of the reactions by which these substances are reached little is known. They represent the very last stages in the process of breaking down complex native protein bodies which still give the characteristic protein tests. Further disintegration leads to bodies which are no longer proteins, but which, as amino acids, are simply constituent groups of the complex protein molecule. The peptone substances represent a more advanced stage of modification than do the albumoses. In both groups of bodies we find the reactions with the alkaloid reagents and with the precipitating metallic solutions in most cases still marked; the biuret reaction is also still present. But for the peptones we find lacking the property on which the salting out processes depend. By adding plenty of ammonium sulphate or zinc sulphate it is possible to throw the albumoses out of solution; the peptones do not respond to this treatment and in other points also they are further removed from the original proteins than are the albumoses.

But it must not be understood that the distinction between the two groups is perfectly simple and clear. Unfortunately much confusion still prevails in the literature of the subject and an elementary presentation which is satisfactory and consistent is not yet possible. In this chapter only a brief outline of the relations now generally accepted among chemists and physiologists will be attempted, while in a following chapter on digestion some of the more practical details will receive consideration.

Basis of Classification. The general classification of these substances commonly recognized is that of Kühne, which was elaborated mainly in conjunction with Chittenden. The scheme has been enlarged and modified somewhat by other workers but in its important features the ideas of Kühne still hold the first place. In the weak acid as well as in the enzymic treatment it is easily seen that the common proteins are not homogeneous or symmetrical bodies. On the contrary they seem to contain two great groups which respond very differently to the action of the digestive agent, whether acid or ferment. A part of the original complex appears to break down rather quickly and go into a soluble form; while a second portion resists this breaking down process pretty effectually as far as weak acid and pepsin fermentation is concerned, at any rate, and in subsequent treatment with the more active pancreatic ferment it yields products different from those derived from the first group. To the first or less resistant fraction, Kühne gave the name *hemi group,* and to the second or more resistant portion, the name *anti group.* It was later noticed that most protein bodies seem to contain a third group which in the subsequent breaking down yields a sugar of some kind. Hence a further or *carbohydrate group* may be assumed to exist in the native protein molecules, or in most of them, at least. But the latest researches seem to show that the amount of this complex present is, in most cases, not large.

Albumoses. In the first stage of the action of the acid and ferment on the protein body a kind of acid albumin appears which passes by continued digestion into the next or *albumose* stage. Different albumoses seem to be derived from the several native proteins, and these may be called, in general, *proteoses.* Names have also been given to them corresponding to their origin. We have, accordingly, *fibrinoses, caseoses, myosinoses, globulinoses,* and so on. Several degrees of albumose digestion are recognized; that is, bodies are produced which behave differently on treatment of the digesting mixture with precipitating reagents, and we have, therefore, *primary* and *secondary* albumoses. The secondary albumose stage represents a more advanced condition of change on digestion than does the primary albumose. Finally, the secondary albumose, by prolonged contact with the digestive agents, passes into the *peptone* stage. Some idea of the existence of these three stages of change may be obtained from the following experiment in which commercial peptone is taken for illustration. This is a substance made by the partial digestion of fibrin, gelatin, serum and other bodies and is not uniform or homogeneous in structure. It contains representatives of the several classes of derived digestion products.

Experiment. Dissolve about 5 gm. of commercial peptone in 50 c.c. of water and use small portions of the solution for these tests: To one portion add some strong nitric acid; this produces a precipitate. To a second portion add a little copper sulphate solution, which gives a light greenish precipitate. To a third portion add a few drops of acetic acid and then some potassium ferrocyanide. This makes a turbidity or may even cause a precipitate. Now to the remaining and large portion of the original solution add an equal volume of a saturated solution of ammonium sulphate. A marked precipitate of *primary albumose* separates and may be filtered off after a time. When the liquid has all passed through the filter note that the precipitate may be easily dissolved by adding fresh water, and further that this new solution is not coagulated by boiling. Note also that the solution gives a good biuret reaction.

Experiment. Use the filtrate from the primary albumose precipitate for a further test. Add to it powdered ammonium sulphate to complete saturation, that is, as long as the powder dissolves on thorough shaking. Then add five to ten drops of a weakly acid solution of ammonium sulphate (which may be obtained by adding to 10 cc. of *saturated* ammonium sulphate solution five drops of concentrated sulphuric acid). This last treatment with the acid ammonium sulphate gives a new albumose precipitate which after a time may be separated by filtration. Save the filtrate and test the precipitate as follows: Dissolve it in fresh water and test portions with copper sulphate, nitric acid and the potassium ferrocyanide. These reagents gave precipitates with the original peptone solution, but yield nothing with the solution of the new albumose, which is called *secondary* albumose.

Experiment. The filtrate from the secondary albumose may finally be tested. Add to it an excess of concentrated sodium hydroxide solution and then a drop of dilute copper sulphate solution. This gives a purple red biuret color, showing the presence of a soluble product not precipitated by ammonium sulphate in excess. This soluble product is the *peptone*, representing the last stage of the true digestion. This peptone gives no precipitation reactions with the reagent used above.

The first of these fractions, or the *primary* albumoses, may be converted by further acid treatment or by digestion into secondary albumoses no longer precipitated by half saturated ammonium sulphate. By solution in water and addition of alcohol it is possible to separate this primary albumose into two sub-fractions which are pretty well characterized. The first of these is known as *heteroalbumose* and is insoluble in weak alcohol, while the second, or alcohol-soluble portion, is called *protalbumose*. The heteroalbumose belongs to the above mentioned anti group and is further changed only with difficulty. The protalbumose belongs to the hemi group. It is quite soluble in water, and in dilute alcohol even more soluble. By prolonged peptic digestion the protalbumose passes into the secondary albumose known as *deuteralbumose A,* and then into *peptone B,* so called.

An enormous amount of labor has been devoted to the study of the various fractions obtained by digestion under different conditions, and a complex nomenclature describing the products has grown up. But the value of much of this is now doubted, as there is no great constancy in the results secured by different observers. This much is true, however, that in the earliest stages of digestion certain amino complexes are very readily split off, while others are not. Tyrosine and tryptophane, for example, separate relatively quickly, and would be considered as belonging to the hemi group. On the other hand glycine, phenylalanine and proline separate slowly and should be referred to the anti group. But from the present point of view the assumption of these two groups is arbitrary and without real justification. The classification based on it need not be further developed in this book, which must be kept within elementary bounds.

Peptones. The amount of real peptone formed by the pepsin digestion is always small; the large amount of peptone produced in the body is a consequence of the action of the pancreas enzyme known as *trypsin.* The peptone of gastric digestion was assumed to be a mixture of products from the hemi and anti groups and was called *amphopeptone.* The term *antipeptone* is generally applied to the final product of the energetic pancreatic digestion. Amphopeptone has been obtained as a yellow powder, very soluble in water and very hygroscopic. It diffuses pretty well through parchment and has a sharp bitter taste. It is not possible to *salt out* the peptone from solution, but the alkaloid reagents give precipitates, which are soluble in excess. Precipitates are formed by solutions of several of the heavy metallic salts also, but not by copper salts.

The two forms of amphopeptone which have been described are known as *amphopeptone A* and *amphopeptone B.* The first is insoluble in 96 per cent alcohol and is further characterized by giving a strong reaction with the Molisch reagent which relates it to the carbohydrate group. The second is soluble in 96 per cent alcohol and does not give the Molisch reaction. Both forms give a strong biuret reaction.

As to the exact nature of the antipeptone referred to above, there is still much uncertainty. This was assumed by Kühne to represent the final product of pancreatic digestion, and it was supposed that even prolonged digestion would not change it further. It was found later, however, that various amino acids appear here in considerable quantity, and that the digestion may be carried so far as to yield a product which no longer gives the characteristic biuret reaction; that is, a product from which everything of a really protein nature has disappeared. This matter will be more fully discussed in a following chapter. The reactions described as characteristic of antipeptone are similar to those for the amphopeptone in the main. A good biuret reaction is obtained if the digestion is not too prolonged, and the alkaloid reagents give precipitates which are soluble in excess. Some of the metallic salts precipitate, but copper sulphate not.

The name *kyrine* has been given to certain kinds of peptones which in a marked degree resist the action of hydrolysis through pepsin and trypsin, and which are basic in character. In some cases the component groups in these kyrines have been determined. For example, a kyrine from casein is apparently made up of 1 molecule of arginine, 1 molecule of glutaminic acid and 2 molecules of lysine.

In the formation of the albumoses and peptones from native protein molecules a large amount of water is added; roughly the action may be compared to the hydration of starch, producing malt sugar and finally glucose. As the original molecule is very large the percentage amount of water taken up in the hydration is much less than is the case in the carbohydrate conversion. It is also very interesting to note that with the progress of the hydration the amount of hydrochloric acid which may be held by the product increases; the smaller molecules in the aggregate resulting from the hydration have a much greater capacity for combining with free hydrochloric acid than the parent substances have. This question assumes considerable practical importance in connection with the subject of gastric digestion and acidity of the stomach, as will be shown later.

It must be remembered that the various commercial products sold as "peptones" may contain many other substances, and may be quite unfit for use as a food or in medicine. While in some cases a considerable portion of real peptone (with albumose) is present, in others the main constituents are decomposition products formed by too long digestion of the meat or fibrin with the acid and pepsin mixture. Some of these commercial peptones appear to be formed by digesting with weak acid under pressure, which results in the formation of bodies of little nutritive value; indeed, it is likely that such products are distinctly harmful when taken into the stomach of man.

It should be mentioned further that many artificial products are known which are related in properties to the peptones of advanced hydrolysis. These may be called *peptides*, or *polypeptides*. Something will be said about them later.

THE PROTEIDS.

The term *proteid*, as already explained, is used to designate a certain group of *protein* compounds. This use is a perfectly arbitrary one as the word was once employed to describe all the bodies discussed in this chapter. It would be perhaps well to drop the term *proteid*, and describe the bodies included under it as *conjugated* or *compound proteins*. According to the generally accepted modern classifications the bodies now called proteids are compound substances in which a true or native albumin is found in combination with some other group which often may be separated as such. In the table given earlier in the chapter three such combinations are mentioned: the *nucleo-proteids*, the *hemoglobins* and the *gluco-proteids*. A brief description of each group will here follow.

NUCLEO-PROTEIDS.

These proteins are important as making up a large part of the cell nucleus. In treating tissues rich in cells with the pepsin-hydrochloric acid digestive mixture it was long ago recognized that a certain portion went easily into solution, while another portion was always left undissolved. This residue was called *nuclein* and was found to contain all the phosphorus of the original protein. If in place of the pepsin mixture some other hydrolyzing agent is used the general result is similar; a separation into two component parts takes place, and one of these parts is a simple native protein substance and the other the nuclein or further and final decomposition product, *nucleic acid*. The nucleo-proteids are therefore described as combinations of native albumins with nucleic acid.

In breaking down the complex nucleo-proteid it appears that several stages must be distinguished, the body described as a "nuclein" containing still some native albumin. Finally, however, the residue or characteristic part, the nucleic acid, is left. Although many investigations have been made there is still much uncertainty about the nature of this acid. Indeed, from different parent substances acids of somewhat different properties have been obtained, so that it is customary to speak of the nucleic *acids*. These will be considered below.

Like the native proteins already described the nucleo-proteids are coagulated by heat and by acids. They are soluble in water, salt solutions and also in alkali solutions. By means of large excess of salt they suffer precipitation. In the last few years nucleo-proteids from different sources have been studied, especially from yeast cells, the thyroid gland, the pancreas and different kinds of spermatozoa. The sperm and spermatozoa of sea urchins and fish have furnished a number of these substances because of their relative richness in cell structures. Thus, characteristic products have been obtained from the spermatozoa of the salmon, the mackerel, the sturgeon and so on. These appear to be distinct bodies, but more exact investigations may show that the apparent differences depend on foreign proteins not completely separated in their preparation.

As intimated above, the nucleo-proteids are found characteristically in the organs rich in cells. The larger part of the solid portion of certain glands and of the heads of spermatozoa consist of these conjugated bodies. The thymus of the calf has been frequently used in the investigations of these bodies, as over 75 per cent of the dried cells of these glands consist of a nucleo-histone. Fish spermatozoa are easily obtainable from hatcheries and have, perhaps, furnished the main material for investigation. The dry, fat-free portion of the heads, which are easily separated, contains 95 per cent, or more, of protamine or histone combinations of nucleic acids. The abundance of these "nucleates" in cell structures, especially in young cells, shows their great physiological importance. The nucleo-proteids of various organs have been investigated in recent years, but the details can not be explained in an elementary book.

Iron seems to be contained in the so-called "masked" or non-ionic condition in the nucleo-proteids. It can not be recognized by the usual qualitative tests, because of its peculiar organic combination. Iron in this form has long been supposed to be important in the formation of red blood corpuscles, which contain hematin. Special methods have been devised for showing the organic iron.

The Nucleic Acids. The occurrence of these important compounds in combination with protamines, histones and other proteins has been referred to several times in the last few pages. They constitute, in fact, the important part of the nucleo-proteids. By different processes of separation a number of these acids have been obtained from various cell structures, and especially from yeast and fish sperm or spermatozoa. The results of analyses lead to formulas of about the following character in nearly all cases: $C_{40}H_{52}N_{14}O_{25}P_4$. These acids have not been obtained in crystalline condition. They are but slightly soluble in cold water, but soluble in weak alkali solutions when they form salts. A number of salts of the heavy metals, which are insoluble in water, have been made and studied. When boiled in aqueous or acid solution the nucleic acids break up, yielding finally the characteristic basic bodies, long known as the purine bases, the pyrimidine bases, phosphoric acid and certain carbohydrates.

Attempts have been made to establish the structural formula of some of the nucleic acids, but without much success, as they are evidently of complex composition. Among the purines the following have been separated:

Xanthine	$C_5H_4N_4O_2$
Hypoxanthine	$C_5H_4N_4O$
Adenine	$C_5H_5N_5$
Guanine	$C_5H_5N_5O$

Three pyrimidine derivatives are known:

Uracil	$C_4H_4N_2O_2$
Cytosine	$C_4H_5N_3O$
Thymine	$C_5H_6N_2O_2$

More will be said later about the relations of these bodies to each other and to the uric acid of the urine. The first are important from that standpoint, as in structure they are closely related to uric acid, and may be forerunners of it.

From the amount of phosphorus and nitrogen found by analysis of the nucleic acids and the amount of the bases secured on cleavage, it has been suggested that they may be complex esters of 4 molecules of phosphoric acid, in which different bases may be combined. This would explain the existence of a large number of closely related acids. It is not necessary to give here the numerous empirical formulas which have been suggested for the acids from different sources. The typical one given above is sufficient as an illustration of the general complexity. Nucleic acids from yeast and other sources have found some application in medicine. The acids from other vegetable sources have been studied, especially by Osborne.

As acids these bodies have rather marked properties; they combine not only with inorganic bases, but also with simple proteins and many toxin bodies. The medicinal uses depend on these facts. The free acids are rather easily hydrolyzed by water and mineral acids, but are stable with alkalies. The salts with sodium and potassium form stiff jellies when dissolved in water by aid of heat, and then cooled below certain temperatures.

HEMOGLOBINS.

The discussion of the important subject of hemoglobins may properly be left to be taken up with the study of the blood in which they are contained. The term is used here in the plural since from different kinds of blood bodies of somewhat different properties have been obtained. Hemoglobin in general must be classed among the compound bodies because it is distinctly made up of two characteristic parts, a histone, already referred to, and hematin.

GLUCO-PROTEIDS.

We have here a group of bodies containing a number of important members about which our knowledge in most cases is not very extended or exact. As the name indicates the proteins here concerned contain a carbohydrate constituent which may be recognized by its reducing properties when the substance in question is warmed with a weak acid and afterwards treated with Fehling's solution in the usual way. The carbohydrate group separated appears to be, in most cases at any rate, glucose amine. Familiar illustrations of these gluco-proteids are found in the mucins and related bodies called mucoids. As a class these substances are characterized by relatively low nitrogen and high oxygen content, due to the presence of the carbohydrate group. The amount of carbon present is also lower than in the common proteins. Of the exact nature of the albumin combined with the carbohydrate little is known, because in separating the two groups by acid or alkali treatment the protein constituent is so changed that no safe conclusion can be drawn as to its original nature.

The gluco-proteids behave as acid bodies. They are not coagulated by heat alone, but heating with acids or alkalies produces a complete alteration. With weak acetic acid a precipitate is in most cases formed which is not easily soluble in excess.

Mucins. These bodies are found in various secretions, especially in the saliva, bile, vaginal fluid, tears, nasal mucus, etc. The amount present, however, is always small and the separation, in pure condition, very difficult. The mucin of the submaxillary gland is probably the best known.

These bodies contain one of the complex protein groups, since they give the reaction with Millon's reagent, the xanthoproteic and the biuret reactions. They are only slightly soluble in water and in presence of alkali produce a viscous stringy liquid which is extremely characteristic, even in great dilution. On warming with dilute alkali the viscous condition disappears through formation of alkali albuminate. On treatment with strong alkali or superheated steam a peculiar body is formed which, from its discoverer, is known as Landwehr's animal gum. This is now known to contain the protein and carbohydrate complexes; after diluting with weak acid and boiling, the sugar reaction may be easily obtained.

The mucins are much more resistant than the nucleo-proteids against the action of reagents or ferments, but they undergo both peptic and pancreatic digestion slowly. In urine the identification of mucin is often a matter of importance, as it is frequently mistaken for albumin. The detection of mucin depends on the behavior with cold dilute acetic acid and also on the solubility in hot water after precipitation with strong alcohol. Albumin is permanently coagulated but mucin not.

Mucoid Bodies. These substances are found in the tendons, cartilage, the vitreous body of the eye, the cornea and elsewhere, and are closely related to the mucins. They have the viscous properties of the latter but in general, in concentrated condition, form stiffer jelly-like masses. The cornea and sclerotic coat of the eye are made up largely of mucoids and collagen dissolved in water.

The mucoids from tendon have been the most thoroughly studied. An extract is made by prolonged treatment with weak lime-water. The solution is precipitated by acetic acid and the precipitate taken up with ammonia. These operations repeated several times give a nearly constant product. The analyses show 48–49 per cent of carbon, 30 of oxygen and below 12 of nitrogen. A small amount of sulphur is present.

In cartilage, along with collagen and albuminoid bodies, a very important mucoid known as *chondro-mucoid* is found. This has a composition not very different from the tendon product just given, but contains over 2 per cent of sulphur, part of which is in peculiar ethereal combination. This ethereal product is separated by cleavage with dilute acids or alkalies and is known as chondroitin sulphuric acid, and, according to Schmiedeberg, has the composition $C_{18}H_{27}NO_{14}.SO_3$. On hydrolysis this acid yields chondroitin, $C_{18}H_{27}NO_{14}$, and sulphuric acid; the chondroitin furnishes acetic acid and chondrosin, $C_{12}H_{21}NO_{11}$; finally further hydrolysis breaks the chondrosin down into glucoseamine and glucoronic acid, according to the same author, but later researches seem to indicate that the cleavage is not as simple as suggested. It has been shown also that this complex acid is not peculiar to cartilage, but is found in many substances belonging to the albuminoid group of proteins as well. Although widely distributed the physiological importance of the body has not yet been determined.

In addition, mucoid substances have been recognized in urine, in blood serum, in white of egg and several pathological transudates in small amount.

ALBUMOIDS OR ALBUMINOIDS.

These substances differ from the real proteins both physically and chemically; the physical differences are, however, the most pronounced and characteristic. The second general classification of proteins places these bodies in the group of *simple proteins*, that is, they are treated as true proteins. The important bodies grouped here contain the different kinds of gelatin or glue-forming compounds, the horn substances, the spongin of the sponge, elastin of the so-called elastic tissues of the body and other substances of less importance. They are all firmer and harder than the common proteins and as a rule quite insoluble in water, and in general resistant against the action of reagents. While by prolonged treatment with superheated steam or acids or alkalies they yield most of the cleavage products described as characteristic of the albumins, some are, however, lacking. The tyrosine group, for example, is absent from gelatin, or present in minute amount at most.

In food value the albuminoids are quite distinct from the other proteins. Most of these substances are so insoluble in the digestive fluids that really no importance as foods could be ascribed to them. Collagen, which yields gelatin, has a limited food value of a peculiar kind which will be referred to below. All these substances serve as supporting, connecting or protective tissues in the body, and they are characterized necessarily by a kind of permanence, which depends on insolubility in the first degree. With increasing age of the body the albuminoid tissues become harder, firmer and less elastic.

COLLAGEN.

The best known of all these albuminoids is the collagen, or glue-forming substance, found as ossein in bone, in cartilage, in the fibrils of connective tissue, in tendons, in fish scales and elsewhere. This substance, wherever found, is insoluble in cold water, but by prolonged heating with water it passes into the soluble form known as gelatin, glutin or, in impure condition, as glue. The change seems to depend on the taking up of a molecule, or more, of water. At the present time it is made in enormous quantities from slaughter house by-products and according to its purity is employed for different purposes. When made by hot water extraction from clean bones or cartilage it is used as an adjunct to food and also in the preparation of emulsions for photographic plates or gelatin paper. The product from common material is used as joiner's glue.

Gelatin softens and dissolves in water at a temperature above 30°.

But this solution point depends largely on the treatment to which it has been previously subjected. By long heating with water, and especially under the action of superheated steam gelatin gradually breaks down into the usual cleavage products of the proteins. As this cleavage progresses a point is finally reached where the mixture no longer solidifies on cooling; a permanent liquid solution is obtained. By hydrolysis with acids this condition is much sooner reached. Many bacteria also have the power of "liquefying" gelatin, which depends of course on their ability to decompose the complex into the more easily soluble amino acids and other compounds.

Among the final cleavage products of gelatin easily recognizable glycocoll and glutaminic acid are probably the most abundant. Leucine, alanine and various other amino acids are found in smaller amount. Like other proteins gelatin yields in peptic or tryptic digestion bodies which have been called gelatoses, gelatin peptones and so on. These resemble but are not identical with the true peptones, which fact has some bearing on the long-discussed question of the food value of gelatin. Gelatin is not converted into true protein in the animal body and for this reason cannot wholly replace the albumins as food. But to some extent it has the power of protecting the so-called circulating albumin from katabolism, by undergoing destruction itself. This sparing or protecting power is limited, however, and the gelatin substances can not permanently replace the native proteins in this way.

EXPERIMENTS TO ILLUSTRATE PROPERTIES OF GELATIN. Dissolve enough gelatin in hot water to make a solution of about one-half per cent strength. Use portions of this for tests:

To some of the solution add a solution of tannic acid; this gives a buff colored precipitate. Gelatin solution is, conversely, employed as a test for tannic acid.

To some of the solution add an excess of strong alcohol; this causes precipitation. This behavior is of importance in the estimation of gelatin.

Use some of the solution with the test reagents. Apply Millon's reagent, the biuret test and the xanthoproteic test.

Prepare a strong solution of gelatin in hot water. To some of this add solution of potassium dichromate and pour the mixture out to cool in a thin layer exposed to sunlight. This treatment produces an insoluble mass which is not attacked by hot water. This property finds application in photo-engraving processes.

To more of the strong gelatin solution add a trace of alkali to neutralize any acidity and then some formaldehyde. On evaporating to dryness a hard mass is obtained which is quite insoluble in water hot or cold and which has found many applications in the arts.

Gelatin to be used in cooking should be nearly white and should dissolve in hot water to form a practically colorless, odorless solution. Inferior gelatin gives off a bad odor when heated with water.

Isinglass is a kind of collagen made from the swimming bladder of certain large fishes. On heating with water it yields a peculiar gelatin which dissolves completely. Common isinglass is largely used in clarifying beer and wine, while the pure white varieties are employed in thickening soups and jellies.

KERATIN.

This is the important insoluble substance in horn, the hoofs of cattle, finger nails, hair and feathers. As can be inferred from the conditions under which it exists, it is not easily attacked by water hot or cold, by weak acids or alkalies, or by digestive fluids. By prolonged action of hot hydrochloric acid, however, it undergoes gradual hydrolysis and cleavage with formation of the usual amino acids and other products. Leucine is apparently the most abundant of these products, as much as 18 per cent of the weight of the horn shavings taken having been obtained by certain investigators. Other important cleavage products found are tyrosine, α-aminoisovaleric acid, aspartic acid, glutaminic acid, phenylalanine, α-pyrrolidine-carboxylic acid, glycocoll, etc.

All keratin bodies contain large amounts of sulphur; some of this is easily split off in the form of hydrogen sulphide. A large part of the sulphur appears to be present in the complex body cystin, $C_6H_{12}N_2S_2O_4$. The xanthoproteic and Millon's reagent reactions are very characteristic with horn substance. The behavior of a drop of nitric acid on the finger nail is well known. The reactions with alkaline lead solutions, yielding lead sulphide, are easily obtained. The use of lead salts in hair dyes depends on this behavior.

ELASTIN.

Elastin differs from keratin mainly in its higher content of carbon and low sulphur content. In their behavior toward reagents they are much alike. Like keratin, elastin can be dissolved only by change in composition. Leucine is produced in large amount by the hydrochloric acid cleavage, and glycocoll, tyrosine and other amino products in smaller amount. Subjected to peptic and pancreatic digestion elastin is slowly dissolved, yielding albumins and a kind of peptone. Most of the protein reactions may be obtained from elastin after bringing it into solution with alkali.

AMYLOID SUBSTANCE.

This is a body which is found in the so-called amyloid degeneration of the liver and kidney. It is particularly characterized by the reddish brown color it assumes when heated with a solution of iodine in potassium iodide. The analysis of amyloid shows a large amount of carbon and some sulphur. It is insoluble in cold water, but partly soluble by long heating. It gives the usual protein reactions when brought into alkaline solution, and contains also a complex group which yields chondroitin sulphuric acid.

FOOD STUFFS.

In the preceding pages the individual substances used as foods or occurring as essential principles of the animal body have been briefly discussed. In nature these compounds do not occur in the pure free condition, but are practically always mixed with other compounds. Before passing to the subject of digestion it will be necessary to have some idea of the general composition of the ordinary foods as used by man. This information will be presented in tabular form, the figures being average values from tables of Atwater. The fuel values are given in so-called large calories.

ANIMAL FOODS.

	Water Per Cent.	Protein Per Cent.	Fat Per Cent.	Ash Per Cent.	Fuel Value in Calories per Pound.
Loin of beef, edible portion	61.3	19.0	19.1	1.0	1155
Flank of beef, edible portion	59.3	19.6	21.1	0.9	1255
Ribs of beef, edible portion	57.0	17.8	24.6	0.9	1370
Round of beef, edible portion	67.8	20.9	10.6	1.1	835
Canned corned beef	51.8	26.3	18.7	4.0	1280
Canned roast beef	58.9	25.9	14.8	1.3	1105
Breast of veal, edible portion	68.5	20.4	10.5	1.1	820
Leg of veal, edible portion	71.7	20.7	6.7	1.1	670
Leg of lamb, edible portion	58.6	18.6	22.6	1.0	1300
Leg of mutton, edible portion	55.0	17.3	27.1	0.9	1465
Lean ham, edible portion	60.0	25.0	14.4	1.3	1075
Fat ham, edible portion	38.7	12.4	50.0	0.7	2345
Loin of pork, edible portion	50.7	16.4	32.0	0.9	1655
Chicken, edible portion	74.8	21.5	2.5	1.1	505
Turkey, edible portion	55.5	21.1	22.9	1.0	1360
Black bass, edible portion	76.7	20.6	1.7	1.2	455
Catfish, edible portion	64.1	14.4	20.6	0.9	1135
Salmon, edible portion	64.6	22.0	12.8	1.4	950
Trout, edible portion	77.8	19.2	2.1	1.2	445
Oysters	83.4	8.8	2.4	1.5	335
Hens' eggs, edible portion	73.7	13.4	10.5	1.0	720
Butter	11.0	1.0	85.0	3.0	3605
Cheese, full, American	31.6	28.8	35.9	3.4	2055
Lard, unrefined	4.8	2.2	94.0	0.1	4010
Oleomargarin	9.5	1.2	83.0	6.3	3525
Gelatin	13.6	91.4	0.1	2.1	1705

In the above table, it will be observed, the animal foods contain all a large amount of water. The solids consist essentially of proteins and fats. In the vegetable foods the water is much less; in most of them carbohydrates are the characteristic principles present. The protein is generally much lower than in the animal foods.

VEGETABLE FOODS.

	Water, Per Cent.	Protein, Per Cent.	Fat, Per Cent.	Carbohydrates, Including Fiber, Per Cent.	Fiber, Per Cent.	Ash, Per Cent.	Fuel Value in Calories Per Pound.
Corn, whole	15.0	8.2	3.8	68.7	1.9	1.4	1610
Cornmeal	12.5	9.2	1.9	75.4	1.0	1.0	1655
Popcorn	4.3	10.7	5.0	78.7	1.4	1.3	1875
Oatmeal	7.3	16.1	7.2	67.5	0.9	1.9	1860
Rice	12.3	8.0	0.3	79.0	0.2	0.4	1630
Rye flour	12.9	6.8	0.9	78.7	0.4	0.7	1630
Wheat flour, entire	11.4	13.8	1.9	71.9	0.9	1.0	1675
Wheat flour, California	13.8	7.9	1.4	76.4		0.5	1625
Wheat flour, general average	12.0	11.4	1.0	75.1	0.3	0.5	1650
White bread, wheat	35.6	9.3	1.2	52.7	0.5	1.2	1205
Whole wheat bread	38.4	9.7	0.9	49.7	1.2	1.3	1140
Crackers	7.1	10.2	8.8	72.4	0.4	1.5	1905
Beans, dry	12.6	22.5	1.8	59.6	4.4	3.5	1605
Beans, dry, Lima	10.4	18.1	1.5	65.9		4.1	1625
Peas, dry	9.5	24.6	1.0	62.0	4.5	2.9	1655
Peas, green, edible	74.6	7.0	0.5	16.9	1.7	1.0	465
Corn, green, edible	75.4	3.1	1.1	19.7	0.5	0.7	470
Cabbage, edible portion	91.5	1.6	0.3	5.6	1.1	1.0	145
Egg plant	92.9	1.2	0.3	5.1	0.8	0.5	130
Potatoes, edible portion	78.3	2.2	0.1	18.4	0.4	1.0	385
Squash, edible portion	88.3	1.4	0.5	9.0	0.8	0.8	215
Apples, edible portion	84.6	0.4	0.5	14.2	1.2	0.3	290
Bananas, edible portion	75.3	1.3	0.6	22.0	1.0	0.8	460
Chestnuts, edible portion	5.9	10.7	7.0	74.2	2.7	2.2	1875
Hickory nuts, edible portion	3.7	15.4	67.4	11.4		2.1	3345
Peanuts, edible portion	9.2	25.8	38.6	24.4	2.5	2.0	2560

FLOUR AND MEAL.

As illustrating the composition of a common vegetable food the following tests may be made:

Experiment. Boil a small amount of wheat flour with Millon's reagent. The red color produced shows presence of proteins.

Experiment. Moisten about 25 gm. of flour with water and work it into a dough. Then hold this under a fine, slow stream of water and by kneading between the fingers slowly work out a portion of the mass as a thin milky liquid. This is largely starch. After some time an elastic residue is left insoluble in water. This is "gluten" and is the chief nitrogenous element of the flour, which has been already referred to. It may be separated into several constituents.

Experiment. To about 5 gm. of flour add 10 cc. of water, shake thoroughly and allow to stand until a nearly clear liquid appears above a white sediment. Filter the liquid and test for sugar by the Fehling solution. Boil some of the residue with water and add iodine solution as a test for starch.

Experiment. To about 5 gm. of fine corn meal in a test-tube add 10 cc. of ether. Close the tube with the thumb and shake thoroughly. Then cork and allow to stand half an hour. Shake again and pour the mixture on a small filter, collect the ethereal filtrate in a shallow dish and evaporate it by immersion in warm water. A small amount of fat will remain.

Action of Yeast on Flour. The following experiment is intended to illustrate the work done by yeast in leavening dough:

Experiment. Crumble two or three grams of compressed yeast into 15 cc. of lukewarm water and shake or stir the mixture until the yeast is uniformly distributed. Then stir in enough flour to make a thick cream and allow to stand over night at room temperature. In this time fermentation of the small amount of sugar in the flour begins and the "sponge" swells up by the escape of bubbles of gas. At this stage mix in uniformly and thoroughly enough flour to make a stiff dough, using for the purpose perhaps 25 gm. Put the dough in an evaporating dish, keep it for an hour or more at a temperature of 30° to 35° C. and observe that it increases very greatly in size, from the continued action of the yeast in liberating bubbles of carbon dioxide. If a good hot air oven is at hand the experiment is completed by baking the leavened mass. The nature of the yeast fermentation will be explained later.

Milk. In milk we have a substance in which all the essential food elements are present. The average composition of cow's milk is given in this table.

Water	87.4
Fat	3.5
Sugar	4.5
Proteins	3.9
Salts	0.7
	100.0

Human milk contains more sugar and less protein than the milk of the cow. Details of this will be given later.

SECTION II.

FERMENTS AND DIGESTIVE PROCESSES.

CHAPTER VI.

ENZYMES AND OTHER FERMENTS. DIGESTION.

In the course of time the conception of fermentation has undergone many changes. The notion was first associated with those processes in which a bubbling or boiling condition without application of heat was observed, and later, as the most familiar kind of fermentation was more closely studied, this phenomenon was found to be due to the escape of gas. This escape of gas came finally to be recognized as the essential feature of fermentation and many operations bearing no relation whatever to alcoholic fermentation were, through confusion of ideas, frequently associated with it.

The real fermentation, which follows when saccharine juices are exposed to the air, had been studied in a way from the remotest antiquity, but no rational attempts at an explanation of the process were made until after the middle of the seventeenth century, when the relation of alcohol and the gas to the destruction of the sugar seems to have been fully recognized. Several other reactions were associated with the alcoholic fermentation; in the leavening of bread the production of a gas was recognized, and it was noticed that in the changes going on in the animal intestine gases were also liberated following the digestion of foods. Along with the alcoholic fermentation there was included under the general name the peculiar change which takes place when the wine formed from saccharine liquids was allowed to stand exposed to the air. To be sure no gas was formed in this action, as in the other, but something in common was recognized. In both cases it was noticed that a scum formed over the liquid and that a small amount of this substance was capable of quickly inciting similar fermentation in more saccharine liquid or wine. The nature of this scum became in time the subject of microscopic investigation (by Leuwenhöck) and we have here probably the beginning of our real study of ferments. It was in 1680 that Leuwenhöck recognized that this scum in the case of beer yeast consisted of minute globules with peculiar properties; the full value of this discovery, however, was not generally admitted and more than a century passed before any great advance was made by others. Lavoisier toward the end of the eighteenth century gave the first explanation of the chemistry of alcoholic fermentation, as he was able to point out the relation of carbon dioxide and alcohol to the parent sugar. But the cause of the action was not much discussed and just what the function of the cells or globules of Leuwenhöck is remained obscure until the time of Pasteur.

Before taking up the important work of Pasteur something must be said of discoveries in other directions. The older conception of fermentation was widened by the addition of new facts. In 1780 Scheele isolated lactic acid from sour milk and later investigators began to look for the agent responsible for the production of this acid. About 1848 the probable nature of an organism which appeared to be always associated with lactic fermentation was pointed out by Blondeau. Various formulas were given for the production of lactic acid in quantity, but it often happened that the final product was an entirely different substance, viz: butyric acid. Butyric fermentation was therefore added to the list of these peculiar reactions, and various speculations were advanced to connect the different phenomena. Meanwhile the situation became still further complicated by the gradual recognition of a new group of reactions which exhibited many of the essential features of the alcoholic and acetic fermentations, and which, therefore, of necessity were classed as ferment reactions. Several chemists had observed the peculiar behavior of a substance produced in germinated barley; this substance possessed the power of converting starch into a sugar which, from its origin, was called malt sugar. Payen and Persoz, in 1833, succeeded in isolating the assumed ferment from the sprouted barley, which they termed *diastase*. About the same time it was recognized that saliva contains a similar starch-converting agent which was later separated and called *salivary diastase* or *ptyalin*. The seeds of the bitter almond were studied by several scientific men and Liebig and Wöhler isolated a ferment body which they termed *emulsin*. This has the property of converting the glucoside called amygdalin into glucose, prussic acid and oil of bitter almonds, or benzoic aldehyde. As the saliva was found to contain a ferment acting on starches, so the gastric juice was recognized as active through the presence of an analogous body called pepsin which acts on proteins.

Most of these discoveries were made before 1840. The ferments in the bitter almond, in sprouted barley, in the saliva and in the gastric juice were all found to be soluble in water. They were therefore called soluble ferments as distinguished from the yeasts and the ferments of acetic, lactic and butyric acids, and from the conditions of their action Liebig was led to formulate the first general theory of fermentation, the *molecular vibration theory*.

THEORIES OF FERMENTATION.

Liebig's Theory. Liebig advanced and maintained for years this view: A ferment is a chemical substance whose particles or molecules exist in a peculiar state of vibration, and in contact with other bodies this ferment is able to set up similar states of vibration which result in the breaking down of the bodies mixed with the ferment. Ferments were considered along with bodies undergoing putrefaction, and many such substances were supposed to be able to bring about real fermentations. According to a somewhat older view ferments were said to act by their mere presence; that is, they exerted what was described as a *catalytic* action. No real attempt, however, was made to define more closely what was meant by this catalytic or contact action.

The Theory of Pasteur. The real nature of yeast as a vegetable growth had finally become established. With this admitted Pasteur advanced the proposition that alcoholic fermentation is a consequence of the life of the organism in contact with sugar and away from the air. Alcohol and carbon dioxide are products of the yeast cell metabolism under these conditions. The cell, according to the Pasteur view, must be furnished with a proper supply of oxygen and this, under the fermenting conditions, it takes from the sugar, giving off carbon dioxide as an oxidation product and producing alcohol at the same time, as a result of the breaking down of the sugar molecule. Fermentation is then to be considered from a purely biological standpoint with alcohol and carbon dioxide as excretory and respiratory products respectively. This Pasteur theory soon found favor with the majority of scientific men and gradually supplanted the mechanical notion of Liebig which could not be brought into accord with experience in other lines. Although the Pasteur view that the yeast produces alcohol only in absence of free oxygen was shown to be incorrect the theory commended itself as otherwise satisfactory and tangible.

Following this a similar explanation was offered for the action taking place in the formation of acetic acid, lactic acid and butyric acid. Here microorganisms are also concerned. These live on certain substances and produce others as metabolic excreta. As to the mechanism of this metabolism we know, of course, nothing; to describe the products formed as *excreta* is perhaps not really warranted by what is actually known. It must be remembered then that the term is used in a broad and general sense only to indicate some kind of a metabolic product.

The work of Pasteur gave an enormous impetus to the study of the common fermentations, but it was evident that this biological explanation was of no value in accounting for the changes produced by the active agents described as diastase, pepsin, emulsin, etc. These, it was pointed out, are as truly "ferments" as are yeast and the mother of vinegar. To avoid confusion it became customary to speak of the *organized* and *unorganized* ferments, or the *insoluble* and *soluble* ferments. The term *enzyme* was later applied to these soluble unorganized agents of change, but this new expression did nothing toward explaining the difficulty or toward relating the two classes of ferments.

Certain scientists from the start, however, refused to admit any fundamental difference between the work of the yeast ferment on the one hand and that of bodies like diastase on the other. Even after the biological theory of Pasteur had become current Berthelot, Hoppe-Seyler and other chemists of prominence maintained that the living cell ferments are active because they secrete soluble or enzymic bodies. In the one case the actual "fermentation" takes place within the cell, as appeared to be the fact with yeast; in other cases enzymes are produced by cells and thrown off to do their work elsewhere. This is true, for example, in the stomach where certain groups of cells produce the active ferment pepsin which, however, does its work of dissolving coagulated protein, or digesting it, outside the cells themselves. In germinating barley the living cells secrete diastase which may be leached out and used to digest starch of other grains. If not leached out the diastase gradually digests the starch of the barley kernel itself, unless the action be checked by heat or other means.

The Work of Buchner. In principle, therefore, the two kinds of ferment action were held to be alike, but although many attempts were made no chemist succeeded in isolating the assumed enzyme from the active cells. Repeated failure in this direction only served to strengthen the belief of the advocates of the vital theory according to which alcoholic and similar fermentations by fungi are processes which cannot be thought of dissociated from the function of life itself. But finally the problem was solved by the German chemist Buchner, who in 1897 succeeded in isolating the active enzyme from yeast cells and in quantity too. This enzyme he called *zymase;* it was found to be as active as the yeast itself and to do all that could be expected of yeast. It has since been produced on the commercial scale in the endeavor to supplant the use of yeast in practice. More recently it has been shown that other ferment cells secrete enzymes and it is possible that all the so-called organized ferments work in this way; but the isolation of the soluble active principle seems to be very difficult in most cases.

All this, however, affords us no real insight into the nature of the ferment process. We have as yet no satisfactory theory as to how these active chemical principles behave in the breaking down of other organic substances. It has not been found possible to prepare any enzyme in a condition of even approximate purity and all analyses made of such substances are doubtless wide of the truth. These analyses appear to show that the enzymes are of protein character, but the impurities in the products analyzed may be responsible for this indication. With this lack of knowledge regarding the chemical composition of the ferments it is naturally impossible to offer a chemical explanation of how they act. It is the effects only that we are familiar with and all our classifications are practically based on what the ferments can do rather than on what they are. It is known that all ferments are destroyed by heat and by the action of even rather weak acids and alkalies. In this they resemble the cells that produce them. While the true ferments or enzymes are apparently complex chemical substances their formation is due in every case, as least it so appears, to cell action. They are organic, but not organized; yet they possess many of the properties of organized bodies. On the other hand certain finely divided metals, especially colloidal platinum, bring about a number of reactions which were long supposed to be characteristic of the true ferments; these reactions are further modified or suspended by the same substances which modify or suspend the ferment actions in question. Based on this behavior it has been attempted to relate the true ferment action to the "catalytic" action of the "inorganic" ferments. All the ferments seem to have the power of decomposing hydrogen peroxide in quantity or catalytically, and this property has been considered as perfectly typical or characteristic. The addition of a number of mineral substances interferes with this catalytic decomposition; in this respect the action of prussic acid is remarkably energetic. It has been found that a minute trace of colloidal platinum in dilute solution decomposes a greatly excessive amount of the peroxide, and further that extremely dilute prussic acid or corrosive sublimate checks the reaction here just as with the true ferment. These analogy reactions are very suggestive, even if they do not explain.

Considering the enzymes as catalytic agents it is to be noted that their distinctive function is, therefore, to hasten certain changes which would take place without them, but, in many cases, with extreme slowness. They have not only the power of hastening or effecting decompositions, as in the alteration of sugars or starches, but also of effecting many syntheses. For example, maltase has the power of bringing about the conversion of glucose into isomaltose. In some cases the enzymes act to *retard*, in place of hastening reactions, and in a great number of instances they seem to act as aids to other catalyzers. In this sense they are spoken of as co-enzymes, kinases or activators.

Enzymes consist of very large molecules which are usually unable to pass through animal membranes or fine porcelain filters. In cases where they can be so filtered the rate of passage is very slow. In addition to large size and probably complex structure the enzymes seem to possess a certain sort of specificity. That is, their activity is exerted in certain directions, or on certain compounds only. The enzyme which aids the conversion of starch into sugar is inactive as far as the digestion of albumin is concerned, although the reactions have much in common. But the specific behavior does not end here. Certain enzymes will effect a decomposition in one form of a glucoside, but not in its optical isomer, and in a large number of reactions it has been found that pancreatic extracts will hydrolyze certain artificial polypeptides, but not their isomers, or related bodies. Such behavior points to a peculiar chemical structure on the part of the enzyme which must bear some relation to the structure of the thing acted upon, or the "substratum," as it is frequently called. Following out this idea there has been no little speculation as to the manner in which the enzymes hasten reactions. It appears that the enzyme, through its structural configuration, unites with the substratum in such a way as to produce a new compound which yields the same end products as the substratum alone would yield, but much more rapidly. In the decomposition to furnish the end products the enzyme group is liberated to combine with a new portion of substratum, and so on, until the reaction is complete, or has reached a condition of equilibrium. It has been suggested that enzymes act as very weak acids or weak bases, or that they combine both properties as do the amino acids, and through this behavior they are able to unite with corresponding groups of the substratum. There is evidence that in a number of such combinations studied the reaction follows the mass action laws with a fair degree of closeness.

CLASSIFICATION OF THE FERMENTS.

With our present knowledge of ferments they are most satisfactorily classified according to the character of the decompositions they effect. Two distinct kinds of action are easily recognized. Many ferment changes are clearly hydrolytic; that is, the reaction follows through the addition of a molecule or more of water to one substance, causing it to break up into smaller groups. In other cases the reaction appears to be in the nature of an oxidation process in which the ferment causes or brings about the addition of oxygen to convert one substance into another. Some authors limit the true ferment reactions to changes which may be referred to one or the other of these heads. But there are a great many decompositions which, while they may not be so clearly defined as those just mentioned, must still be looked upon as of ferment origin. These are produced by bacteria, and in all probability by the enzymes secreted by bacteria. It will be well therefore to add a third general group to make the classification complete.

KINDS OF FERMENT REACTIONS.

We may make three general divisions of the ferment changes, as follows:

   A. Hydrolytic Reactions.
   B. Oxidation Reactions.
   C. Bacterial Decompositions.

A brief discussion of the more important changes coming under each one of these heads will now follow.

A. HYDROLYTIC REACTIONS.

The most important of our ferment reactions, with one exception, perhaps, and at the same time the most thoroughly studied, the changes involving hydrolysis have long claimed the attention of chemists. The true nature of some of these reactions is easily recognized and the earlier workers in this field were able to compare the behavior of the enzymes in question with that of dilute hydrochloric or sulphuric acid. In other very important cases this analogy is far less readily pointed out and it remained for recent workers to satisfactorily establish the true relations. When malt digests starch or when certain enzymes convert the malt sugar formed into glucose the general nature of the changes, as requiring the addition of water, may be shown without difficulty. But with the behavior of pepsin in digesting protein we have more difficulty. Here the reaction is not so easily followed, and the quantitative relations between the original substances and the products formed are more complicated than is the case with the carbohydrate decompositions. However, these reactions likewise have been shown to involve true cases of water addition and therefore may be properly grouped with the carbohydrate reactions as hydrolytic.

This hydrolytic ferment activity is exhibited mainly in the following directions:

1. In the modification of carbohydrates as illustrated by the saccharification of starch and further changes in the sugar thus formed, and in other sugars.
2. In the breaking down of glucosides.
3. In the splitting of fats.
4. In the digestion of proteins.
5. In the so-called fermentation of urea.

Some of these reactions may be represented by definite equations. In general they correspond to the changes produced in the same substances by weak acids with some variations in the details. The salient points will be indicated here, leaving in most cases the fuller discussion for following chapters which deal with the details of the digestion phenomena.

CHANGES IN CARBOHYDRATES.

Amylase or Diastase. Certain enzymes convert starch paste into malt sugar by a reaction which is indicated by the equation:

$$(C_{12}H_{20}O_{10})_n + (H_2O)_n = (C_{12}H_{22}O_{11})_n$$

The enzymes here active are usually described as *diastases* or *amylases*, the terms being employed in the plural, since the action is not confined to a single substance. Of these two terms the word *diastase* is frequently employed in the broad sense to include all the enzymes which act on the starches and sugars formed from them, while the term *amylase* is employed to describe the enzyme which changes starch into malt sugar. In this sense it will be used here. In nature ferments of this character are very widely distributed and serve very important functions. They are active in the changes going on in the vegetable kingdom during the growth of plants and the ripening of fruits, as well as in the germination of seeds. On the commercial scale malt represents the best known diastase-containing substance. In the animal body similar substances are found in the saliva and in the pancreatic secretion. The first of these is called *salivary diastase* or *ptyalin* and the second *pancreatic diastase* or *amylopsin*.

These diastases have never been secured in anything like pure condition. Very active solutions which digest starch quickly may be obtained by extracting ground malt with water, which will be illustrated later. These solutions may be concentrated at a moderate temperature, but the activity of the enzyme is destroyed by heat. A stronger product may be secured by extracting with 20 per cent alcohol and precipitating the solution so obtained by absolute alcohol. This precipitate in turn may be redissolved in water and precipitated again with strong alcohol or with ammonium sulphate, to secure a purer and more active product.

Besides the well-known ferments in malt, in the saliva and in the pancreatic secretion the following may be mentioned here. By many authors the active substance in the liver which converts glycogen into glucose is supposed to be a form of diastase. Others hold the conversion of sugar into glycogen and the subsequent and gradual formation of sugar from glycogen to be specific vital functions performed by the liver cells. The name *cellulase* or *cytase* is given to a ferment body which is found in many vegetable substances and which has the power of converting cellulose into sugar. *Inulase* is the enzyme which acts on the peculiar starch known as inulin found in many vegetable substances, converting it into fructose. Inulase does not appear to act on ordinary starch and on the other hand, malt diastase is not able to convert inulin into sugar. *Pectinase* is another little known vegetable enzyme which converts the so-called pectin jelly substances into a reducing sugar. The original *pectose* in the seed or fruit is first changed into *pectin* by a kind of coagulating enzyme called *pectase*. The true behavior of these bodies is not yet fully known.

Maltase or Glucase. The sugar formed by amylase from starch is known as maltose. This is a primary product and may readily be further converted into glucose by another enzyme occurring in malt and properly known as *maltase*. The nomenclature of these enzymes is unfortunately somewhat confused. An effort has been made to name them systematically, using in each case the name of the carbohydrate or other body on which the enzyme acts, as the first part of the descriptive term, to be followed by the suffix *ase*. Thus *amylase* refers to the enzyme acting on amylose or starch and *maltase* to the enzyme which acts on maltose or malt sugar. But many authors do not follow this system consistently; hence we have as describing the same ferment the term *glucase* in use, since glucose is the product formed. It is preferable to employ the first designation or *maltase*. This enzyme belongs to the class of so-called inverting ferments which convert disaccharides into monosaccharides. In this special case malt sugar yields glucose:

$$C_{12}H_{22}O_{11} + H_2O = 2C_6H_{12}O_6.$$

This maltase is found not only in malt extract, but in various yeasts and elsewhere in the vegetable kingdom. It is also present in saliva, but in small amount, in the pancreas, the liver and in the blood. The formation of glucose in most cases is probably a secondary reaction, maltose being formed first as the primary product. The general importance of this reaction will be pointed out later, as it plays a very essential part in the digestion of the carbohydrate foods.

Lactase. In analogy with the conversion of malt sugar into glucose we have the conversion of its isomer, lactose or milk sugar, into two monosaccharide groups. This is accomplished by the ferment called *lactase* which is found in several kinds of yeast, and which appears to be distinct from the maltase just described. The change of milk sugar is represented by this reaction:

$$C_{12}H_{22}O_{11} + H_2O = \underset{\text{Glucose}}{C_6H_{12}O_6} + \underset{\text{Galactose}}{C_6H_{12}O_6}.$$

Lactose and glucose have nearly the same specific rotation, $[\alpha]_D = 52.5°$ for the first and $53°$ for the second, while for the galactose it is about $83°$. The inversion may be readily followed by the polariscope, therefore.

As to the distribution of this enzyme in nature there is still some dispute. According to some authors lactase is not present in the gastric juice or in the pancreatic secretion, but other investigators have reported finding it in both secretions. It was formerly held that the disappearance of milk sugar in the body is due largely to bacterial actions, as some of these organisms are able to secrete an enzyme which acts on the sugar.

Invertase or Sucrase. One of the most common and important of these enzymic reactions is the inversion of cane sugar, forming glucose and fructose.

$$C_{12}H_{22}O_{11} + H_2O = \underset{\text{Glucose}}{C_6H_{12}O_6} + \underset{\text{Fructose}}{C_6H_{12}O_6}.$$

The name *invertin* or *invertase* has been given to the enzyme which accomplishes this, but *sucrase* would be in better accord with the general nomenclature. The presence of this inverting ferment in many kinds of yeast has been long known. The yeast cell alone is not able to convert cane sugar into alcohol and carbon dioxide; an inversion must first be brought about in some manner. In old yeast or in yeast in which the cell has been destroyed by heat or by mechanical means the inverting enzyme seems to be present in greatest abundance.

Invertase is found in various animal secretions, especially in the intestinal juice. The inverting power of this secretion is marked, while with the pancreatic secretion the inverting power is much less pronounced. In the gastric juice the inverting enzyme is said to be present in some amount and is sufficient to change part of the cane sugar of the food independently of the acid likewise present.

The blood does not appear to contain this invertase, since a solution of cane sugar injected into the veins is eliminated later by the kidneys unchanged. If injected into the portal vein, and thus made to pass the liver, inversion takes place rapidly, as that organ possesses the enzyme in quantity.

Many of the higher as well as the lower plant organisms contain invertase, which accounts for the change of the cane sugar into invert sugar in certain cases. In general this reaction may be easily followed by the polariscope, as the strong dextro-rotation of cane sugar gives place to the levo-rotation of invert sugar. It is possible to make a fairly pure invertase solution from some kinds of yeast, and such a solution has certain practical applications in analytical investigations. By extracting yeast with thymolized water a solution is obtained which rapidly inverts cane sugar, but which is practically without action on malt sugar or milk sugar and which, at the same time, will not induce alcoholic fermentation. This property of the yeast extract is made use of in the determination of cane sugar in presence of the others just mentioned.

GLUCOSIDE REACTIONS.

For our purpose it will not be necessary to go into many details here. A few decompositions only need be mentioned by way of illustration. The glucosides are peculiar compounds which may be looked upon as more or less complex ethers of glucose. They are decomposed in various cleavage processes, with the separation, usually, of glucose as one of the constituent products.

Emulsin. The best known reaction in this group is that which takes place spontaneously in the crushed bitter almond. Along with other substances this kernel contains a characteristic nitrogenous glucoside known as *amygdalin* and the enzyme called *emulsin*. In presence of water the amygdalin breaks up in this way:

$$C_{20}H_{27}NO_{11} + 2H_2O = 2C_6H_{12}O_6 + HCN + C_6H_5CHO,$$

that is, glucose, hydrocyanic acid and benzoic aldehyde are formed. In the uncrushed dry almond this reaction does not take place because the enzyme and glucoside are not in direct contact, but are contained in different cells. The same result is accomplished by distillation of the bitter almond with dilute acids.

Similar reactions are observed with salicin,

$$C_{13}H_{18}O_7 + H_2O = C_6H_{12}O_6 + C_6H_4.OH.CH_2OH,$$
$$\text{Saliginin}$$

and with coniferin, arbutin and other glucosides. A related ferment, known as *myrosin*, converts the potassium myronate found in black mustard into allyl mustard oil, C_3H_5NCS, glucose, and potassium acid sulphate.

THE SPLITTING OF FATS.

The general reactions of fats have been already referred to and it has been shown that in general they may be broken up by the action of water in the form of superheated steam:

$$C_3H_5(C_nH_{2n-1}O_2)_3 + 3H_2O = C_3H_5O_3H_3 + 3HC_nH_{2n-1}O_2.$$

The action of the pancreas in the emulsification of fats was recognized as early as 1834 and in seeking for the cause of this it was finally found to depend on a ferment reaction, and subsequent soap formation.

Lipase or Steapsin. The active principle in the pancreas which accomplishes this fat splitting was first called *steapsin* and afterwards *lipase*. The details of its behavior in the digestion of fats will be explained in a following chapter. Besides its constant occurrence in the pancreas, it has been found in the blood, the liver and the kidney. More recently the existence of lipase in many vegetable substances has been observed and thoroughly studied. It has been found that it hydrolyzes some of the simpler ethereal salts very readily and on this behavior is based a method of recognition of convenient application. Of these ethereal salts ethyl butyrate is possibly the best, as it suffers but very slight change by the action of water alone at ordinary temperatures. The fat-splitting power, or enzyme strength, of various extracts may be compared by noting the amount of the ethyl butyrate decomposed in a given time under the influence of the extract. The extent of hydrolysis of the ethereal salt is determined by titrating the butyric acid liberated with dilute alkali.

PROTEOLYTIC REACTIONS.

While it is not possible to write equations illustrating accurately the absorption of water in the digestion of proteins, as may be done for the carbohydrates and the fats, yet there is abundant evidence to show that water addition is in most cases the characteristic preliminary change here also. The action of superheated steam has been referred to in a former chapter, but at the ordinary temperature certain proteolytic changes take place which are the results of enzyme action. At least three of these changes have been thoroughly studied, and are of great importance in the digestion of foods.

Rennet or Rennin. It has long been known that a certain product found in the stomachs of young animals and especially in the calf's stomach, has the power of clotting milk rapidly, which property has been applied in the manufacture of cheese. The same substance is found also in the pancreas, and the same or a quite similar enzyme in a number of plants. In fact, this curdling or clotting ferment, like others already described, is quite widely distributed in nature. As occurring in the stomach it is mixed with another ferment, which will be described below, known as pepsin. The two substances are apparently quite distinct from each other and may be more or less perfectly separated. Some chemists are, however, inclined to consider them as essentially similar.

Rennet acts on the protein substance casein, throwing it into a coagulated or clotted form. The chemistry of the reaction is obscure and not thoroughly worked out. The essentials of what is known about it will be given later. It is possible to obtain an active extract from the stomach of the calf or young pig which may be kept indefinitely and used for cheese making or other purposes. Formerly in the cheese industry small fragments of the dried calf's stomach, preserved for the purpose, were mixed with the milk and stirred about to induce coagulation. At the present time a liquid extract is made on the commercial scale by the action of an appropriate solvent on the cleaned stomach. Glycerol may be used, or water plus a small amount of salicylic acid to prevent putrefaction. In some European countries certain plants have been employed in the place of animal rennet in the cheese industry. Rennet works well in an acid medium and is easily destroyed by alkalies.

Pepsin. The best known and most thoroughly studied of the proteolytic enzymes is pepsin which has the power of digesting coagulated albumin in an acid medium. It may be obtained best from the mucous membrane of the hog's stomach by extraction with acidulated water or glycerol. In the stomach it appears to exist as a *propepsin* or zymogen, in which condition it is known as *pepsinogen*. The action of acid converts this into the true ferment. Pepsin is very sensitive to the action of alkalies which even in weak solution destroy it or materially lessen its power of dissolving protein. In presence of weak acids, preferably hydrochloric acid of 0.1 to 0.2 per cent strength, it forms from the native or coagulated proteins the derived products known as albumoses and peptones. This change is unquestionably associated with the addition of a number of molecules of water to the original protein group.

Commercial "pepsin" appears in commerce in the form of powder or scales. The latter are obtained by drying the extract from the glands of the stomach on glass plates. The product is far from pure, as it contains a large excess of other extractives. Yet, as now made, one part by weight of the scale or powder is capable of digesting or rendering soluble two to four thousand parts of coagulated albumin in the form of hard-boiled eggs. In an experimental way products of enormously greater activity have been prepared; it is said that one part of a dry pepsin may be made to dissolve three hundred thousand parts of coagulated egg albumin. The relative strengths of pepsin products are always compared by noting the amount of egg albumin or washed fibrin which they will digest in an acid medium of definite concentration.

Pepsin, like most of the enzymes, is precipitated by alcohol. In aqueous solution with a little acid it is most active at about 40° C., and loses its power at about 56°. In the dry condition it withstands perfectly a much higher temperature. While hydrochloric acid is usually employed as an aid to pepsin digestion, other acids may be used with equally good results. Oxalic acid, lactic acid and formic acid work well, but the action with acetic and propionic acids is weak. In presence of alkalies there is no activity and certain salts also interfere with the digestive power.

Through fractional precipitations and by other means many attempts have been made to obtain a "pure" pepsin. The strongest, that is, the most active, products so secured have been analyzed. The results do not differ greatly from those found on analyzing the proteins, yet with some of these products it is not possible to obtain the ordinary protein reactions. We have no clew to the real composition of the substance. It is not at all diffusible through parchment and must have a high molecular weight.

It is stated above that pepsin and rennin are possibly identical substances. The view commonly held by the majority of chemists and physiologists has been that they are distinct ferments produced possibly by different regions of the stomach, but in late years a mass of evidence has been accumulating which appears to throw doubt on this notion and suggest the perfect identity of the two proteolytic enzymes. The chemists of the Pawlow school have been particularly active in advancing this theory. According to them an extract which shows the digesting power will also show the milk curdling action. If it is strong in the one case it will be found strong in the other if the conditions are made right. This amounts to saying that the same enzyme does the two kinds of work, but the conditions of reaction, concentration, salt content, etc., must be different in each case. A commercial rennet, for example, if largely diluted with 0.2 per cent hydrochloric acid, will show a strong proteolytic reaction, while without such dilution it may appear quite inactive. It is held further that the milk curdling ferment of the pancreatic extract is probably identical with the trypsin to be now described.

Trypsin. One of the most active and important of the body ferments is the substance which occurs in the pancreas and known as *trypsin*. It may be extracted from the minced organ in a variety of ways and in crude form is a commercial product. In its action it bears some resemblance to pepsin, but works under different conditions. While pepsin digests protein compounds in dilute acid medium trypsin is most active in presence of weak alkali, preferably sodium carbonate. Action may be observed, however, in neutral solution and even in presence of a trace of acid. In its hydrolysis of proteins trypsin goes farther than pepsin. The action of the latter, under ordinary conditions, ends with the production of albumoses and peptones, while the pancreatic enzyme carries the splitting process to the extent of producing a number of comparatively simple amino acids, and the hexone bases. In this respect the behavior of the trypsin is comparable with that of weak sulphuric acid when heated with the protein bodies. As already pointed out in a former chapter this acid resolves the proteins into complexes which may be considered as the constituent groups of the large molecule. The trypsin digestion may be carried far enough to leave products which fail to show more than a very faint biuret reaction. This reaction, it will be remembered, persists as long as anything having the characteristics of the original protein substance remains. From all this it is evident that the trypsin is a hydrolyzing agent of marked power.

Of the real nature of the enzyme nothing is known. It has never been isolated in a condition of even approximate purity. The pancreatic extracts of the market contain the enzymes acting on fats and carbohydrates as well, in addition to a very large amount of other matter. The active ferment is very soluble in water and in dilute glycerol or dilute alcohol, but in strong alcohol or glycerol it is insoluble. In presence of weak hydrochloric acid trypsin is quickly digested or destroyed by pepsin, and at temperatures much above 50° C. it soon becomes inactive. The temperature optimum is probably about 40° to 45°, in weak alkaline solution, but the statements in the literature on this point are somewhat discrepant.

Erepsin. In this connection another peculiar ferment, which in some respects resembles trypsin, must be mentioned. Erepsin is found in the walls of the small intestine and is concerned in the final splitting of protein derivatives. It acts on proteoses and peptones, essentially, and carries the hydrolysis to the formation of comparatively small amino acid groups, that is, to practically complete hydrolysis.

THE UREA FERMENTATION.

Urease and Other Ferments. Urine exposed to the air soon becomes alkaline and the presence of ammonia is easily shown. This behavior is due to the formation of ammonium carbonate from the urea by a reaction which may be expressed in this way:

$$(NH_2)_2CO + 2H_2O = (NH_4)_2CO_3.$$

The agency concerned in this addition of water molecules was for a long time in doubt, but investigation finally showed it to be a case of ferment action. In all urines undergoing this change numerous bacteria are present and by separating and making pure cultures of these, several species have been found which are capable of decomposing pure solutions of urea. The name *micrococcus ureæ* has been given to one of the most active of these bacterial organisms. It has been found, however, that the action is certainly due to a soluble product or enzyme secreted by the bacteria, since it may be brought into solution. This solution, after the most careful filtration even, is very active when properly made and will quickly induce the ammoniacal decomposition in urea solutions.

The name *urease* has been given to this soluble ferment, which must belong to the hydrolytic group. It is active up to about 50° C. and is much more stable in presence of alkalies than with acids, as would be supposed from the reaction it produces. The enzyme is not readily extracted from the living bacterial cells; these must first be destroyed or allowed to die out in process of making strong cultures. The cells holding this enzyme are very widely distributed in nature, being found in the air, in most soils and in river water. This accounts for the usually rapid fermentation of urine.

B. OXIDATION REACTIONS.

Under the head of oxidation reactions it is very easy to include some in which the essential phenomenon is clearly one of addition of oxygen to the decomposing substance. This is certainly the case in the production of acetic acid from weak alcohol. In other cases, however, the actual nature of the chemical change which occurs is more obscure and the classification of such reactions as oxidation reactions is possibly open to doubt, as will appear below. In some cases the classification appears very arbitrary, as the grouping of the alcoholic fermentation among the oxidation processes illustrates. But there is sufficient reason for this to justify the place the reaction is given.

ALCOHOLIC FERMENTATION.

As mentioned at the outset the phenomena of alcoholic fermentation were the first to claim attention and many of the fundamental conditions were empirically established long before the part played by yeast in the process was recognized. After the investigations of Pasteur our knowledge in this field rapidly widened.

Yeast. The common agent of alcoholic fermentation is known as *yeast*, but under this term are included a very large number of really distinct species. In fact several different genera may be and actually are employed in practice. The following table gives an idea of the relations of the commoner organisms classed among the alcoholic ferments. The yeasts with many other cells are classed in a group of the *budding fungi*, or *Eumycetes*, as distinguished from the *fission fungi* or *Schizomycetes*.

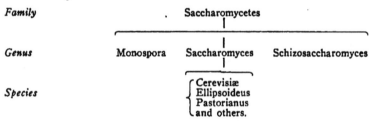

A few molds, also, bring about alcoholic fermentation. We have included here *Mucor mucedo*, *Mucor racemosus*, *Mucor Rouxii* and others which are not in any way technically useful.

Ordinarily, however, we take as the type of a yeast the common beer yeast *Saccharomyces cerevisiæ,* which is a cultivated species employed in fermentation by brewers and distillers. In the natural wine fermentation other species seem to be the most active. These are found on the skin of the grape and hence find their way into the juice after crushing. *S. apiculatus* and *S. ellipsoideus* are the names of two of the most important of the species active in this way. The common beer yeast appears in the form of nearly spherical cells having a diameter of 8 to 9 μ. It is active through a comparatively wide range of temperature. In practice the fermentation of malt wort to produce beer is carried out at a very low temperature, while a grain mash to produce common alcohol or whisky is fermented at a high temperature.

In the one case, however, weeks are required to complete the change, in the other one or two days.

Like most similar reactions brought about by living cells a limit to the quantity of product formed is soon reached. With ordinary glucose the reaction follows approximately according to this equation:

$$C_6H_{12}O_6 = 2C_2H_6O + 2CO_2.$$

The mechanism of the reaction is not known, but it may be considered as one of internal oxidation, since the carbon of the liberated gas is in the fully oxidized condition.

As the alcohol formed accumulates a point is reached where the activity of the yeast cell is impeded and the fermentation finally stopped. This occurs when about 20 per cent of alcohol has accumulated as a reaction product. Very strong sugar solutions do not ferment at all. Indeed some of the common uses of cane sugar in preserving fruits depend on this fact. Some of the conditions of fermentation may be readily illustrated by simple experiments.

Experiment. Make a strong cane sugar syrup, by boiling or heating together 10 gm. of sugar and 10 cc. of water. Allow to cool and add about a gram of crumbled compressed yeast, and then set aside for several days. The solution should be found free from any signs of fermentation.

Experiment. Prepare a 20 per cent solution of commercial glucose and pour 50 cc. of it into a small flask. Add some yeast and allow to stand two days in a moderately warm place. At the end of this time it should be found in active fermentation, as shown by the escape of gas bubbles and the odor of alcohol. If allowed to stand several days longer in the ordinary atmosphere the liquid in the flask usually becomes sour from acetic fermentation.

Experiment. Prepare a tube with sugar solution and yeast as in the last experiment. Close it loosely with a plug of absorbent cotton and heat to boiling, allowing steam to escape through the cotton. If the tube is now left to itself for several days it will be found that fermentation has not taken place, showing that heat destroys the characteristic property of the yeast cell.

Experiment. Prepare another tube with sugar and yeast and add 10 cc. of strong alcohol. Shake the mixture and allow to stand. No fermentation appears, as the activity of the yeast cell is destroyed by alcohol. We have good familiar illustrations of this in the self-preservation of certain "heavy" wines, as ports, sherries and malagas, while "light" wines, which contain 10 to 12 per cent of alcohol usually, must be kept tightly bottled for preservation.

Experiment. TEST FOR ALCOHOL. We have many tests by which the presence or formation of alcohol may be shown. The fermentation of a saccharine liquid is followed by a lowering of the specific gravity as may be easily found by experiment, and a practical quantitative test is based on this fact. A simple chemical test for the presence of alcohol, which in most cases is sufficient, is the following: Add to the clear liquid to be examined a few small crystals of iodine, warm to about 60° C., and then add enough sodium hydroxide or sodium carbonate to produce a colorless solution. An excess of the alkali must not be used. In a short time bright yellow crystals of iodoform precipitate, easily recognized by their color and odor. Certain other liquids give the same test.

Ordinary yeast contains the soluble ferment called invertase, which has been already referred to. This may be shown by experiment, as follows:

Experiment. Crush some yeast, add water and wash by decantation or on a filter thoroughly. Then rub up the washed yeast with some water in a mortar and add the mixture to a solution of pure cane sugar which has previously been treated with a few drops of a strong alcoholic solution of thymol. 50 cc. of a 5 per cent sugar solution will answer. The thymol prevents the action of the yeast cell fermentation, but does not prevent the action of the invertase. The mixture should be kept about 24 hours at a temperature of 40° to 50° C. At the end of this time it is filtered and the filtrate tested for invert sugar by means of the Fehling solution. Ether and chloroform are sometimes employed in place of the thymol; the latter must be removed by heating before making the Fehling test.

Zymase. It has been intimated already that the activity of yeast as an alcoholic ferment is due to the presence of an enzyme. This fact, long suspected and much debated, was finally demonstrated by E. Buchner, as explained above. Buchner rubbed the yeast with fine, sharp sand and water and then subjected the mixture to great pressure. The liquid pressed out was carefully filtered and was found to be as active as the original yeast. The enzyme in it he called *zymase*. It clings tenaciously to the yeast cell, hence the necessity of destroying the structure by grinding with sand, and employing great pressure.

Zymase is not a very stable ferment and in the solution obtained is soon destroyed by other ferments present. The yeast extract may, however, be concentrated at a low temperature and obtained in dry form which is more stable. Extracts made from yeast by simple treatment with water may contain invertase but no zymase. It seems probable that the ferment is not confined to the yeast cell. It has long been known that many overripe fruits produce a small amount of alcohol, even when the possibility of the presence of yeast cells is entirely absent. This formation of alcohol was finally ascribed to the cell activity of the fruits themselves, but since the work of Buchner it seems more rational to refer the appearance of alcohol to the presence of an enzyme in the ripe fruit.

It should also be said that sugar may be made to yield alcohol by a much simpler process. It has been found that a sugar solution mixed with a little potassium hydroxide and placed in bright sunlight yields some alcohol and carbon dioxide. This is of course a purely chemical decomposition, and suggests the possibility of chemical reactions in other cases. The old notion as to the necessity of the presence of living cells to break down the sugar is thus completely disproved.

ACETIC FERMENTATION.

In this a true oxidation takes place, the oxygen of the air being employed to convert weak alcohol into the acid according to this reaction:

$$C_2H_6O + O_2 = C_2H_4O_2 + H_2O.$$

The active agent concerned in the fermentation oxidation is the cell found in "mother of vinegar."

Mother of Vinegar is an old name given to the slimy scum or sediment which forms in weak alcoholic liquids that turn sour, in wine or cider, for example. Microscopic examination shows this substance to consist of minute cells which have received the name of *Micoderma aceti;* more recently the name *Bacterium aceti* has been given to the plant organism. Thus far it has not been found possible to isolate a soluble enzyme from the cell ferment. One may be present, but attempts to obtain it have failed.

Besides this Bacterium aceti several other vinegar ferments are known. Most of them float in the air, and when lodged in a weak alcohol containing certain mineral substances produce a fermentation quickly. A dilute aqueous solution of pure alcohol will not ferment in the same way; the presence of various salts and organic matters in addition is necessary. An experiment may be made to illustrate vinegar or acetic acid fermentation.

Experiment. If available a fruit juice, freshly expressed and left in contact with the skin, should be allowed to undergo alcoholic fermentation. Or, a sugar solution, as described some pages back, may be allowed to ferment. The weak alcoholic liquid obtained in the case of the fruit juice will next turn sour from the production of acetic acid by the action of the germs on the skin. In the case of the alcohol from the sugar it may be necessary to add a little "mother of vinegar" from a vinegar factory to induce the fermentation. Presence of the air is necessary to complete the change. The acid strength of the product may be finally tested by means of a standard alkali solution and phenol-phthalein.

THE OXIDASE ENZYMES.

We come now to a very brief consideration of an obscure but interesting subject about which our knowledge is of comparatively recent origin. In certain vegetable substances reactions occur which are ascribed to the presence of a class of oxidizing enzymes called *oxidases*. These changes are illustrated by the blackening of an apple, potato or beet which is cut and exposed to the air. The cut surfaces soon turn dark. If the same substances are thoroughly heated before the cutting the color change does not follow. Potato or apple pulp speedily darkens in the air, but if previously cooked the natural light color persists. To account for these and many similar reactions it has been assumed by many chemists that the fruits or vegetables in question contain an oxygen-carrying enzyme and at the same time some chemical substance on which this can act with the production of color, the oxygen necessary for the change being taken from the air. The action of this enzyme or oxidase may be shown in other ways, especially by the use of hydroquinol and pyrogallol, which substances yield very dark solutions when oxidized. It is simply necessary to make an aqueous extract of certain vegetables and fruits and add this to the aqueous solution of the hydroquinol to produce the dark color. Here the enzyme appears to be active enough to carry oxygen to the hydroquinol.

Laccase and Tyrosinase. These are the names which have been given to two of these oxidases. The first was originally found in the sap of the Japanese lac tree, which when expressed and exposed to the air darkens and produces the well-known lacquer. The same laccase is said to be one of the agents which brings about the darkening in many other saps and juices. Tyrosinase acts on the phenol derivative tyrosine which is found in traces in many vegetables and causes its oxidation.

These two reactions may be taken to represent a wide range of changes in which phenol bodies are concerned. In another group of reactions aldehyde bodies are turned into acids, as happens to salicylaldehyde. It is possible that many of the obscure oxidative changes of the animal body are brought about by enzymes of this type, but our knowledge here is not very definite. It is known that extracts from the liver and spleen have the power of changing hypoxanthine and xanthine to uric acid, but of the more profound oxidations of the body much less is known. Cohnheim has described a glycolytic ferment, active in the combustion of sugar, when aided by a co-enzyme, or activator, but the nature of the change is not one which can be clearly explained.

Peroxidases. Recently the term *peroxidase* has been introduced to describe a peculiar enzymic ferment which occurs in animal and vegetable cells, the striking feature of which is to induce the oxidation of a great variety of substances through hydrogen peroxide. Milk, for example, when fresh has the power of bringing about the oxidation of phenol-phthalin to phenol-phthalein by hydrogen peroxide, and the same behavior has been observed in other animal secretions. The intensity of the oxidation is much increased by the presence of various other substances which serve as accelerators. As hydrogen peroxide is very readily formed by a wide range of reactions, it is possible that it is produced in living cells, to undergo immediate destruction through the activity of the ferment bodies. This may have some bearing on the explanation of animal oxidations, but as yet our knowledge on this point is scarcely beyond the speculative stage.

C. BACTERIOLYTIC PROCESSES.

The term bacteriolytic is applied to such fermentation-splitting processes as may be carried out by bacteria without the addition of oxygen. In the acetic acid fermentation, which is likewise a bacterial process, the presence of oxygen is necessary, but there are several somewhat analogous reactions in which oxygen is not required and these are included in the present group. It must be admitted of course that the division is a perfectly artificial one based on convenience rather than on marked differences in agents or products. Some of the reactions classed here have long been described as fermentations and have been studied in connection with the other common ferment changes. These will be taken up first. But we have, in addition, further changes which are certainly of the same general character and call for like treatment.

LACTIC AND BUTYRIC FERMENTATIONS.

Why milk turns sour spontaneously in warm weather is an old question, but it was not satisfactorily answered until after the time of Pasteur's pioneer labors. Following his work on the yeasts Pasteur took up other problems of fermentation and pointed out the general nature of the reaction by which the sour substance present, lactic acid, is formed. He found the production of lactic acid to depend on the ferment activity of certain microorganisms, which have later been more fully described by bacteriologists.

Lactic Acid Bacteria. It was found that lactic acid is formed from the simple sugars by a splitting process which for a long time was illustrated by an equation supposed to represent the facts quantitatively:

$$C_6H_{12}O_6 = 2C_3H_6O_3.$$

It was also recognized that not merely one, but many species of bacteria are capable of decomposing sugar solutions in this way. Of these the form known as *Bacillus acidi lactici* has been perhaps the most thoroughly studied; it appears to be always present in milk which has soured spontaneously, and can be found in the air, especially of pastures or cowsheds. Many soils also contain the organism. In no case, however, is the reaction a perfectly sharp one; along with the lactic acid other products are formed, acetic acid, alcohol, formic acid, carbon dioxide and hydrogen being the most common. In some cases the proportion of lactic acid is relatively small. The formation of lactic acid may be illustrated by a laboratory experiment.

Experiment. To 100 cc. of 20 per cent cane sugar solution add an equal volume of aqueous malt extract and 10 to 15 grams of precipitated chalk. Inoculate this mixture with a culture of lactic acid bacteria and keep at a temperature of about 40° C. for some days. The chalk is necessary to take up the acid as fast as formed; without it the fermentation soon ceases, as the ferment is extremely sensitive to the action of free acid. The mixture must be shaken from time to time. As the fermentation progresses the slightly soluble calcium lactate begins to separate. In a good fermentation enough of this forms to fill the fermenting vessel with a mass of crystals. These crystals are redissolved in hot water, and the solution filtered. The filtrate on concentration deposits crystals of calcium lactate, $Ca(C_3H_5O_3)_2.5H_2O$, which may be collected and dried between folds of filter paper. The free lactic acid may be obtained by decomposing the calcium salt with sulphuric acid in the proper amount and shaking out with repeated small portions of ether. The lactic acid dissolves in the ether and is left when this is evaporated. Zinc oxide may be employed in place of calcium carbonate to neutralize during the fermentation. In this case zinc lactate forms, from which the acid may be separated by dissolving the crystals in hot water and decomposing the solution by means of hydrogen sulphide.

Several pure cultures of lactic acid bacteria can now be obtained for technical use. For the rapid production of the acid Lafar recommends *Bacillus acidificans longissimus*.

Pure lactic acid as prepared by fermentation is a thickish liquid, with marked acid taste and but slight odor. It is optically inactive, but may be resolved into active components by treatment with strychnine, which crystallizes with the levo modification. This common fermentation acid is employed for several purposes in the industries and is now comparatively cheap since the introduction of methods of fermentation with pure cultures.

Lactic acid fermentations are concerned in many common operations. In the leavening of bread along with yeast fermentation there is usually a bacterial fermentation with production of acid. In some kinds of bread this is extremely important. In the preparation of sauerkraut and many pickles a lactic acid fermentation is the characteristic feature. Several well-known beverages produced from milk are fermented in such a manner that they contain lactic acid; kephir and kumyss are illustrations. Yeasts and the lactic acid bacteria work together in many instances and symbiotic products are the rule, perhaps, rather than the exception in fermentations. In the milk industries these mixed fermentations are apparently essential in the ripening processes, and in certain distillery fermentations with yeast a lactic acid fermentation is encouraged to prevent the growth and action of the bacteria. This fermentation lactic acid is found also in the stomach and the intestine. In the stomach the formation of any large amount is usually impossible because of the presence of hydrochloric acid. About 0.1 per cent of free hydrochloric acid is sufficient to impede the lactic fermentation. Free mineral acids are not present in the intestine; the organic fermentation acids may therefore be formed in appreciable quantities. Fermentation lactic acid must not be confounded with the isomeric sarcolactic acid found in the muscles.

Butyric Acid Fermentations. Another very important kind of acid fermentation is that which results in the formation of normal butyric acid:

$$C_6H_{12}O_6 = 2H_2 + 2CO_2 + C_4H_8O_2.$$

As in the case of lactic acid this butyric acid fermentation is not the result of the action of one organism only, but it may be produced by several, and furthermore several by-products are always produced in quantity. The above reaction is then merely a limit reaction, which is approached but never absolutely realized.

In the milk fermentation the lactic acid or calcium lactate formed may be further changed to butyric acid, the necessary ferment entering from the air. Most river waters contain butyric acid bacteria, which bring about the characteristic fermentations when the water is mixed with some sterilized milk, as in one of the common tests carried out in the sanitary examination of water. Garden soils are also rich in some of these butyric acid-producing organisms, and may be used in starting a fermentation, as may be illustrated in this way:

Experiment. Mix 100 cc. of a 5 per cent glucose solution with four or five grams of fibrin and heat to boiling. To the hot solution add a few grams of garden loam and allow the liquid to cool rapidly. The bacterial spores resist the heat while other forms succumb and are thus disposed of. Keep the mixture at a temperature of about 37° to 40° C. Fermentation begins in about two days and is assisted by neutralizing with a little sodium hydroxide from time to time. After several days the presence of butyric acid may be shown by warming some of the liquid with sulphuric acid, or with sulphuric acid and alcohol. In the latter case the odor of ethyl butyrate formed is very characteristic.

Butyric acid in pure condition is a strongly acid liquid possessing a rather disagreeable odor. It is frequently present in the stomach, but its occurrence there is really abnormal. If the gastric juice contains the proper amount of hydrochloric acid a butyric acid fermentation is not possible. With diminished hydrochloric acid, however, bacterial fermentations can take place. In the arts, while lactic fermentation is desirable frequently, and encouraged, the butyric fermentation is usually considered very objectionable and is prevented if possible.

Other Fermentations. It will not be necessary to explain at length any other cases of bacterial fermentations, as these two just given are sufficient for illustration. What is known as the mucous fermentation sometimes takes place in saccharine liquids or in wines which have not been completely fermented. A slimy mucilaginous product is formed here which contains a kind of gum. Certain microorganisms have the power of decomposing cellulose and the operation is called the cellulose fermentation. The products of this reaction with certain bacteria are mainly gaseous, hydrogen and marsh gas predominating. Certain organisms are able to produce fatty acids also. In the intestines of the herbivora changes of this character take place, and the acids produced are doubtless of value in aiding in some of the other digestive processes which take place there.

CHAPTER VII.

SALIVA AND SALIVARY DIGESTION.

It has already been said that the saliva contains an enzyme known as ptyalin, the function of which is to begin the digestion of starchy foods. It remains now to look into the nature of this process a little more closely, and to study the conditions of this kind of fermentation. The saliva as secreted by the three large pairs of glands of the mouth is a thin liquid with slightly alkaline reaction. Because of the constant presence of mucus and epithelial cells it is never clear but presents always an opalescent appearance. The amount secreted daily varies between 1 and 2 liters. In the last few years Pawlow has shown how a normal saliva may be collected from animals.

In the older literature several complete analyses of saliva are given, but less importance is now attached to these than formerly, since a great degree of exactness is not possible in such tests and besides the composition of the secretion cannot be a constant one. In the mean the amount of water present is 99.5 per cent. In the 0.5 per cent of solids about 0.2 per cent consists of inorganic salts and the remainder of organic substances, including the ferment. Among the salts there is a minute trace of potassium thiocyanate, KSCN, which may frequently be recognized by the test with ferric chloride. It is not known that this substance exerts any specific function, and in different individuals it is present in different amounts. Some of the important properties of saliva may be illustrated by simple experiments.

Experiment. After washing out the mouth thoroughly with water chew a piece of rubber or other neutral insoluble substance to stimulate the flow of saliva. Collect 25 to 50 cc. in a clean beaker and after diluting with an equal volume of distilled water allow to stand a short time to settle. Then filter through a small filter paper into a clean vessel and use the filtrate for the following tests:

To a few cc. of the clear saliva in a test-tube add several drops of a dilute solution of ferric chloride. This gives a more or less marked red color from the formation of ferric thiocyanate. A very strong reaction must not be expected. Make a comparative test by adding a like amount of ferric chloride to dilute solutions of potassium thiocyanate.

The addition of solution of mercuric chloride discharges the color. This test is of value in distinguishing between a thiocyanate and a meconate, which sometimes has value in medico-legal work.

Test the reaction of saliva with neutral litmus paper. It will be found slightly alkaline. Now add two or three drops of dilute acetic acid and note that a stringy precipitate of mucin separates. Filter off this precipitate and test the filtrate for proteins by boiling with Millon's reagent or by the xanthoproteic reaction.

Make a thin starch paste, about a gram to 200 cc. of water, and observe that it does not respond to the Fehling sugar test already described. Mix 10 cc. of this paste with 5 cc. of the filtered saliva and warm to a temperature not above 40° C. for about 15 minutes. At the end of this time apply the sugar test again. A yellow or red precipitate will appear now, showing that the starch has been converted, in part at least, into sugar.

The saliva alone fails to reduce the copper solution, as should be shown by trial.

Pour about 5 cc. of the clear saliva into a test-tube and boil a few minutes; add the starch paste and allow to stand as in the above experiment. On testing with the copper solution no sugar will be found, showing that heat destroys the activity of the ferment.

The digesting power of the saliva is destroyed also by the addition of a small amount of strong acid or alkali solution, which the student should prove by experiment.

Saliva is practically without action on raw starch, as may be shown in this way. Stir a small amount of uncooked potato starch into 5 cc. of saliva, and allow to stand 15 minutes at 35°–40°, and filter. Now apply the Fehling test, and note the absence of precipitated copper suboxide.

THE CONVERSION OF STARCH.

The action of ptyalin on starch is a complicated one and in all details cannot be satisfactorily described. In many respects the digestive behavior of the enzymes of the saliva and of malt is similar to that of weak acid. The complex insoluble starch molecule is in some manner broken up and partly soluble bodies result. This change is at first unaccompanied by hydration, but later the normal enzymic reaction of water addition follows and the dextrin bodies first produced become sugars. Malt sugar is formed first, and in the case of acids this gives rise finally to glucose by further conversion. But with ptyalin the main action seems to end with the production of maltose; at all events no large amount of the hexose sugar is formed. A little maltase is said to be present. Furthermore the whole of the starch is not brought into the sugar condition; a portion remains in the form of a dextrin. In an earlier chapter something was said about the character of these dextrins.

In most respects the behavior of ptyalin is very similar to that of malt diastase, which can be shown by a simple experiment with commercial malt. This substance is usually made by germinating barley and permitting the growth to continue some days, the barley in moist condition being spread out on a so-called malting floor to encourage the growth and prevent overheating. In the germination the enzyme is developed, probably from a portion of the protein substance present. When the action has gone far enough, which the malster recognizes by the appearance of the rootlet thrown out, the action is checked by quick drying, leaving the diastase in permanent stable condition. This malt is made in enormous quantities for use in breweries and distilleries. In the germinating seed in the ground the same enzyme is formed which converts starch into soluble food for the young plant.

Experiment. Mix about 10 gm. of pale ground malt with 50 cc. of lukewarm water, and allow the mixture to stand a short time, with frequent stirring and shaking. Then filter and add the clear, bright filtrate to a thin starch paste made of 10 grams of starch with 250 cc. of water. The starch paste must be cool when the malt extract is added. Place the mixture on the water-bath and warm to 50°–60° C., and maintain this temperature. Note that the liquid gradually becomes thin and transparent. From time to time remove a few drops by means of a pipette, and test with iodine solution. At first a deep blue color appears, but this grows weaker, giving place to violet, then to yellowish brown and finally no color is obtained, indicating completion of the reaction. The starch paste is first converted into dextrin and finally into maltose. Evaporate the solution to a very small volume and observe the taste and appearance of the residue. In the end product there is usually about 80 per cent of maltose and 20 per cent of dextrin when made at the temperature of this experiment.

It has been found in practice that the amount of malt sugar formed depends on the temperature and duration of the digestion with diastase. At a lower temperature with longer action the conversion of the dextrin becomes more perfect. This corresponds with the behavior of the pancreatic diastase which is active through a longer period usually than is possible with the saliva.

BEHAVIOR OF THE DIASTASE.

The question of the identity of the malt diastase with that from saliva is still a disputed one; while some writers describe them as identical, others apparently find characteristic points of difference. The behavior of saliva with various reagents has been pretty thoroughly studied; stronger acids and alkalies have, of course, a destructive action, but experiments seem to show that very weak acids favor rather than retard the digestive power. When the acid strength is gradually increased up to that of the gastric juice, the effect of the ptyalin on starch paste grows weaker and finally becomes zero long before the maximum acidity is reached. In the mouth the action of the saliva is certainly largely mechanical, since the time for any other action is entirely too short, but with the passage of the food into the stomach it does not follow that all diastatic digestion ceases because of the acid condition of that organ. After the beginning of a meal some time is required for the commencement of hydrochloric acid secretion, and a further time before enough has accumulated to seriously interfere with the activity of the diastase. The effect of the acid is dependent on its *concentration,* not on the gross amount present. Up to a concentration of about 0.01 per cent the acid seems to have but little inhibiting action. Therefore while this amount of free acid is accumulating we may suppose the salivary digestion to go on in the stomach. Later, with increase in acid, the ptyalin disappears, possibly through gastric digestion.

Many salts exert an influence on the rate of diastatic digestion. Usually this is to retard the action, but sodium chloride and other neutral salts in small amount have a beneficial effect. With other substances the action is generally unfavorable. Small amounts of protein matter, or preferably the syntonin or acid albumins formed from the proteins by combination with traces of hydrochloric acid, seem to increase slightly the activity of the salivary diastase. This is a point of considerable importance in explaining possibly the continuation of the ptyalin reaction in the stomach. Acid combined with protein behaves as *free* acid toward certain indicators, while with other indicators it does not show. Starch digestion with saliva in a mixture containing protein and hydrochloric acid, as indicated by dimethyla-minoazobenzene, cannot continue, but if the indication is merely by phenol-phthalein the ptyalin action may still go on, since in this case the acid shown may possibly be wholly or largely combined with protein substances. Recent investigations have shown that under such conditions, which are probably duplicated in the stomach, the digestion of starch may go on at practically the normal rate, the hydrochloric acid being rapidly combined with protein, and therefore comparatively inert with ptyalin. The alkalinity of human saliva is usually referred to as due to the presence of sodium carbonate, but soluble phosphates are present which may account for the reaction as shown by certain indicators, especially by litmus. With phenol-phthalein the reaction appears neutral ordinarily or even slightly acid. With the latter substance as indicator it is generally necessary to add a little alkali to secure neutrality. With litmus as indicator the average alkalinity, expressed in terms of Na_2CO_3, is 0.15 per cent. This reaction seems to vary with the time of day and is strongest before breakfast. Although carbon dioxide is present in saliva, it probably occurs as bicarbonate rather than as carbonate, which would account for the reactions noticed.

Many soluble substances introduced into the blood in any way soon appear in the saliva. This may be shown by an experiment which illustrates also the rapidity of absorption.

Experiment. Swallow about a gram of potassium iodide in a gelatin capsule. In this manner the salt is gradually dissolved in the stomach without having come in direct contact with the mouth. After a few minutes begin testing the saliva for iodine. At first the tests are all negative, but in time a reaction appears on treating the saliva with something to liberate the iodine in presence of starch paste. Solution of sodium hypochlorite may be used for this purpose. The time required to exhibit this absorption and secretion with the saliva varies greatly in different individuals.

CHAPTER VIII.

THE GASTRIC JUICE AND CHANGES IN THE STOMACH.

The gastric juice free from saliva and particles of food is a thin liquid with specific gravity ranging from 1.001 to 1.010. It contains besides certain enzymes some small amounts of protein matters, a little sodium chloride and traces of other salts and free hydrochloric acid. Lactic acid is also frequently present in traces. The older analyses of human gastric juice, which have been frequently quoted, are misleading, as they were made with material containing saliva and food products. By aid of a fistula it has been possible to obtain a fairly normal secretion from certain animals, especially from the dog, and much of our knowledge of the conditions of secretion and variations in composition has been secured in this way. In this direction the work of Pawlow has been of the greatest importance, and his experiments have given us new ideas on the subject of the gastric secretion. The physiologically important substances in the gastric juice are *free hydrochloric acid, pepsin, rennin,* and a *lipase.*

The secretion is furnished by two kinds of glands known as the *pyloric glands* and the *fundus glands.* Both groups of cells yield the two enzymes, but the pyloric cells do not seem to furnish an acid secretion. It is probable that certain of the fundus cells only are concerned with the acid secretion. The gastric secretion is promoted by two kinds of stimuli. Certain chemical substances when taken into the stomach have the power of exciting a flow of the juice from the mucous membrane, and are themselves not subject to gastric digestion. Small amounts of alcohol, ether, spices and meat extracts act in this way. But more important than this is the so-called "psychic" stimulus, depending on the desire for food and the satisfaction derived from partaking of it. The amount of the secretion varies with the nature and kind of food.

THE DIGESTIVE AGENTS.

Origin of the Free Hydrochloric Acid. The material from which the fundus cells produce the enzymes and the acid is the blood. But this is always slightly alkaline and to account for the secretion of a characteristic acid from such a source has long been a puzzle to physiologists. Several hypotheses have been advanced, but these are all more or less faulty. The difficulty is not with the liberation of hydrochloric acid, which is a purely chemical question, and one which may now be explained, but with its secretion.

The blood contains always a small amount of sodium chloride and an excess of carbonic acid. In a solution containing these two things some double decomposition must take place with production of a little free hydrochloric acid and sodium carbonate. According to the older view hydrochloric acid is so much stronger than carbonic acid that the liberation of the former from a chloride by the action of the latter is impossible. But this view leaves out of consideration the effect of a much greater mass of the weaker acid through which in fact a dissociation of the chloride is to a certain extent accomplished. But, granting this kind of a double decomposition, it is still beyond our powers to explain how the free acid formed in the cells is able to pass in one direction into the stomach, while the sodium carbonate produced at the same time wanders in the other direction into the blood.

This acid is not liberated in constant amount at all times but its flow is subject to the influence of the various stimuli referred to above. The quantity present then in the stomach may vary from a mere trace, or zero even, to a maximum. This maximum may be 0.5 or 0.6 per cent of the liquid contents. It has usually been given as much lower. How it is measured will be shown below. Just what is meant by the term *free hydrochloric acid* will be presently explained.

The Enzymes. In an earlier chapter the general nature and behavior of the three gastric ferments, the *pepsin, rennin* and the *lipase* was pointed out. Whether the first two bodies are always secreted simultaneously and in corresponding amounts is not definitely known, but that this is the case is often assumed; it will be recalled that the followers of the Pawlow school consider the enzymes identical, as referred to above. In fact, one of the clinical methods in use for the estimation of "peptic" activity is based on the measurement of the rennet action through milk coagulation. The process seems, however, of doubtful value. In speaking of gastric digestion in adults we are concerned mainly with what takes place through the action of pepsin, which will now be discussed. A briefer discussion of the other ferments will follow.

PEPTIC DIGESTION.

In presence of free acids of the so-called "stronger" type pepsin has the power of effecting remarkable changes in protein substances, which have been the subject of numerous investigations. In the stomach hydrochloric acid only comes into play and it first gradually converts the proteins present into acid-albumins or syntonin bodies. This is the preliminary stage in the digestion of these food substances and must be accomplished before the other steps in the stomach are possible.

In this reaction the hydrochloric acid enters into a kind of combination with the protein. The product has just been spoken of as acid albumin, but it is evidently through the basic character of the protein complex that the combination can take place. The protein here is in effect a very weak base. The amount of acid which may be so held is considerable, and may in fact amount to 5 per cent or more of the weight of the protein. With certain of the derived protein products the weight of hydrochloric acid combined is even larger, at times as much as 15 per cent of the protein weight being so held. These derived products are hydrolysis products with smaller molecular weight and evidently more available amino groups to hold the acid.

It is generally held, as just stated, that this acid fixation is the first step in the gastric digestion, although some authors claim to have recognized the albumose stage as the primary one in some cases. While this acid reaction may take place in pure aqueous-acid solution, it is much more quickly reached in presence of pepsin, as is the case in the stomach. Experiments with artificial mixtures show that the combination then is almost immediate, as is made evident by the loss of " free " acidity, to be explained below. Then the hydrolysis goes on and the various derived products mentioned in a former chapter are produced. In the gastric digestion it is likely that the cleavage does not usually extend beyond the production of the secondary albumoses; that is, not much real peptone is formed in the time naturally consumed in normal digestion. In practice the larger part of the peptone production is doubtless left for the trypsin conversion.

Hydrolytic cleavage beyond the acid albumin stage is favored by abundance of free acid, but in absence of this it still goes on. In actual digestion the whole of the hydrochloric acid may be combined with albumin, leaving some of the latter in excess even, yet primary and secondary albumoses will appear, leaving the remaining native albumin to begin the reaction later. In other words, it is not necessary that one stage of the digestive process must be complete before the following may begin. All these reactions may be in progress simultaneously, and if needed hydrochloric acid will be taken from the advanced products to hasten the beginning hydrolysis of the protein yet to be digested. It has in fact been shown that hydrochloric acid in combination with leucine and other amino acids, which it will be recalled are advanced products of proteolysis, will still digest fresh albumin rather rapidly, but not as well, of course, as the equivalent amount of free acid.

The amount of acid taken up by an original native protein substance during gastric digestion has been referred to already. Starting with a given weight of pure protein, hydrochloric acid may be added until a distinct reaction is shown by dimethylaminoazobenzene. This indicator behaves as a very weak base and will show no free acid until the protein, considered as a basic body, is saturated. As digestion proceeds more and more acid must be added to complete the saturation. The amount of acid which may be so added is to some extent a measure of the advancing cleavage. With phenol-phthalein, which is a very weak acid, the whole of the hydrochloric acid behaves as "free" acid. The acid joined to the protein is "combined" acid as far as the dimethylaminoazobenzene is concerned and this indicator may be used to show the excess of free acid in examinations of stomach contents. More will be said about this below.

As hydrolytic digestion goes on the amount of water combined becomes appreciable, and finally may reach three or four per cent, as has been determined by direct experiment. In a series of investigations carried out in the author's laboratory with casein the water addition amounted to 3.6 per cent, and the acid addition, at the same time, to 7.2 per cent. The water and acid are added in molecular proportions, therefore. The analysis of the albumose and peptone products shows practically the same thing; these substances are always lower in carbon than are the original proteins since oxygen and hydrogen have been taken up in the cleavage. These products of diminished molecular weight pass from the stomach in the condition of hydrochloride salts into the small intestine, where they undergo a new order of changes.

THE ISOLATION OF PEPSIN.

It has been stated already that not one of the enzymes is known in even approximately pure condition. Very strong active extracts of the secretion of the gastric glands of animals may be made by the use of various solvents. Such extracts naturally contain much besides the pepsin, but they are suitable for experimental and other purposes. A good process originally suggested by Wittich is illustrated by the following experiment.

Experiment. Separate the fresh mucous membrane of the hog's stomach from the outer coatings and mince it fine in a meat chopping machine. To 10 gm. of the minced membrane add 200 cc. of glycerol to which a little hydrochloric acid has been added. The acid should amount to about 0.1 per cent of the weight of the glycerol, and may be added in the form of the "normal" volumetric acid of which 5 cc. will be sufficient. Allow the mixture to stand a week with frequent shaking, then filter it by aid of the pump. This extract, bottled, will keep many months. For use 5 cc. of it may be diluted with 100 cc. of water containing the right amount of hydrochloric acid, generally 0.1 to 0.3 per cent.

For many laboratory experiments a fresh aqueous extract is preferable which may be secured in this manner:

Experiment. The washed mucous membrane of the hog's stomach is chopped fine and then rubbed up in a mortar with sharp sand or powdered glass. Water is added (plus 0.1 per cent HCl) in amount ten times as great as the weight of the minced membrane, the mixture is thoroughly stirred, and is allowed to stand over night. It is then filtered and is ready for use. Such an extract is relatively strong.

A much purer product may be secured by the following process as worked out by Kuehne and Chittenden:

Experiment. Remove the mucous membrane of a hog's stomach, wash it thoroughly with water and spread it out on a plate of glass. Scrape the membrane with a knife or piece of glass and mix the scrapings with hydrochloric acid of 0.2 per cent strength. About half the membrane should be reduced to the form of scrapings and for this mass 500 cc. of the acid may be used. Allow this to digest at a temperature of 40° C. for about two weeks in order to convert as much as possible of the protein present into peptone. The mixture is filtered, and to the filtrate powdered ammonium sulphate is added to complete saturation. The object of this is to throw down the pepsin and some albumose, the peptone formed in the digestion being left in solution. This precipitate is collected on a filter, washed with saturated ammonium sulphate solution and redissolved in a little 0.2 per cent hydrochloric acid. The solution so obtained is placed in a tube dialyzer with a little thymol water and dialyzed in running water until the sulphate is all removed. The pepsin solution left is mixed with an equal volume of 0.4 per cent hydrochloric acid and kept at 40° C. 5 days to complete peptonization of albumose still present. Then precipitation with ammonium sulphate to saturation is again effected, the precipitate collected and washed as before and taken up with 0.2 per cent hydrochloric acid. This solution is dialyzed in running water for the removal of all sulphate. The liquid remaining in the dialyzer is a comparatively pure pepsin solution. It may be concentrated in shallow dishes or on glass plates at a temperature not above 40° C., and leaves finally a scale residue. It may be evaporated perhaps better in shallow dishes placed in a large vacuum desiccator with sulphuric acid. The flakes or scales resulting may be kept in dry form almost indefinitely and will be found extremely active.

Commercial Pepsin. What is commonly known as pepsin is a product prepared on the large scale from the hog's stomach and preserved in dry form. Sometimes the mucous membrane is cut into shreds, dried at a low temperature and ground to a powder, in which condition it keeps very well. In presence of weak hydrochloric acid this powder becomes active and is able to digest a large amount of albumin. Usually, however, the mucous membrane is extracted in some manner as illustrated by the first steps described in the Kuehne-Chittenden process. In the commercial processes the following steps are much simpler however. As carried out in the United States, they aim to furnish a finished dry product, one part of which will digest 3,000 parts of egg albumin prepared in a certain way. Several different methods are in use by manufacturers for purifying and concentrating the extract from the stomach glands.

Some Reactions with Pepsin. The behavior of peptic extracts may be easily shown by experiment. For this purpose an extract made by the use of glycerol, as described above, is very convenient. An aqueous extract will answer if freshly prepared.

Experiment. Boil an egg until it is hard, take out the white portion and rub it through a clean wire sieve with fine meshes, by means of a spatula. Add about five gm. of this egg to 100 cc. of 0.2 per cent hydrochloric acid in a flask, and then add 2 cc. of the glycerol extract. Keep the flask at a temperature of 40° C., with frequent shaking. In time the egg albumin will dissolve, forming an opalescent liquid. Unless the flask is very frequently shaken the solution of the albumin will be slow. Use the solution for experiment to be described.

Experiment. To 2 cc. of the glycerol extract in a test-tube add a little water and boil a few minutes. Now add this boiled liquid to albumin and 0.2 per cent hydrochloric acid, as in the last experiment, and note that under the same conditions digestion does not take place, the heating having destroyed the active enzyme. In like manner it may be shown that the enzyme is destroyed by alkalies or stronger acids.

Experiment. Tests for proteoses and peptones. Some instructive experiments may be made with the digesting mixture just described. Some hours after the beginning of the digestion pour or filter off as much of the liquid as possible and use it in this way. Divide the filtrate into several small portions. Boil one portion in a test-tube and observe that it does not coagulate. On cooling the contents of the tube the addition of alcohol produces a rather voluminous precipitate. With other small portions, a few drops is enough, try the xantho-proteic and the Millon's and other color tests. These all give good reactions. Then neutralize the remainder of the liquid with ammonia, exactly, and add powdered ammonium sulphate to saturation. In this way we secure a precipitate of the proteose fraction. After a time filter off this flocculent precipitate and test the filtrate for the more advanced digestion product, the peptone. The biuret reaction may be employed, adding enough sodium hydroxide to cause a separation of sodium sulphate first. Because of the extreme solubility of the peptone bodies a real separation is extremely difficult, but by concentration, and crystallization of the greater part of the ammonium sulphate after cooling, followed by precipitation by alcohol, it is possible to secure a solution in which the peptones may be more clearly recognized.

In practice pepsin is always valued by the amount of protein matter it will digest in a given time. Hard-boiled white of egg is generally employed with hydrochloric acid of 0.3 per cent strength. Sometimes well-washed fibrin is used, with a somewhat weaker acid. As an illustration of practical pepsin testing the following may be given, which is essentially the process of the U. S. Pharmacopœia:

Pepsin Valuation. A. Prepare $N/10$ hydrochloric acid such as is employed in volumetric analysis. B. Dissolve 66 milligrams of pepsin in 100 cc. of water. Mix

175 cc. of A with 25 cc. of B, giving a solution of about 0.32 per cent acid strength.

Boil a fresh hen's egg fifteen minutes, then cool it by placing in cold water. Separate the coagulated white part and rub it through a clean sieve having 40 meshes to the linear inch. Reject the first portions which pass through. Weigh out exactly 10 gm. of the clean disintegrated substance, place it in a 100 cc. flask and add 40 cc. of the acid-pepsin mixture last described. Put the flask in a large water-bath or thermostat kept at 50° C. and let it remain three hours, shaking gently every fifteen minutes. At the end of this time the albumin should have practically disappeared, leaving at most only a few insoluble flakes. Much depends on keeping the temperature constant, and shaking at regular intervals.

In the above test if the albumin is all digested it shows that the pepsin has a converting power of 3,000 or over, which meets the practical requirement of the Pharmacopœia. The relative digesting power of stronger or weaker pepsin may be ascertained by finding through repeated trials how much of a pepsin solution mixed with the acid and made up to 40 cc. will be required to dissolve the 10 gm. of disintegrated white of egg under the same conditions. The process, although not thoroughly satisfactory, is a good one for practical purposes.

THE EXAMINATION OF STOMACH CONTENTS.

From the clinical standpoint the examination of the contents of the stomach at any given time is a matter of considerable importance. The examination may extend to the detection or recognition of the nature of various solid products present, but ordinarily it is confined to the detection or estimation of the acid and the enzymes on which the functional activity of the organ depends. For such examinations it is necessary to collect the liquid contents of the stomach by the aid of some kind of stomach tube. Vomited matter may be used for the same tests. In any event it is preferable to have as much of the solid contents as possible along with the liquid.

Inasmuch as the secretion of the gastric juice does not take place all the time, as was pointed out above, but depends largely on the action of certain stimuli, of which the passage of food down into the stomach is the common and most important one, it is customary to encourage the flow of the secretion by giving what is called a "test-meal" some time before introducing the stomach tube. Unless this is done it might be possible to collect from the stomach a liquid practically free from either acid or enzyme. The Ewald test-meal consists of wheat bread and water or tea without sugar; 50 gm. of bread to 400 cc. of water is an average meal. The content of protein in this would amount to less than 5 gm. ordinarily, and in the normal stomach enough hydrochloric acid to more than combine with this would soon be secreted. After about an hour therefore "free" acid should be detected by the tests given below. With a meal richer in proteins more time would be consumed in producing an excess of hydrochloric acid. In such a case two or three hours might elapse before it would be possible to detect the free acid. The Riegel test-meal consists of a plate of broth or soup, 200 gm. of beefsteak, 50 gm. of wheat bread and 200 cc. of water. The protein in this would amount to about 60 gm., which would require 2 to 3 gm. of hydrochloric acid for preliminary saturation. Some hours would therefore be consumed in producing this. The detection of free acid, then, in such a case would be evidence of relatively high secreting power.

The Detection of Free Acid. In the early digestion stages of a meal rich in carbohydrates organic acids, especially lactic acid, may be formed by bacterial fermentation. But the amount so produced is usually very small if the normal secretion of hydrochloric acid begins in the proper time. The organic acids produced are in amounts ordinarily below 0.1 per cent. Pathologically, when the bacterial fermentation goes on unchecked by the production of hydrochloric acid, the organic acid may accumulate far beyond this and may then be readily detected by the processes given below. At present the detection of the free hydrochloric acid will be considered. Some of the gastric secretion collected by a tube or otherwise is filtered, and the filter (always a small one) is washed with a very little water. The mixed filtrate and washings is used for the following tests:

DIMETHYLAMINOAZOBENZENE TEST. To a few cc. of the gastric filtrate add a drop or two of this reagent used in weak alcoholic solution (about 0.2 per cent). Free hydrochloric acid present strikes a pink or even red color with the indicator. Combined acid and the traces of organic acids which may be present have no such action.

CONGO RED TEST. This substance in aqueous solution is turned blue by very dilute hydrochloric acid. Organic acids do not give the test, except when present in relatively much stronger solution.

The reaction is most conveniently carried out by means of test papers made by dipping filter paper in a solution of the coloring matter and drying. These strips are dipped in the gastric filtrate and allowed to dry spontaneously.

METHYL-VIOLET TEST. A dilute violet-colored aqueous solution of this substance, when mixed with weak hydrochloric acid, turns blue. The reaction with gastric juice is faint, but when care is observed, characteristic. Organic acids, even when present in quantity, do not give the test, which was first successfully used for the detection of traces of mineral acids in vinegar. Use a few drops with 2 cc. of the gastric filtrate.

GUENZBERG'S REAGENT. This is a well-known solution and is made as follows:

Phloroglucin	2 grams
Vanillin	1 gram
Alcohol	100 cubic centimeters

To make the test for free hydrochloric acid, mix 5 cc. of this solution with 5 cc. of the gastric filtrate and concentrate in a glass or porcelain vessel on the water-bath. In presence of free hydrochloric acid the liquid gradually becomes red as the concentration proceeds.

BOAS' REAGENT FOR FREE HYDROCHLORIC ACID.

Resorcinol	5 grams
Cane sugar	3 grams
Alcohol, 50 per cent	100 grams

Add a few drops and evaporate as above. Color appears as in the other test.

Total Hydrochloric Acid. By the use of the above tests the excess of hydrochloric acid beyond that which the proteins and bases will hold is recognized. At one time this acid was supposed to be all that could have any physiological value. The importance of that combined with the proteins in the form of acid albumin was not considered. From the explanations given above it is evident that in some stages of the digestive process the hydrochloric acid may be largely or wholly in combination and therefore not in a form to be recognized through the aid of the tests just given. From experiments made under such conditions it would be wrong to conclude that the stomach is secreting no acids. It has been found that by making the test in a different way, employing phenol-phthalein instead of the reagents mentioned above, the combined acid may be readily recognized. To do this we must make practically a quantitative analysis, and the method employed depends on the proper use of certain indicators. This will be taken up presently.

The Organic Acids. Under normal conditions, as already stated, these are present in the stomach contents in very small amounts only. As their formation depends on bacterial fermentation processes, they appear only when hydrochloric acid is absent, or present in relatively small proportion. Mineral acids arrest bacterial fermentation quickly, from which it follows that in the healthy stomach there is never opportunity for the accumulation of much lactic or other acid of like origin. These acids are never products of secretion as is hydrochloric acid; they are not formed in the cells of the walls of the stomach, but in the food contents. If from some pathological cause the fundus glands fail to secrete hydrochloric acid or secrete it in traces only, then the fermentation bacteria can work unhindered on the carbohydrates in the stomach and produce relatively large amounts of acid. Lactic acid is usually the most abundant of these fermentation products, but butyric acid is occasionally formed and also acetic acid.

The recognition of these organic acids is not difficult if they are alone or mixed with only a little mineral acid. These are of course the cases of practical importance; much hydrochloric acid and much lactic acid could not be found together. Among the simpler reactions employed the following with iron salts are the most useful.

TEST FOR LACTIC ACID. Prepare a dilute solution of phenol by dissolving 1 gm. of the pure crystallized product in 75 cc. of water. To this add 5 drops of a strong solution of ferric chloride, which produces a deep blue color. Five cc. of this mixture suffices for a test. Add to it a few drops of the liquid containing lactic acid, and note the change from blue to yellow. (Uffelmann's test.)

A weak, colorless solution of ferric chloride serves also as a test substance, as its color becomes much deeper by addition of a trace of lactic acid. (Kelling's test.)

This reaction is not influenced by the presence of small amounts of hydrochloric acid, as can be readily shown by adding some to the liquid to be tested. The color change depends on the peculiar behavior of ferric salts with organic acids in general. These acids are relatively weak and with ferric iron tend to form "undissociated" salts which all have a deeper color than have those with the stronger acids.

Both of these tests are much more delicate if applied to the product obtained by shaking out the gastric juice or stomach contents with ether. About 10 cc. of the filtered juice may be shaken with 100 cc. of ether in a separatory funnel through half an hour. When the ether is drawn off and evaporated slowly the lactic acid, if present, is left as a residue. This residue is taken up with a few cc. of water and used for the tests.

The Amount of Acid. It has just been shown how we are able to recognize the free acids existing in the gastric juice, and also, under certain conditions, that in combination with the protein. An equally important problem is the determination of the proportions in which these fractions of the total acid exist. Several different schemes have been proposed by which these degrees of acidity may be measured. The total acid not combined in the form of inorganic salts may be most accurately found by the methods of ordinary quantitative analysis. The total chlorine is found by precipitation or by the Volhard titration. The total bases are found by the usual gravimetric methods. On calculating the amount of chlorine necessary to combine with these bases an excess is left over which must be considered as existing in the form of free acid. In very exact work the traces of phosphates and sulphates present must be also determined and these first combined with bases. The method is one which requires great care in manipulation, and besides does not distinguish between free acid and that held as acid albumin, and this is a very important point.

The principle of another general method may be illustrated in this way. Three portions of the gastric juice of 5 cc. each are measured off. The first is mixed with a little pure sodium carbonate, evaporated and ignited. The total chlorine is so retained and may be found by the Volhard titration. The second portion is evaporated slowly to dryness at a low temperature, mixed with sodium carbonate and ignited. The chlorine is determined in the residue. This represents the fractions combined to proteins and to inorganic bases, as the free hydrochloric acid is lost in the original evaporation at low temperature. Finally, the third portion is evaporated and ignited without any addition. The chlorine now found in the residue is that which originally existed in inorganic combination. With these three operations, as is at once apparent, it is possible to measure the element in the three kinds of combination. The process has been modified and improved so as to be fairly exact.

Attempts are now made to determine the acid accurately volumetrically by the aid of indicators, and here, it may be said, if we can neglect the lactic acid present, pretty good results are possible. But if the lactic acid is present in amount more than traces, as suggested by the qualitative tests above, the process becomes more difficult. Before describing the details of a method something must be said about the indicators themselves, as an understanding of their nature and behavior is necessary for much that is to follow.

Theory of Indicators. The indicators employed in acidimetry and alkalimetry are all weak acids or weak bases themselves, and in general much weaker than the acids or bases in the determination of which they are employed. These indicators, as acids or bases, form salts with the bases or acids to be titrated; it is on the peculiar properties of these salts that the value of the indicators depends. As is well known the change in " reaction " in employing an indicator is accompanied by a change in color. This change in color is accounted for in two general ways. According to one view, which is usually described as the "chromophoric theory" substances which may be employed as indicators must be capable of existing in two modifications, one of which, at least, must possess a so-called chromophoric group. By change of reaction one of these modifications must pass over into the other practically instantaneously, and by the addition of the smallest excess of alkali or acid. Hundreds of substances show this phenomenon in a general way, but to be of use as indicators the change must be both rapid and delicate.

Phenol-phthalein, for example, may be assumed to exist in two forms, one of which is an extremely weak carboxylic acid, and weaker than the acids which are to be titrated by its aid. The acid itself is not stable, but it forms more stable red salts with alkalies. By addition of acids the phenol-phthalein passes over into the other form, which is a lactone and colorless. The value of the indicator depends on the fact that these changes are extremely sharp. The salt form has a chromophoric complex which appears to be of quinoid structure.

In methyl orange, or its related substances, we have evidently two chromophoric groups, one of which is found in the yellow salt form, given with alkalies, and the other in the red isomer produced when combined with acids. The stable yellow form is produced by even very weak alkalies, while the weakest acids are not able to effect a transformation into the red isomer. This property has its advantages, in the titration of mixtures containing both strong and weak acids.

The other, and perhaps more commonly accepted, view of indicators is based on the ionization theory. Phenol-phthalein is assumed to possess a red ion which does not appear in the acid form because of its slight dissociation, but when combined with alkalies the salt dissociates and the red ion then shows itself. With extremely weak alkalies this change does not follow, but the weakest acids are able to suppress the ionization and with it the color. Hence the value of the indicator in titrating weak organic acids. As weak bases do not form stable salts with very weak acids, the whole of the hydrochloric acid combined with protein may be titrated, as illustrated by this equation:

$$\text{Prot. HCl} + \text{NaOH} = \text{Prot.} + \text{NaCl} + H_2O.$$

This reaction will be studied more fully later.

Methyl orange exhibits the opposite behavior, and is assumed to act as a weak base which in the undissociated form is yellow. The ion of the salts, formed with acids, is red. With weak acids it forms extremely unstable salts and therefore cannot be used in the titration of such acids. Carbonic acid is practically inert with it. But bases, even very weak ones, are able to displace it from its combinations with acids, just as weak acids displace phenol-phthalein. Weak ammonia, for example, which combines imperfectly with phenol-phthalein, is strong enough to react with the acid combinations of the dimethylaminoazobenzene or methyl orange. Protein in the so-called acid albumin combination, in which the protein is really basic in character, is stronger than the indicator and able to displace it from its salts. If we add a weak alkali to a solution of the red salt of methyl orange the color changes immediately on the neutralization of any free acid which may be present. The yellow color of the undissociated base takes the place of the red of the salt or free ion. With the very weak solutions of the indicator used the merest drop of alkali should be sufficient to bring about the change in the indicator salt alone. Assuming in solution a mixture of free hydrochloric acid, protein and hydrochloric acid, and the red methyl orange-hydrochloric acid salt, addition of weak sodium hydroxide would produce a change in color immediately after the neutralization of the last trace of free hydrochloric acid. Any excess of alkali added would separate the protein-acid combination, but the protein would behave itself as a base and furnish hydroxyl ions to decrease the dissociation of the indicator and produce the characteristic yellow. Hence the "neutral" point is reached with the disappearance of the actually uncombined HCl. With a weak acid, like lactic acid, present in small amount the condition would be practically the same. Such an acid is but slightly ionized and not able to form stable salts with the indicator.

Illustration. Before taking up the actual titration of the stomach contents a practical illustration of the steps may be found useful. To this end a mixture of about 10 gm. of finely divided coagulated egg albumin with 10 milligrams of powdered pepsin and 100 cc. of 0.4 per cent hydrochloric acid should be made up to 200 cc. with water. This will give an acid strength at the very outset of 0.2 per cent.

Immediately after diluting measure out three portions of 25 cc. each of the thoroughly shaken mixture. Filter one portion (A) at once and wash the residue on the filter with water several times, adding the washings to the filtrate. Add a few drops of phenol-phthalein solution and titrate this liquid with $N/10$NaOH, preferably after warming. Warm the second portion (B) of 25 cc. and titrate with the alkali and phenol-phthalein without filtering. In general the result here will be higher than in the first case. It represents the total acidity and corresponds to one-eighth of the acid taken. In the titration of A the result will be lower because a part of the acid combined at once with the albumin and is left in a form not yet soluble. To the third portion (C) of 25 cc. measured off add 2 drops of a weak dimethylaminoazobenzene indicator and titrate directly with the $N/10$ alkali. The result will be found distinctly lower than that with A or B, even in this beginning stage of the process.

The remainder of the albumin and acid mixture in a loosely stoppered flask is placed in a water-bath and kept as exactly as possible at a temperature of 40° C. through five or six hours. The mixture is shaken frequently as in the pepsin test described above. At the end of two or three hours measure out two portions of 25 cc. each; titrate one with phenol-phthalein addition and the other with dimethylaminoazobenzene. The result with phenol-phthalein present should be nearly the same as before, while with the other indicator it will probably be a little less than in the first case and not much more than half the acidity shown by the phenol-phthalein. After the digestion has continued six hours, or until practically complete, test two further portions of 25 cc. each in the same way. The *total* acidity as measured by the aid of phenol-phthalein will be found but slightly changed, while with the dimethylaminoazobenzene the "free" acid should be found still further lowered probably, and not over half the total acid. The exact relation of the free to the total acidity depends on the strength of the pepsin and the amount of albumin taken. In a long-continued artificial digestion or in presence of much pepsin the acid is gradually combined more and more completely because the basic digestion products formed have relatively lower molecular weights and combine with the acid more or less perfectly, and as shown below the total acidity as measured by phenol-phthalein will be increased. Exact numerical relations here have not yet been established by sufficiently numerous or detailed experiments.

Titration of the Gastric Juice. The illustration given above shows about how this should be carried out. In general as large a volume as there used will not be available, but 5 or 10 cc. should be collected by the tube or otherwise for each test. In testing for the total acidity the mixture should not be filtered, unless the digestion is far advanced, for the reason just pointed out. A part of the hydrochloric acid may be held in the insoluble residue. In testing for the free acid, however, the measured portion should be filtered and the residue on the filter washed with a little water. The whole of the free acid will then be found in the filtrate. As the color change with the indicator here used is not as sharp as in the other case a clearer liquid is essential for the test.

These two titrations give us the total acidity and the free hydrochloric acid, but do not measure the organic acid which may possibly be present. Attempts have been made to estimate this by aid of another indicator. Sodium alizarin sulphonate has been used for this purpose, but the reaction is not as sharp as desirable. This substance appears to behave as a weak acid, but one not as weak as phenol-phthalein. Lactic acid may be titrated with it, but protein separated from HCl behaves as a base toward it. Theoretically the three indicators are related in this way, as illustrated by diagrams, in which H Pht represents phenol-phthalein, HAl alizarin sodium sulphonate, Or Cl the hydrochloric acid salt of dimethylaminoazobenzene and HL lactic acid:

$$\left.\begin{array}{l}\text{H Pht}\\ \text{HCl Prot.}\\ \text{HL}\\ \text{HCl}\end{array}\right\} + \text{NaOH.} \qquad \left.\begin{array}{l}\text{HAl}\\ \text{HCl Prot.}\\ \text{HL}\\ \text{HCl}\end{array}\right\} + \text{NaOH.} \qquad \left.\begin{array}{l}\text{Or Cl}\\ \text{HCl Prot.}\\ \text{HL}\\ \text{HCl}\end{array}\right\} + \text{NaOH.}$$
$$\qquad\quad 1 \qquad\qquad\qquad\qquad 2 \qquad\qquad\qquad\qquad 3$$

It has been shown above how phenol-phthalein and the methyl orange bodies act. The alizarin sulphonate as standing midway between them in properties is influenced by the protein which may be separated from acid albumin in titration with NaOH. Therefore the difference between the titrations in schemes numbers 2 and 3 must measure the lactic or similarly acting organic acid.

Under some conditions this appears to be true. When there is relatively much lactic acid present and not much of the digestion products a fairly good end reaction is obtained. This corresponds of course to a practical case and the indicator then has some value. But as digestion goes on the products formed are more or less basic. While not strong enough to affect the phenol-phthalein they do appear to act on the alizarin compound in such a manner as to diminish the alkali required for titration; the free hydrochloric acid is thus made to appear low. It is plain that the indicator has but limited value.

The Amount of Pepsin. Thus far the detection and estimation of the acid in the stomach contents has alone been considered, but the question of the amount of pepsin present may be of equal importance. We have no very satisfactory tests to determine this amount, but approximate values may be obtained by observing the action of a filtered portion of the gastric juice on some albumin solution to which weak hydrochloric acid has been added. Comparative tests may be made in this manner:

Prepare some egg albumin solution of about 2 per cent strength (2 per cent of dry albumin) and mix this with 0.4 per cent hydrochloric acid in equal proportions; that is for every cubic centimeter of the albumin solution take one cubic centimeter of the acid solution. The resultant mixture has an acid strength of 0.2 per cent and an albumin strength of 1.0 per cent. Measure out 20 cc. of the acid-albumin mixture and add to it 5 cc. of the filtered gastric juice or stomach contents. To another 20 cc. of the mixture add 5 cc. of a 0.2 per cent pepsin solution. Incubate both mixtures through 24 hours at a temperature of 40° C., with frequent shaking. At the end of the time examine both the incubated liquids for digestion products. To this end neutralize a few cc. of each portion with very weak alkali, using phenol-phthalein, and observe whether a precipitate forms or not, directly or on warming gently. If no precipitate forms in either fraction the digestion has gone beyond the acid albumin stage, which should be the case of course in the comparison sample with the pepsin. Next test 5 cc. portions of each mixture in the Esbach albuminometer, adding the usual Esbach reagent (10 gm. picric acid and 20 gm. citric acid with water to 1 liter). This reagent precipitates proteoses but not peptones, when used in excess, and from the extent of the reaction in the tube some conclusion can be drawn as to degree of digestion. A similar test should be made with potassium ferrocyanide and acetic acid in place of the picric and citric acids. Ferrocyanhydric acid does not precipitate peptones.

In another general method the action on solid coagulated protein is observed. White of egg is drawn up into narrow glass tubes having an internal diameter of about 2 mm., and coagulated by heat. The tubes are then cut into lengths of 1 centimeter, thus exposing the ends of the coagulated columns of albumin. These prepared tubes are then immersed in the filtered gastric juice and in standard pepsin solution to be compared and kept at a temperature of 40° some hours. The change in length of the albumin column is taken as the measure of the peptic activity. The filtered gastric juice must be largely diluted with water before making the test, as salts and carbohydrates present interfere with the normal solution of the end of the coagulated mass. The amount of albumin dissolved under these conditions is said to be proportional to the square root of the ferment strength, but the rule is far from exact.

It has been explained above that according to late researches pepsin and rennin are believed by many chemists to be identical substances. As the milk coagulating behavior seems to be much more easily followed and measured than the proteolytic, the ferment strength is frequently determined by observing the extent of the coagulating power. The test may be made in a number of ways and has already found clinical application, but the real value of the process remains to be demonstrated.

PRODUCTS OF PEPTIC DIGESTION.

Frequent reference has already been made to the question of what is produced from the food proteins during peptic digestion. In answering this question it is necessary to distinguish between what *may* be formed under the influence of pepsin and hydrochloric acid, with sufficiently long time afforded for the action, and what actually *is* formed in the few hours in which food, under normal conditions, remains in the stomach. On this subject the views of physiological chemists have undergone various and marked changes. The stomach has long been considered, popularly, as the chief organ of digestion, and this view appeared to be confirmed by the results of the earlier experiments carried out with artificial digestive mixtures. The gradual disappearance of coagulated egg albumin or of fibrin, with the simultaneous formation of soluble products, is a phenomenon easily observed. Various precipitation reactions served to recover from the mixture the products formed and these were early spoken of as "peptones."

At this time, however, the distinction between real peptones and proteoses was not thought of. It remained to be shown that all this abundant mass of material formed in the course of a few hours' digestion consists actually in the main of products preliminary to peptones. This later knowledge came somewhat slowly and led to a radical view of just the opposite order from the early popular one of the function and importance of the stomach. If the stomach is not the principal organ of digestion, it was asked, what is its real value? If the operations carried out there may be accomplished as well later in the intestine, if its work is wholly preliminary and if in turn these preliminary stages are not really essential, what are the functions for which the presence of the stomach appears to be "practically" necessary? A number of remarkable experiments made with animals threw some light on the question. It was found that dogs were able to live and thrive without the stomach, mixed foods of various kinds being almost, if not quite, perfectly digested in the intestine. One of these dogs was kept under observation several years after complete removal of the stomach, and in other cases dogs have been fed through long periods by direct injection of food into the small intestine, the connection with the stomach being meanwhile completely broken by ligature. The feces of these animals were found to be practically normal in most cases.

With such facts in view a school of chemists following Bunge have come to the conclusion that the main use of the stomach is in the destruction of bacteria taken in with the food. The acid usually present in the gastric juice is assumed to be sufficiently strong to destroy most of the ferment organisms, which if allowed to live and pass into the alkaline intestine would certainly work great harm. It must be granted that this view appears plausible; the protection of the intestine through the sterilizing action of the acid is beyond question of prime importance and that the stomach actually accomplishes this to some extent must not be forgotten in any discussion of the relations of the one organ to the other. It is well known what happens in the human stomach when, from some cause, the hydrochloric acid is temporarily absent or greatly diminished. A great development of organic acid-producing bacteria follows, and the products of these are a source of much discomfort without being at the same time strong enough to check the growth of certain pathogenic bacteria.

But after all these facts have been given due weight we must still admit that the peptic digestion if not actually "essential" is in practice really important. Some peptone, although not a large amount, is formed in the stomach and this is ready for immediate absorption, or for the further conversion by erepsin. The proteoses are ready for the final conversion into peptones, or they may be attacked directly by the erepsin. This preliminary work saves, therefore, much work in the intestine. In studying the products of pancreatic digestion it will appear that some of them are identical with those formed in the stomach, or by pepsin-hydrochloric acid action in general. Others appear at first sight quite distinct and their existence leads to the long-accepted notion that the peptic action is incapable of carrying the conversion of proteins through to the final stages. The conclusions which may be drawn from the most recent of the long investigations which have been carried out on this question are somewhat conflicting, but in the main they show that with a sufficiently long time allowed the end products of peptic and tryptic digestion are essentially the same.

Some idea of the extent of the changes taking place in reasonably prolonged peptic digestion may be obtained from a study of the rapidity of combination with hydrochloric acid which has been already referred to in speaking of digestion experiments. A digestive mixture was made with 90 gm. of coagulated and finely divided white of egg, 900 cc. of approximately 0.2 per cent hydrochloric acid and 150 mg. of commercial pepsin. Two portions of this mixture, of 25 cc. each, were titrated at once, one with use of phenol-phthalein and the other with dimethyl-aminoazobenzene. The remainder of the mixture was poured into a large flask which was maintained at a temperature of 40° in a thermostat through a number of days. Titrations were made from time to time with the following results, 25 cc. of the mixture being always taken. The phenol-phthalein titration was made warm, the other cold.

Time.	Cc. of $N/10$ NaOH with Phenol-phthalein.	Cc. of $N/10$ NaOH with Dimethyl-aminoazobenzene.
at once	14	9.0
10 hours	14.5	8.5
24 "	14.5	8.0
40 "	14.7	7.5
96 "	16.0	6.9
168 "	16.6	6.5

It appears, therefore, that the "total" acidity as measured in the phenol-phthalein titration undergoes a slight increase. The hydrochloric acid remains, and added to it are some digestive products of amino-acid character, and strong enough to show in this way. On the other hand in the course of the week's digestion there is a decrease in the "free" hydrochloric acid as measured by aid of the dimethyl-aminoazobenzene indicator. The titration here is not as sharp as with phenol-phthalein, but close enough to indicate the facts. An amount of acid corresponding to 9 cc. of the $N/10$ alkali was "free" immediately after mixing. About 5 cc. had evidently combined with the egg albumin to form the acid albumin. As digestion progressed, and smaller molecules were formed, more acid was required to unite with these. Finally the perfectly uncombined acid amounted to the equivalent of 6.5 cc. of alkali only. Before the end of the digestion bodies were formed which acted as both acids and bases with the proper indicators. The amino acids are of this character.

Some similar results were obtained in the author's laboratory in a very prolonged digestion of casein with pepsin and hydrochloric acid. A mixture was made containing in 1000 cc. 9.6 grams of pure casein, 2.33 grams of hydrochloric acid and 500 milligrams of commercial pepsin. This was incubated at 38°, and from time to time portions were withdrawn for titration with $N/10$ NaOH. The following data were obtained. The original acid was of such strength that 25 cc. required 16 cc. of the alkali for titration with phenol-phthalein or methyl orange. The first titration after mixing with the casein was made at once.

Time.	Cc. of $N/10$ NaOH with Phenol-phthalein.	Cc. of $N/10$ NaOH with Methyl Orange.
at once	16.2	14.6
24 hours	19.4	13.4
48 hours	19.8	13.2
13 days	20.0	12.5
29 days	20.1	11.8
38 days	21.2	11.5
54 days	21.5	11.5

As the digesting mixture leaves a residue of so-called pseudo-nuclein the titration is not quite as sharp as in the other case.

THE MILK CURDLING FERMENT. As has been intimated in this and earlier chapters, two distinct views are held concerning the coagulation of the casein of milk. Hammarsten studied this reaction very carefully and ascribed it to the presence of a peculiar ferment which he called rennin. It has been explained that the adherents of the Pawlow school consider this phenomenon as merely one of the varied manifestations of peptic digestion, in which casein, as the substratum, becomes first coagulated and then dissolved in part. The small portion which is left in this peptic digestion is known as *paranuclein*.

It has frequently been observed that a coagulum is formed when a rennin solution is added to a crude proteose product from peptic digestion. This coagulum is called a *plastein*. But as the precipitate is formed in other ways, as well, it can no longer be referred to as a true rennin reaction. This plastein was at one time assumed to be a step in the synthesis of larger groups from the proteoses, in other words a reversed digestive process. How the product is formed, or what it actually is, is not yet clearly known.

THE DIGESTION OF FATS. It is a discovery of comparatively recent date that a lipase of considerable power is found in the gastric secretion, but as to the extent of the action of this ferment in the normal stomach but little is yet known. The ordinary lipase is an intestinal enzyme which works in a slightly alkaline medium, whereas this lipase is said to be active in a weak acid medium.

ABSORPTION FROM THE STOMACH. Among other newer observations on the functions of the stomach, it has been shown that the absorption of certain digestive products and soluble salts follows to an appreciable extent. Water is not absorbed here, but under certain conditions sugars, peptones, alcohol and bodies soluble in alcohol. It has usually been assumed that no absorption of any importance is possible before the chyme passes into the intestines, but the later investigations on dogs seem to show that this view can no longer be maintained, and that the stomach may, in large measure, play the part in digestion which the earliest investigators ascribed to it.

CHAPTER IX.

THE PRODUCTS OF PANCREATIC DIGESTION.

After leaving the stomach where the food is subjected to the influences described in the last chapter it passes into the small intestine, where it comes in contact with other agents of change. The work in the stomach is largely preliminary and serves to bring the food into a finely divided homogeneous semi-liquid condition, in which it may be readily attacked by the new digestive enzymes. As explained, the chemical actions in the stomach are comparatively simple, and, leaving out of consideration the continuation of the salivary digestion, are due essentially to the combined effect of pepsin, hydrochloric acid and protein substances. In the upper part of the small intestine, however, the work of the pancreatic enzymes is much more complicated; at least three kinds of reactions take place here, due to the three distinct types of ferments in the pancreatic secretion. The protein digestion begun in the stomach is completed, the carbohydrate digestion begun by the saliva is continued or completed, while the fats, not yet attacked, are brought into a condition for absorption through the intestinal walls. These three groups of changes will be taken up in detail, but something must be said first about the pancreatic juice as a whole.

COMPOSITION OF PANCREATIC JUICE.

For obvious reasons it was not possible to give any fair analysis of the gastric juice. But something more is possible in the case of the pancreatic secretion which may be collected by means of a fistula. Most of the experiments have been made with dogs, and the flows, collected under conditions to give a secretion as nearly normal as possible, show that it contains in the mean over 95 per cent of water, and solids consisting of salts and organic substances. Our knowledge of the pancreatic secretion has been greatly increased by the work of Pawlow, who, by specially devised surgical methods, succeeded in securing a product with little disturbance to the animal, and which probably represented the normal liquid sufficiently well. As in the case of the stomach secretion it was clearly shown that the pancreatic flow is excited by certain stimuli, some of which may be clearly followed, while others are beyond explanation, at present. The entrance of the acid chyme from the stomach into the intestines seems to be the most important of these stimulating factors; the hydrochloric acid apparently aids in the production of a peculiar ferment which enters the blood and finally reaches the pancreas. Other theories have been advanced to explain the manner of action of the acid, but the fact is clear that the acid is the most active of all the stimulants. Various foods have also marked effects, and the character of the secretion following the consumption of milk, bread and meat has been reported by Pawlow and his pupils. In one set of experiments these figures are given for the percentage amounts of organic and inorganic solids in the juice:

	Inorg.	Org.
Meat	0.907	1.558
Milk	0.869	4.399
Bread	0.925	2.298

In general, it has been noticed that the amount and nature of the pancreatic flow seem to be adjusted to meet the requirements of the peculiar chyme furnished by the stomach. Of the mechanism of this adjustment practically nothing is known. The juice is always alkaline, and the alkalinity is about that of a sodium hydroxide solution of 0.5 per cent strength. Of the volume of the secretion in man little is accurately known; from fistulas several hundred cubic centimeters daily have been collected, but this may not represent the normal flow. The important enzymes present are trypsin, lipase and amylopsin.

THE BEHAVIOR OF TRYPSIN.

In an earlier chapter a few words were said about the function of this important pancreatic enzyme and it remains to discuss its practical relations to food digestion. In view of the discoveries of Pawlow, it seems probable that the trypsin is not secreted by the pancreas as such, but in the form of trypsinogen, which is activated by the intestinal ferment, to be later described, known as enterokinase. The acid chyme from the stomach passing into the intestine is neutralized by the alkaline pancreatic fluid and the bile. In this neutralized condition the trypsin is able to continue the breaking down process begun by the pepsin, and the proteoses formed in the stomach are carried further to the peptone stage and made ready for absorption or further cleavage by erepsin. From what was said in the last chapter it is evident that the trypsin could effect the preliminary changes also; that is, it is not really necessary that the food proteins should be brought into the proteose condition before the action of trypsin may begin. This enzyme is able to effect the complete digestion from the beginning, and rather rapidly too, which may be illustrated by experiments, using either the minced gland from some animal or an extract made by the aid of a proper solvent. Such an active extract may be secured in several ways. The following methods answer very well.

DIGESTIVE EXTRACTS.

Experiment. Mince a hog's pancreas fine and weigh out about 10 gm., which cover with absolute alcohol in a small bottle. Cork and allow to stand over night. Then pour off the alcohol, which is added to remove water, and squeeze out the residue. Return to the bottle, add 10 cc. of glycerol and allow the mixture to stand about a week with frequent shaking. At the end of this time pour or strain off the glycerol which is now a fairly strong pancreatic extract, and able to act on the three classes of food stuffs. It has been found by experience that the extracts from beef and hog glands are not quite the same in digestive activity, but the hog's pancreas yields a product suitable for all practical purposes, and which keeps a long time when made in this manner.

Experiment. An active pancreas powder which keeps indefinitely is also very useful and may be made in this way. Remove the adhering fat as carefully as possible from a hog or beef pancreas and mince it fine in a meat chopping mill. The disintegrated substance is treated as above with an excess of absolute alcohol to remove water. The alcohol is poured off and the residue pressed dry. This residue is mixed with ether, allowed to stand an hour, and then freed from ether by pouring, pressing and air evaporation. This treatment removes practically all the water, traces of fat remaining and other substances soluble in alcohol and ether. What is left is thoroughly air dried, ground to a fine powder and sifted through gauze with 20 to 30 meshes to the inch. The powder so secured may be kept in a stoppered bottle. In digestion experiments the powder may be used directly, or an extract may be employed. This is best obtained by soaking a few grams of the powder with fifty times it's weight of thymol water through 24 hours.

Some of the conditions of pancreatic digestion may be illustrated by very simple experiments.

Experiment. Pour 25 cc. of a 1 per cent solution of sodium carbonate (crystallized salt) into each of several small flasks or test-tubes. Add to each half a cc. of the glycerol extract of pancreas and about a gram of finely divided hard boiled white of egg. (The white of egg can be easily prepared according to the method given under the pepsin test.) Make one of the tubes slightly acid by the addition of dilute hydrochloric acid, enough to amount to 0.2 or 0.3 per cent. Now place all of them in water kept at 40° C. At the end of half an hour remove one of the alkaline tubes, and note that it still contains unaltered coagulated albumin. Test the liquid for albumoses and peptones as given above. After another half hour, test a second tube (after filtration). It will be observed that as the coagulated protein disappears peptones become more abundant.

Allow one of the alkaline tubes to remain several hours at a temperature of 40° C. In time it develops a disagreeable odor, due to the presence of indol formed. The tube containing the hydrochloric acid kept several hours at 40° C. does not show the effect of digestion, indicating that an acid medium does not suffice for the converting activity of the pancreatic ferment. A minute trace of acid, below about 0.05 per cent, does not appear to check the action.

To readily recognize the final products of the pancreatic digestion of proteins it is necessary to start with larger quantities of materials than are given above.

THE PRODUCTS OF PANCREATIC DIGESTION.

An experiment made as above shows at first digestion and finally bacterial putrefaction as disclosed by the indol odor. A better idea of some of the products formed in digestion may be secured by operating as follows:

Experiment. Mince 50 gm. of fresh fibrin and 25 gm. of pancreas, mix and cover the mixture with 250 cc. of alkaline thymolized water, the thymol being added to check too rapid putrefaction. Keep the mixture at 40° two or three days in a closed vessel, the mass being frequently shaken or stirred. At the end of the digestion the alkali of the mixture is neutralized with a faint excess of acetic acid, after which it is boiled in a porcelain dish and filtered. Some of the fibrin may remain and there will always be some fat to separate by the filtration. The filtrate is used for the identification of important products, some of which are readily recognized, while others are not.

PRODUCTS OF DIGESTION.

Of the albumose stage in pancreatic digestion little is known, as peptones seem to be the first recognizable products. The formation of peptones is greatly facilitated by the previous activity of the stomach ferments. The peptones of trypsin formation are speedily followed by other products, the most important of which are amino acids. Different proteins break down with very different degrees of readiness, some quickly, others very slowly. Among the important cleavage products the following may be referred to. Tryptophane and tyrosine seem to split off in a very early stage, and may appear even with the peptones.

Tryptophane. This name is given to a peculiar product or mixture of products found in a pancreatic digestion like the above. It is characterized by giving a marked violet red color when mixed with a little chlorine water or bromine water. The composition of the tryptophane is not yet known, but on treatment with alkalies at the fusion temperature a mixture of several complex aromatic products, including indol and pyrrol, is obtained. Quite recently the name tryptophane has been given to one of the constituents of this mixture which has the formula $C_{11}H_{12}N_2O_2$ and which has been shown to be indol amino propionic acid.

In a concentrated solution the addition of bromine or chlorine produces a precipitate. This may be redissolved only in a very considerable excess of water. The solution does not yield the protein reactions at all, from which it follows that the body is an advanced decomposition product. The substance is sometimes called protein chromogen.

Experiment. To recognize the chromogen or tryptophane use two or three cc. of the above filtrate from the digestion experiment. Add to the liquid some bromine water, drop by drop, shaking after each addition. Finally the desired color appears.

Tyrosine and Leucine. These important amino acids have already been referred to when the decomposition products of proteins were described. They are formed abundantly in a prolonged digestion like the above and may be easily recognized. Tyrosine is paraoxyphenyl-α-aminopropionic acid,

$$C_6H_4\begin{cases}OH\\CH_2CH(NH_2)COOH\end{cases}$$

and is formed from most of the protein bodies on digestion. It is not formed in appreciable quantity from gelatin. Leucine is regarded as a caproic acid derivative, or possibly as α-aminoisobutylacetic acid $(CH_3)_2CH.CH_2.CH(NH_2)COOH$. It is one of the most common of the protein cleavage products, and is formed from gelatin also. Both of these substances are but slightly soluble in cold water and may be easily separated in crystalline form.

Experiment. To recognize the two amino acids in the digestion mixture proceed as follows: Concentrate the bulk of the liquid to a volume of 25 cc. and allow it to stand in a cold place several days. At the end of this time filter through fine muslin or a coarse filter paper. The granular mass so collected contains some tyrosine while the bulk of the leucine remains in the filtrate. Examine the residue first. Wash it into a beaker with a little cold water, allow to settle, decant and wash again by decantation. Then add a large volume of water and enough ammonia to give a marked odor. Heat to boiling and filter hot. The tyrosine dissolves in the alkaline liquid. Concentrate the filtrate until the odor of ammonia has disappeared and allow to cool; crystals of tyrosine separate.

Examine some of these under the microscope. The appearance is that of bunches or sheaves of fine needles. These needles may be dissolved in alkalies and also in hydrochloric acid on the slide, which behavior distinguishes them from other somewhat similar crystalline deposits.

MILLON'S TEST. A very distinctive test is by the use of Millon's reagent, which has been already illustrated. Mix a little of the crystalline deposit with some water and Millon's reagent in a test-tube and apply heat. A red precipitate forms after a time if much tyrosine is taken. With only a minute amount a red color only may result. It will be remembered that this reaction is not confined to tyrosine alone, but is given by many benzene derivatives containing a hydroxyl group attached to the nucleus. Hence phenol gives the test distinctly.

By heating a little of the crystalline residue with 2 or 3 cc. of strong sulphuric acid solution follows. On adding a drop of formaldehyde solution a red color is produced which becomes green on heating further with addition of some glacial acetic acid.

The solution left after filtering off the tyrosine is concentrated still more, which finally causes a separation of leucine. The concentration is continued until a volume of about 5 cc. is reached. Crystals which have separated may be examined under the microscope. To the concentrated liquid about 20 cc. of alcohol is added, the mixture heated on the water-bath to the boiling point and then allowed to stand until cold. It is then filtered. The filtrate contains most of the leucine present. In the precipitate there is peptone. Evaporate the alcoholic liquid slowly to dryness, take up the residue with water, add some lead hydroxide (lead oxide with a little alkali), boil and filter. From the filtrate remove the excess of lead by means of hydrogen sulphide, filter again and concentrate the liquid to a small bulk for the crystallization of the leucine.

Examine some of the leucine crystals under the microscope. They appear as spherical bunches of very fine needles. Often the needle structure is not visible. Hydrochloric acid and weak alkali solutions dissolve the needles on the slide.

Leucine gives some marked chemical tests. Dissolve some of the crystals in water, add sufficient sodium hydroxide to give a good alkaline reaction and then a *few drops* of copper sulphate solution. The precipitate of copper hydroxide which forms at first redissolves, giving place to a blue solution containing a compound of leucine and copper.

Leucine may be oxidized to yield valeric acid. On this behavior a test is based. To some of the crystalline residue containing leucine add 3 drops of water and 2 or 3 grams of solid potassium hydroxide. Heat in a test-tube until the alkali melts. The leucine decomposes, giving off ammonia. Allow the mass to cool, add enough water to dissolve the residue and then enough dilute sulphuric acid to give a sharp reaction. On applying heat the odor of valeric acid becomes evident. Through the alkaline oxidation carbon dioxide is split off.

The Hexone Bases and Other Bodies. In recent years much attention has been paid to the more complex residues left on tryptic digestion. In this mixture the hexone bases, arginine, lysine and histidine, are important components. These are all amino acids with six carbon atoms, and, because of their constant occurrence in digestive mixtures and other products of protein decomposition, they must be looked upon as essential factors in the protein structure. Leucine and tyrosine always seem to accompany the hexones in these decompositions.

Although by prolonged digestion products are reached which do not give the biuret reaction, it is shown by Fischer and others in recent work that residues remain which are still relatively complex. The name *polypeptides* has been given by Fischer to such residues, and their relations to chemical substances of definite composition pointed out. But even these may be finally broken down into simpler amino acids.

Synthesis of Polypeptides. In the last few years a number of these polypeptides have been produced by several synthetic processes. Among such bodies described by Fischer the following may be cited as illustrations:

DIGLYCYLGLYCINE, $NH_2CH_2CO \cdot NHCH_2CO \cdot NHCH_2COOH$. This is a *tripeptide* and, as the formula shows, is formed by a condensation of three groups of aminoacetic acid.

ALANYLGLYCYLGLYCINE, $CH_3CHNH_2CO \cdot NHCH_2CO \cdot NHCH_2COOH$. In this compound alanine, α-aminopropionic acid, is one of the groups brought into the combination with glycine.

PHENYLALANYLGLYCYLGLYCINE, $C_6H_5CH_2CHNH_2CO \cdot NHCH_2CO \cdot NHCH_2COOH$. This body is of interest because of the occurrence of phenyl alanine among the commoner protein cleavage products, where reagents are used. Residues containing this group appear to be much more resistant toward tryptic fermentation.

LEUCYLPROLINE. Proline = α-pyrrolidine carboxylic acid.

$$\begin{array}{c} CH_3 \\ CH_3 \end{array}\!\!>\!\!CH \cdot CH_2 \cdot \underset{NH_2}{CH} \cdot CO \cdot N\!\!<\!\!\begin{array}{c} CH_2 - CH_2 \\ \underset{COOH}{CH} - CH_2 \end{array}$$

In this case the synthesis of leucine and the pyrrolidine carboxylic acid has been made. In trypsin digestion residues containing the latter body along with phenyl-alanine seem to be characteristic, especially where casein is used. But these residues are easily decomposed by hydrochloric acid with separation of the constituent amino acids.

These four artificial polypeptides are among the earlier products of laboratory synthesis. In recent studies by Fischer and his coworkers the number has been greatly extended.

Besides the hexone bases many simpler amino acids are always found in the digestive residue; glutaminic acid, aspartic acid, alanine, amino valeric acid, glycocoll and others have been separated. The hexone bodies as end products of definite composition are of great theoretical importance because of their relation to the protamines referred to in a former chapter. Some of these protamines break down almost quantitatively into arginine and the other hexones, so that the latter may well be looked upon as nucleus structures which unite, with loss of water, to form the more complex molecules. These diamino acids seem to bear about the same relation to the peptones and proteins that sugar bears to dextrin and starch. As in the hydrolysis of starch the nature of the end product depends on the nature of the agent of cleavage, so in proteolysis the same thing is true; acids and enzymes work *nearly* in the same way, but not absolutely.

In this connection it should be pointed out that Siegfried has separated by a somewhat peculiar method of treatment a number of bodies which he calls trypsin-fibrin peptones and pepsin-fibrin peptones which may be represented by the following formulas:

$$\begin{aligned}
\text{trypsin antipeptone } \alpha &\quad C_{10}H_{17}N_3O_5 \\
\text{trypsin antipeptone } \beta &\quad C_{11}H_{19}N_3O_6 \\
\text{pepsin peptone } \alpha &\quad C_{21}H_{34}N_6O_9 \\
\text{pepsin peptone } \beta &\quad C_{21}H_{36}N_6O_{10}
\end{aligned}$$

The pepsin peptone α seems to be related to the antipeptones in this way:

$$C_{21}H_{34}N_6O_9 + H_2O = C_{10}H_{17}N_3O_5 + C_{11}H_{19}N_3O_5$$

It is urged by Siegfried that the constant optical rotation of these various products is a satisfactory evidence of their constant composition. In connection with these formulas the formulas of the hexone bodies may be recalled:

$$\begin{aligned}
\text{histidine,} &\quad C_6H_9N_3O_2 \\
\text{arginine,} &\quad C_6H_{14}N_4O_2 \\
\text{lysine,} &\quad C_6H_{14}N_2O_2
\end{aligned}$$

These compounds are relatively much simpler than the Siegfried peptones and might readily be derived from them, with separation, at the same time, of still smaller molecules.

The conception of "end product" in tryptic digestion is evidently a somewhat indefinite one. In the last edition of this book the following sentence occurs: "Certainly in the animal body the digestive cleavage cannot extend to the production of these small molecules which would doubtless be useless for nutrition. What is obtained in artificial digestions depends largely on the time given and the activity of the enzyme employed; the term 'end product' is therefore wholly relative."

In a few short years our views have been materially changed, and largely through the results of the investigations of Cohnheim and Abderhalden. It has been shown that these advanced cleavage products are sufficient to maintain the body in nitrogen equilibrium through long periods, and that they may play a very important part in nutrition. This point will be taken up again presently.

At one time a great deal was written about the toxicity of these digestive products. A toxic effect was certainly observed on injection of the commercial peptones into the circulation, but this action seems to be due to the presence of impurities, and to residues of the ferments left, rather than to anything inherent in the amino acids themselves. Since their behavior in nutrition has been shown the notion of toxicity has been abandoned.

Indol and Skatol. In a prolonged pancreatic digestion, especially in the absence of the protecting thymol or chloroform, these bodies are always formed. Their appearance has nothing to do, however, with the enzymic fermentation which gives rise to the other products. They are always products of bacterial decomposition and seem to be produced by the bacteria from some of the enzymic products, most probably from tryptophane. Indol has the composition,

$$C_6H_4\underset{NH}{\overset{CH}{\diagup\diagdown}}CH.$$

Skatol is the methyl derivative,

$$C_6H_4\underset{NH}{\overset{C}{\diagup\diagdown}}\underset{CH.}{\overset{CH_3}{\diagup\diagdown}}$$

Pure indol is a crystalline substance melting at 52°. Skatol melts at 95°. Indol is oxidized in the body to indoxyl, which appears in part in the urine as indican or potassium indoxyl sulphate,

$$\underset{K}{\overset{C_8H_6N}{\diagdown\diagup}}SO_4.$$

Skatol suffers a similar change. More will be said about these reactions later. Although these bodies are not true pancreatic products, it may be well to illustrate their production in this place, since they frequently appear in pancreatic digestions. An experiment will show this.

Experiment. Chop fine 500 grams of meat and 25 grams of pancreas and allow the mixture to stand exposed a day. Then mix with 2 liters of water and 50 cc. of a saturated solution of sodium carbonate, place in a flask and keep at a temperature of 40° through about 10 days. Then transfer the whole mass to a large tin or copper can and distil off most of the liquid. For a complete separation 500 cc. of water should be added at this stage and this distilled also. The whole of the distillate is now acidified with hydrochloric acid and divided into portions of 300 cc. each, which are shaken out thoroughly in a separatory funnel with ether. For the first 300 cc. of acid liquid about 200 cc. of ether should be used. The extracted aqueous layer is drawn off and a new portion of 300 cc. added to the same ether. About 50 cc. of fresh ether must also be added. The mixture is thoroughly shaken, separated as before, and the operation repeated until all the acidified distillate is extracted. The ether is mixed with an equal volume of water and enough sodium hydroxide to give a strong reaction. The alkali combines with and holds the volatile acids which are present while indol and skatol remain in the ether layer. Separate as before, transfer the ether to a flask and distil at a low temperature. Drive off three-fourths of the ether and allow the remainder to evaporate spontaneously. It will not be necessary to purify the residue in any way. Dilute it largely with water and apply the following tests:

Transfer 10 cc. of the dilute indol solution to a test-tube and add 1 cc. of a dilute sodium nitrite solution, mix thoroughly by shaking and then pour carefully a few cc. of strong sulphuric acid down the side of the tube so as to form a layer below the other liquid. At the junction of the two liquids a purple red color is formed, which changes to bluish green on neutralization with alkali. This test is similar to the one commonly employed in water analysis to detect the presence of indol-producing bacteria. The nitrite solution used must be very weak, preferably not over 0.02 per cent in strength.

Another test is performed in this way. A splinter of soft pine wood is moistened with strong hydrochloric acid and then dipped in a weak aqueous solution of indol. The wood gradually becomes red. With much indol the color becomes deep and characteristic.

A characteristic test of value depends on the formation of a salt of nitroso-indol. Acidify the indol solution to be tested with nitric acid and then add a few drops of a 2 per cent solution of sodium nitrite. The nitrate of nitroso-indol, $C_{16}H_{12}(NO)N_2HNO_3$, forms and produces a red precipitate if much indol is present. If the indol solution is weak a red color only forms. By adding some chloroform and shaking, the indol may be concentrated in the junction layer between the two liquids.

By adding a weak solution of sodium nitroprusside to an indol solution a yellow color is first obtained. The addition of weak sodium hydroxide changes this to violet, which, in turn, becomes blue by acidifying with acetic acid. This is known as Legal's test.

Skatol fails to give the above tests.

THE CARBOHYDRATE DIGESTION.

The pancreas furnishes an enzyme called amylopsin or pancreatic diastase which acts on starch or dextrin to form sugar. Beginning with starch we have the gradual formation of maltose by hydrolysis.

THE PRODUCTS OF PANCREATIC DIGESTION. 153

It has been already pointed out that this is not a simple process but one which takes place in several stages, various kinds of "dextrins" coming in between the original starch and the final sugar. In addition to the enzyme which forms the malt sugar the pancreas furnishes, in small amount, a "maltase" which converts this malt sugar into glucose. The action may be very well shown by means of the glycerol extracts of pancreas described some pages back under the head of tryptic digestion.

Experiment. Prepare a starch paste with 5 gm. of starch to 100 cc. of water. Mix 10 cc. of this paste, after cooling, with 5 cc. of the pancreatic extract, warm to a temperature of 35°–40° C. and notice that the paste soon becomes thin and nearly clear. After a time test for sugar. Repeat the experiment, using pancreatic extract which has been boiled before mixing with the starch. The sugar reaction now fails to appear, showing that high temperature destroys the activity of the enzyme, as in the case of saliva. Note in the solution of the starch the dextrin stages which may be followed by the iodine test. For the complete conversion of the amount of starch here taken some hours may be required. This depends on the strength of the pancreas extract.

It is well to vary the experiment by employing some fresh minced gland in place of the extract. The pancreas powder may also be used.

We have then the two principal reactions here, the formation of malt sugar and the inversion of the same. It is not possible to isolate the enzyme which produces the one reaction from that which gives rise to the other, but that both are products of the cells of the pancreas has been satisfactorily shown against the view that the inverting enzyme is furnished by the so-called intestinal juice. It may be recalled that both reactions are hydrolytic.

It is possible that the living gland does not contain the active ferment itself, but a pro-ferment or zymogen, which becomes active after the secretion has passed into the intestine, but a zymogen action has not been clearly proven as in the case of trypsin. In the minced gland the change appears to take place through the agency of air and moisture. There is a marked difference in the activity of the glands of different animals, which fact is practically recognized by the manufacturers of the commercial products. The pancreas of the hog furnishes an enzymic mixture richer in the starch digesting agents, while the beef pancreas seems to be most active in the digestion of proteins. At the present time some very active "pancreatic diastases" are prepared by several firms in this country.

As the proteins are prepared practically for final absorption from the intestine by the action of trypsin, so the remains of the carbohydrates are brought into the proper final condition by the amylopsin and maltase; at any rate the starches are so prepared, and maltose from any source also. But as to cane sugar and milk sugar there appears to be some little doubt, several authors claiming that the pancreas does not contain lactase or invertase, but that the changes in these substances, when not already accomplished by the acid gastric juice, take place through the agency of the enzymes of the intestine.

THE ACTION OF THE PANCREAS ON FATS.

In the general discussion of the subject of enzymes it was shown that a certain product of the pancreas called steapsin or lipase is active in splitting neutral fats into glycerol and acid. This is a true change by hydrolysis and in effect is similar to the splitting by water alone at an elevated temperature. In the pancreas the reaction may not be complete, but may extend only to the separation of one-third of the acid as illustrated by this equation for stearin:

$$C_3H_5\begin{cases}C_{18}H_{35}O_2\\C_{18}H_{35}O_2\\C_{18}H_{35}O_2\end{cases} + HOH = C_3H_5\begin{cases}OH\\C_{18}H_{35}O_2\\C_{18}H_{35}O_2\end{cases} + HC_{18}H_{35}O_2$$

This amount of liberated acid combining with the sodium carbonate of the intestinal juices produces a soap which in turn aids in the emulsification of the rest of the fat and thus prepares for its passage through the intestinal walls. On the other hand, certain writers maintain that the fat must be essentially all split before absorption is possible. The fatty acid and glycerol pass through the intestinal walls directly and recombine. The change in this case would follow through the typical equation:

$$C_3H_5(C_{18}H_{35}O_2)_3 + 3H_2O = C_3H_5(OH)_3 + 3C_{18}H_{36}O_2$$

The behavior of the pancreas may be shown by experiment, but for this purpose it is much better to use the fresh pancreas than to depend on extracts. The lipase seems to be soluble in glycerol to some extent, but unless the fresh gland is employed for the extraction the result may be unsatisfactory. These points may be tested by the student:

Experiment. Rub up a part of a pancreas with some fine, *clean* sand in a mortar to bring it into the condition of a creamy paste. Add some water, mix thoroughly, and use this for the tests. Next melt some butter to allow the curd and salt to settle. Collect the clear butter fat. Mix a few grams of the fat with an equal volume of the pancreas paste, add some water and a few drops of chloroform or toluene as preservative. Keep the mixture at a temperature of 40° through a period of several hours or over night, and observe that it gradually becomes acid through the liberation of butyric acid from the butyrin. This may be shown qualitatively by means of the reaction with rosolic acid, a slightly alkaline solution with red color changing to yellow on addition of some of the pancreas-fat mixture. It may also be shown by adding a few drops of phenolphthalein to some of the mixture and then gradually very weak sodium hydroxide until the alkaline reaction is secured. With a standard alkali solution the volume used becomes a measure of the amount of acid set free.

In this form of the experiment the butter fat is more readily decomposed than are the more solid neutral fats. Indeed the lighter esters, such as ethyl butyrate, are frequently used to detect vegetable lipase through the same general reaction. If in the experiment the butter fat used is not neutral to begin with, it is best to add a few drops of rosolic acid and then very weak alkali until the color just changes to red. Lipase is destroyed by heat as are the other enzymes.

Experiment. The emulsifying power of the pancreas may be shown also. Grind some fresh pancreas to a thin paste with a little water. Add several grams of this mixture to some perfectly neutral refined cotton-seed oil, about 10 cc., in a warm mortar and rub thoroughly with a pestle. After a considerable time an emulsion forms which will bear dilution with much water. With common oil containing a little free fatty acid the emulsion forms more rapidly, but in this case the reaction may be largely due to the formation of soap first, from the combination of the fatty acid and alkali of the pancreatic secretion. The experiment would therefore fail to show the presence of an enzyme as fat splitter. For success here a fresh pancreas is necessary.

A pure neutral fat suitable for such experiments may be obtained by adding a little caustic soda solution to some refined cotton-seed oil and then ether. On shaking thoroughly the neutral fat dissolves in the ether, leaving the soap formed and excess of alkali undissolved, practically. The fat-ether layer is poured off, shaken several times with water for the removal of traces of soap or alkali, and then slowly evaporated. Neutral fat is left after the volatilization of the ether.

In these emulsification reactions the pancreatic secretion is assisted by the alkaline bile. According to the theory of the formation of soaps, as a preliminary to absorption from the intestines, the bile must act as a very important factor, as its alkali would be needed for the purpose. The bile acids have been shown to be activators for the steapsin, and to assist materially in the cleavage of the glycerol esters. In addition the bile has a distinct solvent action on fatty acids which may be of help in the ultimate passage of the fat products from the intestine. The general nature of the bile products will be discussed later.

THE FUNCTION OF THE INTESTINAL JUICE.

Closely related to the action of the pancreatic diastases is the behavior of certain enzymes entering the small intestine from other sources, especially from the glands of Lieberkühn. As these enzymes seem to follow up and complete the pancreatic digestion, they may be briefly mentioned here. It should be said first, however, that any specific digestive action due to ferments in the secretion of these glands was for a long time denied, but there appears now to be no further question as to the actual behavior of the secretion in this respect.

Character of the Secretion. The flow into the intestine from the Lieberkühn glands consists of a thin serum-like liquid, holding in solution protein bodies and salts. The reaction is strongly alkaline because of the presence of sodium carbonate. The amount of this is sufficient to give rise to an evolution of carbon dioxide when an acid is added to the secretion collected by a fistula. This alkali is doubtless important in two ways; it aids in the emulsification of fats, and also helps in the neutralization of the remaining hydrochloric acid from the gastric juice carried into the intestine with the chyme current.

The ferments appear to have little or no action on fats or proteins, but work on the residues of carbohydrates only. Some chemists claim to find in the intestinal juice a slight starch-digesting power; others deny that such a behavior is possible and limit the activity of the secretion to the inversion of certain sugars, especially cane sugar and malt sugar. Indeed some authors go so far as to urge that all of the inversion processes taking place in the intestine are brought about in this way, while the pancreas can produce malt sugar only. Investigations of this kind are attended with considerable difficulty, which fact must be kept in mind when attempting to draw conclusions from apparently contradictory statements, such as are quoted above. All recent investigations have shown this, that while the intestinal juice may not be the *sole* agent of inversion, it is certainly an *important* agent in this direction. The ferments present are evidently of two types; one resembling the invertase of cane sugar already described, while the other is of the maltase type.

The Secretion from Brunner's Glands. The collection of the product from these small glands offers considerable technical difficulty, and until recently no very clear statements were found in the literature as to the exact nature of the secretion. By taking special precautions, however, Glaessner succeeded in securing the secretion free from other fluids, and has found that it possesses marked proteolytic properties in solutions of all reactions. The digestion of protein is carried to the stage where tryptophane may be easily recognized. The name *pseudopepsin* may be given to the active enzyme. A lipase and an inverting enzyme are also present.

Erepsin. Comparatively recently a ferment called erepsin has been described by Cohnheim in the intestinal juice. It does not digest the true proteins but has the power of splitting albumoses and peptones as far as the mono and diamino acids. While trypsin has this power, it is very weak as compared with erepsin, which seems to be the important agent concerned in the last change in the proteins. Under the old view of protein digestion there was no place for a ferment of this character, as extensive cleavage of proteins was assumed to take place only in artificial media. But the newer views of protein metabolism and protein synthesis are perfectly consistent with the profound hydration of the digesting material, which erepsin is able to effect. In all discussions, then, of protein splitting in the body erepsin must be considered as one of the most active factors.

Enterokinase. This name has been given to a ferment-like body which occurs in the intestinal juice and which has the power of activating trypsinogen. Without the presence of this activator, it is held by some recent writers, trypsin is not formed and therefore cannot digest protein. The enterokinase is not a digestive agent itself, but a co-ferment of great importance. This ferment is one of a class much discussed recently. In many reactions two enzymes seem to be concerned, or perhaps, better, an enzyme and an *activator*. These activators are sometimes called *kinases,* and in some cases they are not actual enzymes themselves.

In the brief discussions of the last few chapters it has been shown in a general way how the important classes of food stuffs through the action of enzymes in different parts of the body are gradually brought into a condition suitable for assimilation and absorption. They have undergone digestion and are ready to be carried through the intestinal walls into the blood stream to be used as food for the building up of the body or as oxidation material for the production of mechanical energy and heat. These various digestive processes differ in many ways, but they have this important element in common which must be kept in mind: they are essentially hydrolytic in character, the addition of water by the enzymes being the essential feature in all of them. We have next to consider some changes in the intestines in which hydrolysis does not play an important part.

CHAPTER X.

CHANGES IN THE INTESTINES. THE FECES.

In an ideally normal condition of the alimentary canal after the completion of the digestive processes described in the last chapters, there should be practically nothing left finally in the intestines but residues of non-nutritive value, along with broken down products from the digestive agents themselves. Every trace of sugar or starch should have been brought into the form of a monosaccharide and absorbed; every particle of fat should have been hydrolyzed or emulsified and then carried into the lacteal circulation; while the proteins should have reached the form of higher albumoses or peptones and have been likewise absorbed. The actual situation approaches this ideal condition only approximately. In the first place the foods we consume are not absolutely pure fats, carbohydrates or proteins. They all contain some mineral matter which may escape the various digestive actions, and they usually contain certain organic substances which are only partially digestible. Some vegetable foods, for example, contain relatively large quantities of cellulose, which is a body related to the carbohydrates but which is not attacked by the weak digesting enzymes. In the foods of animal origin there are likewise substances which are very difficult of digestive hydrolysis. This is true of some of the albuminoids; horn-like substances, for example, are practically not attacked, while the cartilaginous and similar bodies are but slowly changed. From foods containing portions of such compounds a residue would always be left therefore, and in the case of poor, cheap meat this residue might be considerable.

OTHER FERMENTATIONS.

Bacterial Processes. But the case is complicated by other considerations. Our foods carry hosts of acid and putrefactive ferments with them; and some of these at times work through the stomach into the intestine, where they start reactions of their own. Just what changes take place in the small intestine depends on the character of the food. Following the alkaline zone where the pancreatic secretion, the bile and intestinal juice rapidly effect the changes already described, there is a zone of acid or neutral reaction where certain fermentation processes of bacterial origin take place. If the food is rich in carbohydrates this fermentation may be considerable, resulting in the formation of appreciable quantities of lactic, butyric and acetic acids. The liberation of these acids at this stage is a matter of very considerable importance, since it prevents the breaking up of not yet absorbed protein by bacterial putrefaction. If the acids were not present bacteria would reach the small intestine in enormous numbers from the large intestine and greatly modify the conditions there. While, along with the acid-forming bacteria, a few others are always present in the small intestine, the real putrefactive changes do not begin to a marked extent until later, when what remains of the food passes down into the large intestine. Ordinarily the small intestine is devoid of disagreeable odor, showing the absence of putrefactive changes.

It is evident therefore that the chemical nature of the food is a factor of great importance in determining the character of the complex reactions which follow the real digestive processes in the upper part of the intestine. Here we have normally the work of enzymes, and this is always followed by bacterial destruction of what is left. But we must distinguish between fermentation changes and putrefactive changes, the former being characteristic of carbohydrate food and the latter of protein food. As one or the other of these predominates, the chemical processes taking place must vary. Throughout the length of the small intestine, and in the beginning of the large intestine active absorption takes place, but between the enzymes and the bacteria a struggle for the possession of the field is in progress all the time. Theoretically, without the bacteria, the foods would undergo complete digestion and be practically all absorbed, but before this ideal condition can be reached the parasitic bacteria begin their work and rob the body of part of its food.

Acid Fermentation. Just when this competition on the part of the acid-producing bacteria begins is hard to say. Through the upper third of the small intestine the reactions are essentially those of true pancreatic digestion, and there is at no time a sharp line of demarkation between this zone and the following one. The point in the intestine where the acid fermentations begin is a fluctuating one and must vary with the time which has elapsed since the beginning of the digestion as well as with the character of the food. The enzymic and acid fermentation zones must besides overlap each other; that is, in the central part of the tract the two kinds of changes must go on simultaneously. Lactic and butyric fermentations are favored by a nearly neutral medium, and this is for a time secured by the slow neutralization of portions of these acids formed through the alkali of the pancreatic, the intestinal and the bile secretions. As the foods push farther down the neutralizing action of the alkali becomes less and less marked, and finally the characteristic acid decomposition becomes the principal feature.

In some animals this acid fermentation, to the almost complete exclusion of putrefactive changes, is easily recognized. The food of the herbivora contains an excess of pentoses, starches and other carbohydrates, and these produce sugar enough to furnish a large portion for intestinal fermentation. The feces of these animals have not the disagreeable odors of those of carnivorous animals, where the putrefactive reactions are very marked and the fermentations of very minor importance. In animals with a mixed diet this condition can be very largely changed at will by causing a variation of the food given them.

With the disappearance of the larger part of the carbohydrates through absorption and acid fermentation, the products of fermentation being themselves partly absorbed, the activity of the putrefactive organisms gains the upper hand and large numbers of complex reactions follow. The nature of some of the bodies produced in this way has been already referred to and further facts may now be given. In laboratory experiments on pancreatic digestion it will be recalled that two general results are obtainable. In working with the pancreas or pancreatic extract plus fibrin or casein we add thymol or chloroform water if it is desired to secure the maximum enzymic effect, but if, on the contrary, the bacterial as well as the enzymic decompositions are desired this protective addition is omitted and putrefaction soon becomes apparent. In the animal body the acid fermentation products take the place in a measure of the chloroform or other substances used in the laboratory experiments. Indol and skatol have been already referred to as characteristic disintegration products resulting from the action of bacteria on proteins; there are many others in addition to these, and most of them are compounds of the aromatic group. Under the conditions of their appearance in the intestines these disintegration residues must be largely formed from the albumoses, peptones, leucine and tyrosine of the previous enzymic digestions; in comparatively rare cases it is possible that the putrefaction may take place with portions of left-over original proteins which for some reason escaped digestion proper. Phenol, paracresol, phenylacetic acid, phenylpropionic acid, para-oxyphenylacetic acid, glycocoll, methyl mercaptan, hydrogen sulphide, marsh gas and still other substances, including various volatile fatty acids and carbon dioxide, have been found here along with the indol and skatol. These various products are produced mainly in the

CHANGES IN THE INTESTINES. THE FECES.

large intestine, and here again we find certain limitations to the extent of the reactions. Through the small intestine the contents have remained very soft and liquid, but in the large intestine normally a marked absorption of water takes place, from which the contents become thick and at times almost hard. This loss of water interferes greatly with the progress of putrefaction. In addition to this the work of the bacteria is hindered by the accumulation of the products of their own production. Some of these products have to a certain degree a bactericidal action and tend to check the more rapid bacterial development. It follows therefore that when the rectum is reached in the downward progress of the intestinal contents there may still be present remains of putrescible matters which might have been broken down if all the conditions had been favorable.

This brings us to a consideration of the final remains or the feces, but first a word must be said about the absorption of certain products in the lower stretches of the intestine. Not only are the normal digestive products taken up from both intestines, from the small intestine mainly, but various products of the bacterial reactions referred to follow the same course. The importance of this fact is very great from two directions at least. The excessive production of such a body as indol is always a consequence of increased bacterial activity which is often a pathological phenomenon. The indol may escape partly with the feces, but a large portion is always absorbed and is oxidized in the tissues, or in the liver mainly, from which it passes into the blood and later into the urine, where it is recognized in the form of indican. The amount of indican and certain similar substances detected in the urine is a measure then of the extent of putrefactive changes going on in the intestine. But it must be remembered that other putrefactive processes, besides those of the intestine, may furnish a *small portion* of the indol. In this oxidation an atom of oxygen is taken up and indoxyl is formed:

$$C_8H_7N + O = C_8H_6(OH)N.$$

This indoxyl, like other basic substances, always finds sulphuric acid to combine with to yield indoxyl sulphuric acid or a salt of the form

$$\begin{matrix}C_8H_6N\\K\end{matrix}\Big\rangle SO_4$$

or indican.

Phenol is another product of intestinal putrefaction and in great part passes also into the circulation from the lower intestine to reach the urine finally in the form of ethereal sulphate. Certain aromatic oxy-acids are also formed in the putrefactive processes, and are likewise absorbed. Other substances referred to above as putrefactive products follow the same course; in part they escape with the feces, and in part they suffer absorption to be more or less changed and finally eliminated by the urine. The importance of the two substances for our purpose is mainly diagnostic. They are not absorbed in sufficient quantity to be poisonous, but if found easily in the urine this points to a more than normal intestinal disintegration of protein substances or their derivatives. If the lower intestine becomes for any reason clogged with fecal products which prevent the easy downward passage and escape of the contents of the small intestine, time is given for the more prolonged action of the bacteria, resulting in the accumulation of these disintegration products. In nearly all conditions of high fever the same thing is observed. The urine test is frequently therefore a suggestion of an approaching pathological condition, or of an aggravated condition.

In another direction these bacterial products have interest and importance. While the traces of indol, phenol, etc., found may be quite harmless, it does not necessarily follow that other things produced in the same way may be equally harmless. On the contrary, some of the putrefactive products found in the intestine are violent poisons and their absorption constitutes an element of danger to the body as a whole. In laboratory experiments it is an extremely simple matter to obtain from certain bacterial cultures soluble products which are very toxic. These are the toxins formed by the bacteria and when injected into the circulation of animals are capable of producing poisonous effects. Similar bodies are undoubtedly formed in the intestines if the bacteria there present become excessive in number. Sometimes the microorganisms themselves penetrate the intestinal walls and pass to other parts of the system, being collected finally by the urine. But the peculiar poisons produced by them are much more likely to be absorbed into the circulation and give rise to special symptoms at points far removed from the infected intestine. No one of these intestinal toxins has been isolated and definitely recognized, but with them other bodies are formed which are readily detected in the urine, as is the indican referred to above. In the urine of typhoid fever, and of other pathological conditions also, certain complex aromatic products are always present which give rise to the well-known reaction designated as the diazo reaction of Ehrlich. When a mixture of weak solutions of sodium nitrite and sulphanilic acid is added to this pathological urine under certain conditions it strikes a carmine to garnet red color, due to the formation of an azo compound of some kind. The urine must add an aromatic body, different from those normally present, to aid in the formation of this azo coloring substance. With normal urine an orange color is usually obtained, but this deeper red is characteristic of some bacterial product probably, of the exact nature of which we are still in ignorance.

THE FECES.

Composition. In the lower intestine the absorption of water is one of the most important of the changes taking place and this leaves what remains in a semi-solid condition ready for final discharge. The amount of water left in the feces is quite variable, and although the fecal mass may appear hard the water content is usually 70 to 85 per cent. In the thinner pathological discharges it may be much higher.

At first sight it might appear that the feces should consist mainly of undigested residues, and this was long held to be the case; but we now know that such substances may make up the least important part of the discharge. The several kinds of products present may be roughly divided as follows:

1. Bacteria.
2. Products formed by bacteria.
3. Remains of the digestive ferments.
4. Epithelium and mucus from the intestinal walls.
5. Food residues partly or wholly undigested.

In the normal fecal discharge all these groups are represented, but incidentally there may be many other things present. The bacteria may make up one third of the whole weight of the feces at times. There are often substances which become accidentally mixed with the food and which are not attacked by the digestive secretions. There may be remains of various substances taken into the body as remedies, for example, oxide or sulphide of bismuth from bismuth subnitrate, or chalk or other insoluble substance taken in the same way.

NORMAL FECES.

For comparison it is necessary to have something as a standard, and as such a fecal discharge from a condition approaching starvation might be taken. In such feces there are no food residues, but the other things are abundantly represented. Many analyses of feces have been made from persons who for a period of several days had consumed no food and these give some idea of the character of the discharges which might be expected when the minimum of food is consumed and no more. It has been calculated in this way that about 10 to 12 grams daily is the average normal discharge from a man of 70 kilograms weight, due to other sources than the remains of food. Numerous attempts have been made to find the average composition of feces from a diet which contains just enough protein, fat and carbohydrate to keep the body in normal condition. Some of the results are given in the table below, in per cent.

	Water.	Dry Subst.	Fat.	Nitrogen.
Mixed diet	76.5	23.5	6.2	1.0
Mixed diet	85.0	15.0	4.0	0.9
Milk diet	71.2	28.8	4.8	1.4
Milk diet	77.0	23.0	2.7	0.9

A better idea of the composition of normal feces from a full and varied diet is shown in the following table, obtained from the analyses of the feces of six men under observation in the author's laboratory through a period of four months. The total feces were collected daily, mixed, in periods of about a week each, and analyzed. The figures below are the means of sixteen analyses, and give a good view of the general composition. The results are given in grams for each 24 hours, and the so-called "crude fat" refers to the ether extract of the dry feces, not including the fat combined as soaps.

Subject.	Moist weight.	Dry weight.	Nitrogen, per cent. in moist.	Crude fat, per cent. in moist.	Nitrogen in grams.	Crude fat in grams.	Crude fat, per cent. in dry.
I	178	33.5	1.4	2.7	2.49	4.81	14.4
II	140	37.3	1.7	3.9	2.38	5.46	14.6
III	234	40.9	1.17	2.32	2.74	5.43	13.3
IV	112	25.6	1.20	3.04	1.34	3.37	13.2
V	197	32.1	1.17	1.98	2.30	3.90	12.1
VI	157	33.8	1.34	3.25	2.10	5.10	15.1

The moist feces in the adult may weigh from 50 grams about to 400 or 500 grams daily in health, or even more. The average weight is about 150 gm., as illustrated by the above table. The variations in the values from which these averages were calculated were between 70 and 309 grams daily. The variations depend on the individual and also largely on the character of the food. This last is illustrated in the following table from König's "Nahrungsmittel," where for certain foods the daily consumption is given, and also the weight of the moist and dry feces in grams.

	Food.		Feces, Grams.	
	Fresh.	Dry.	Fresh.	Dry.
Roast beef	884	366.8	65.3	17.7
Eggs, boiled	948.1	247.4	42.7	13.0
Milk	2438.0	315.0	96.3	24.8
Milk	4100.1	529.7	174.0	50.0
Milk	2291.0	296.0 }	98.3	25.3
Cheese	200.0	123.8 }		
Milk	2209	285.4 }	273.7	66.8
Cheese	517	320.0 }		
Meat	614	135.9 }		
Bread	450	303.3 }	299.1	46.5
Bacon	95.6	— }		
Cornmeal (mush)	750	641.4	198.0	49.3
Potatoes	3077.6	819.3	635.0	93.8
Rice	638.0	551.9	194.6	27.2
Flour (as bread)	500	438.8	95.2	23.5
Carrots	2566	351.6	1092.6	85.1
Peas	959.8	835.6	927.1	124.0

It will be noticed here that the highest weights of feces correspond to the high weights of certain vegetable foods which are rich in cellulose. Meat and milk in proper amount yield feces which are not excessive, but with milk and cheese in excessive amounts the weight of feces becomes large.

The mixed diets consumed by the subjects of the experiments from the author's laboratory contained considerable amounts of the so-called breakfast foods, as well as fruits and vegetables, and the presence of these always tends to hinder the complete utilization of protein. The apparently high loss in protein is not all waste, however, as suggested above. It is not possible to consume a diet which is satisfactory through a long period, and still show no loss of nitrogen.

In what may be called normal feces certain relations exist between the nitrogen, the fat and the ash, if we understand by the term "normal feces" a product containing no excess of unabsorbed food, as explained above. Such feces contain as nitrogen compounds only those substances that are left over from the digestive secretions or bacterial ferments, or are produced from the intestines themselves, while the "fats" are ether-soluble products of similar origin, rather than the original complex glycerides. In some cases recently reported the following figures were obtained which will serve as illustrations. Three dogs of similar character were fed on a meat diet through a number of days, two receiving just enough to keep them in nitrogen equilibrium, while the third received an excess of meat. The results of analyses of the feces were, from the dried substance, in per cent amounts as follows:

Dog.	N.	Fat.	Ash.
I	8.59	13.18	19.24
II	8.85	11.46	22.09
III	10.56	10.12	14.14

The high nitrogen of No. III indicates an excess of protein; this being high, the fat and ash must be correspondingly low. The difference in the two kinds of feces becomes more apparent when it is remembered that a much higher factor must be used in multiplying the N values to obtain original substance in III than is probable for I and II. Since in III an excess of protein is known to be present the factor approaches 6.25, while for I and II it is probably not over 4.5 or 5.0. Something will be said below about the general character of the important fat and nitrogen substances present in the feces.

THE ANALYSIS OF FECES AND INTERPRETATION OF RESULTS.

This may extend to the recognition of a large number of products, but usually includes the detection and determination of a few important ones only. Fat of some kind is always present, hence, a qualitative test is of little value. The total amount of fat must be determined by some kind of an extractive process. Nitrogen is determined generally by the Kjeldahl method and special tests are made for proteins or their more immediate derivatives. Occasionally unchanged starch and other carbohydrates are present which may be recognized by methods given below. The amount and character of the ash is sometimes of value, likewise the reaction. Below a few details will be given about some of these tests.

Separation. Ordinarily tests are made on feces of mixed diets, but frequently it is desirable to observe the character of feces following special diet and not modified by the product from a previous or later diet. To separate the feces for such tests several schemes have been proposed. It is best to give some inert substance at the beginning of the period which will pass through the stomach and intestines unchanged, and at the conclusion of the period of dieting the same substance may be given. Fine precipitated or floated silica, powdered charcoal, carmine and other things have been used in this way. The feces to be examined are collected between the discharges of the inert and insoluble limiting substances. Various devices are also used to collect the feces apart from the urine, which is essential for exact tests.

Reaction. In health the reaction of feces with litmus is practically neutral in most cases, but at times either acid or alkaline reaction may be found. With excessive putrefaction in the lower intestine various aromatic products and ammoniacal compounds are formed which may show alkaline behavior with indicators sensitive in this direction. On the other hand free fatty acids may occasionally be present in sufficient quantity to give a distinct acid reaction. In speaking of the reaction in the intestine the conditions for the formation of light fatty acids were explained. Putrefaction on the one hand and fermentation on the other are the important factors in this connection, and these depend in turn largely on the diet. Meat diet gives usually neutral or alkaline reaction; with excessive carbohydrates the reaction may turn to acid. With infants on mother's milk the reaction is commonly acid, while with cow's milk it is neutral or alkaline. These tests refer to the behavior with litmus, but with phenol-phthalein, which is not very sensitive with weak alkalies, an acid reaction due to carbonic acid is often obtained. This fact should be kept in mind, since the question of reaction is often a question of indicators.

In testing for the reaction some of the mixed feces is spread on one side of the test paper by means of a glass rod; the color effect is looked for on the other side of the paper. If the feces are not quite moist it will be necessary to rub up with a little water.

Dry Residue or Solids. As explained above, the larger part of the fecal discharge is always water. The amount of solid matter is best obtained by drying a weighed portion, at a relatively low temperature, in a current of hydrogen or air. By evaporating over a water-bath there will be always some loss of volatile substances besides water. It is very difficult to obtain a perfectly dry product on the water-bath in most cases, especially if fat is present. For most purposes it is safest to evaporate a relatively large amount to moderate dryness on the water-bath, after mixing the weighed feces with a little alcohol. This air-dry product is weighed and finely powdered and a new portion is weighed out for the final complete drying, at a temperature of 105° in the air bath. There will be some little loss by volatilization of light acids and other substances.

For this kind of work a vacuum drying apparatus which can be heated to a moderate temperature renders good service. It is also possible, where time is not an object, to finally dry the air-dried product in a desiccator *under* sulphuric acid; that is, in the form of drying apparatus in which the acid is above and the substance below. For this purpose the air-dried feces must be thoroughly powdered, or distributed in a very thin layer. In some pathological stools there is an abundance of fat, even to one-half of the total solids. In such cases a perfect drying is always difficult with any process.

Specific Gravity. An exact determination of this datum is not easily made, as the occluded gases interfere greatly with the test. The normal specific gravity is about 1.045 to 1.070, but may be much lower pathologically. Fatty stools may have a specific gravity as low as 0.935.

THE TOTAL FATS.

In the analysis of feces a number of substances are included under the term "fat." In the extraction of dried feces with some solvent everything which goes into solution is classed as crude fat, to be more fully identified by special tests later. Besides the fats proper feces may contain fatty acids and their soaps, traces of lecithin, cholesterol, cholalic acid and other bodies soluble in ether or chloroform. In the acidified feces these substances go into solution, the acids of the soaps being taken up also. In the feces of adults the fatty acids combined as soaps may make up 30 to 40 per cent of the total "crude fat," obtained after acidification.

For this extraction it is customary to add enough acid to impart a faint acid reaction to the feces and then evaporate to dryness with addition of sand or other inert insoluble substance. The dry residue may be transferred to a paper tube and extracted with anhydrous ether in the Soxhlet apparatus. A better plan is to spread a weighed portion of the mixed and acidified feces over paper such as is employed in the well-known milk fat extraction process. The test is completed by drying the paper and extracting in the Soxhlet tube, as in the case of milk. The results so obtained are higher than those from the ordinary process and the time required for extraction much shorter. But it is not easy to obtain in this way enough fat for further study, as not much more than 10 gm. can be easily worked.

The amount of crude fat in the dry feces is variable, but may make up in the mean about 25 per cent if the acids combined as soaps are included. Much of it under normal conditions must be derived from other sources than the unutilized original fat of the foods; a portion is always derived from residues from some of the intestinal secretions, and from organized elements thrown off from the walls of the intestines. The extent of the utilization of the food fat depends largely on its physical character, especially its melting point. The solid fats with high melting point are but poorly utilized as the following figures illustrate, in which the amount of loss in the feces from different kinds of fat is given. (v. Noorden.)

	Melts.	Loss.
Olive oil	liquid	2.3 per cent.
Goose fat	25°	2.5
Lard	34	2.5
Bacon	43	2.6
Mutton tallow	49	7.4
Stearin plus almond oil	55	10.6
Pure stearin (and palmitin)	60	90.0

The free fatty acids do not appear to be as well absorbed as are the neutral fats, and in general from mixed vegetable foods the fat loss in the feces is much greater than from the animal foods.

CHANGES IN THE INTESTINES. THE FECES.

Pathologically there may be a very great increase of fat in the feces, so great in fact as to be readily recognizable by the eye. This is especially true in cases where the flow of bile into the intestine is diminished or altogether hindered. The fat in the dry feces may then amount to 50 per cent of the whole. Any derangement of the normal pancreatic functions leads also usually to an increase of fat in the feces. In the last case protein and carbohydrates would suffer also in absorption.

ANALYSIS OF THE CRUDE FECES FAT.

The method of separating or extracting the crude fat has been briefly referred to above. An extract obtained in this way may be used for a number of tests after having been weighed. The recognition of all the substances in it is practically out of the question, but there is no difficulty in making an approximate separation if enough fat be used. A simple heating test will show the presence of light and volatile fats, which, however, are not usually present in more than traces. The following scheme will be sufficient for the recognition of the more important constituents. The extraction is completed in the Soxhlet apparatus, the ether distilled off and the residue dried and weighed.

Cholesterol. Cholesterol is not a fat chemically and therefore does not undergo saponification. This behavior makes it possible to recognize it. Add to the crude fat some alcoholic potassium hydroxide, for 1 gm. of fat about 1.5 gm. of the stick alkali in 25 cc. of alcohol. Boil under a reflux condenser half an hour and then drive off the alcohol. From the dry residue extract the unchanged cholesterol by use of an excess of ether. The substance is rather slowly soluble and a little soap may be dissolved at the same time. The result is fairly accurate. The ethereal solution of the cholesterol is evaporated and the residue weighed. By evaporating a little of an ethereal solution of the substance on a glass slide a residue is secured which will serve for microscopic identification. Cholesterol crystallizes in large thin plates.

Cholalic Acid. Cholalic acid from the bile is an important constituent of the crude fat, provided this is obtained from slightly acidified feces. This acid may be detected in the soap left after extraction of the cholesterol as just explained. The soap is mixed with a little water and acidified with dilute sulphuric acid to free all the organic acids. The mixture is extracted with ether in a separatory funnel. After completed extraction the ether is evaporated, leaving the free fatty and other acids. To these a slight excess of barium hydroxide solution is added and heat applied to form barium soaps. While warming, the mixture must be well stirred or shaken. The separation to be made depends on the fact that the barium soap of cholalic acid is soluble in about 25 parts of hot water while the true fatty soaps are not. Therefore on washing with plenty of hot water the bile acid soap dissolves. By evaporating the solution to a small volume, acidifying with dilute sulphuric acid and shaking with ether the cholalic acid will pass again into ethereal solution, from which it may be recovered on evaporation. The acid may be recognized by mixing with strong sulphuric acid. A yellow solution results which soon shows a green fluorescence. The acid may be identified also by mixing with a small amount of water and a little cane sugar, and adding then a few drops of strong sulphuric acid. A red color develops which becomes purple. The sulphuric acid must be added in just sufficient quantity to warm the mixture to about 70° or 75° C. This test is the Pettenkofer bile test.

Fatty Acids Proper. After separating the cholesterol and cholalic acid as just described the true fatty acids are left in the form of insoluble barium soaps. By acidifying with a little hydrochloric acid and shaking with ether these acids go into solution and may be recovered by evaporation of the ether. The fatty acids of any lecithin originally present are included, as the lecithin would be decomposed in the first saponification. The acids of soaps as well as of neutral fats are also included if the original extraction was made, as assumed, on acidified feces.

Lecithin. The separation of lecithin as such from feces is not practicable but the amount may be estimated from the phosphoric acid separated in the saponification. In the above tests the glycerophosphoric acid would go as barium salt along with cholalic acid into the hot water solution. The phosphate could be recognized or determined in this. But it may be estimated much more accurately by using some of the original crude fat. This is mixed with some sodium carbonate and ashed carefully. Then a little saltpeter is added to complete oxidation; the fused mass is dissolved in water. The phosphoric acid may be determined by titration with uranium nitrate, or, better, with molybdic acid by the Pemberton method. In this way it is usually possible to secure enough phosphate to make an accurate titration, in most cases, by starting with a gram of crude fat.

Soaps. The above tests give the total acids. It may be desirable to measure the amount present in the form of soaps. For this purpose a double extraction is necessary. In one case the feces are dried and extracted with ether without preliminary acid treatment. The soaps, not being ether soluble, remain behind. Then a second extraction, after acidification, is made; the result gives the total fats and acids and the difference between the two extractions shows the acid due to soaps.

For those extractions the paper coil method is very satisfactory as sufficient extract may be obtained from about 10 gm. of moist feces for satisfactory weighing.

CARBOHYDRATES.

Under this term starch, sugar, gums and cellulose must be included. With a vegetable diet the last named is always present, while on a purely animal diet not one of the group can be found in the feces.

Starch. This substance is found commonly in feces, under normal as well as pathological conditions, and especially when the diet has contained starch in the form of coarse meal, which is difficult of digestion. This is readily shown with corn meal and other products containing much cellulose. On the other hand with fine, well-sifted flours in which the starch granules are easily turned into paste by boiling or baking, the utilization of the starch is usually much more complete. For the identification of starch several methods have been applied, some direct, others indirect.

The recognition of starch by the microscope usually fails, as the outline of the granules is destroyed by the process of cooking. When the cooking has been imperfect, however, the granules may be intact and in a condition suitable for identification. The iodine test is frequently satisfactory. For this the feces are boiled with water and filtered. In the filtrate the iodine solution is added as for other tests. To facilitate the separation of the starch from cell structures a little hydrochloric acid may be added to the boiling water.

The amount of starch is determined after conversion into sugar. This may be accomplished by prolonged heating with a little hydrochloric acid, which changes the starch into glucose. This may be measured by one of the copper reduction processes, but not without some difficulty as the solution is always highly colored. The necessary precautions can not be given here.

The normal starch content of the feces is always small, but in disorders of digestion, especially with diminished activity of the pancreas, more starch may be found. In very young children the consumption of starchy foods is very often followed by the appearance of starch in the feces.

Sugar. The presence of traces of sugar in the normal feces has been frequently affirmed and as often denied. The larger part of the more recent evidence on the question goes to show that sugar is not normally present even in traces. All forms of sugar are very soluble and easily absorbed from the intestine. The ability of sugar to escape absorption through the whole length of the intestinal tract would therefore appear very problematical. Even in disease sugar is of rare occurrence in the feces as far as has been determined by experiment. But the detection of traces is not an easy task.

Pathologically sugar seems to pass through the intestine and escape with the feces only when the conditions for absorption through the intestinal walls are reversed. This is the case in diarrhœa because of the more rapid movement downward in the intestine and because of the diminished or interrupted flow from the intestine to the blood. There may be an increased loss of proteins at the same time.

The detection of traces of sugar calls for a preliminary extraction, and purification of the extract from substances which might interfere with the copper or analogous tests. A fermentation test is sometimes made and without preliminary treatment. This depends on the spontaneous decomposition of the sugar by bacteria with liberation of gas which is collected and measured. Starch present gives the same result, however, but not so rapidly.

Cellulose. This is a common constituent of feces after the consumption of vegetable foods. Digestion of cellulose takes place in the small intestine of man, but in the large intestine there is sometimes a bacterial destruction. The detection of cellulose is not difficult, although the methods are somewhat complicated. The best of the methods depend on the solution of other carbohydrates by treatment with weak acid and then with alkali and finally with water. The mixture is filtered and the residue washed with alcohol and ether; it is crude cellulose contaminated with a little ash and protein substance, both of which may be determined. The appearance of cellulose in the feces has of course no pathological significance.

Gums. As these are not common articles of food they do not occur usually in the feces. When they are consumed in pastry and confectionery they may be found later in the feces since they are not digested with readiness in many cases. Some of the gums are but slightly soluble and undergo pancreatic digestion slowly. Experiments have shown that gum tragacanth and gum arabic may be found in considerable quantity after their consumption in bon-bons.

NITROGEN AND THE PROTEINS.

Nitrogen is found in the feces in many combinations. Some of these represent residues from the digestive operations and some are found in secondary products formed by bacterial or chemical action. Some of the molecular combinations are large, while others are relatively small. No conclusion, therefore, as to the weight of the nitrogenous bodies can be drawn from the nitrogen found, but the datum has value from other standpoints.

Total Nitrogen. The total nitrogen in the feces may be accurately determined by a combustion process, but most readily by the Kjeldahl process which is now everywhere employed. This depends on the conversion of the nitrogen into ammonia by prolonged heating with sulphuric acid to which a very little metallic mercury is added. Often a mixture of pure sulphuric acid and potassium sulphate is employed. At the end of the digestion the mixture is made alkaline with a slight excess of ammonia-free sodium hydroxide and distilled. The ammonia formed is collected in standard acid and measured in this by titration of the excess of acid with standard alkali.

Even in condition approaching starvation, when no food proteins can possibly be present, the feces always show some nitrogen, which, as pointed out above, must come from the secretions thrown into the intestines and from the remains of bacteria and their products. A part of this nitrogen therefore has once been absorbed to be later thrown back into the intestine, which fact must be kept in mind in making deductions from the nitrogen found as to the loss of nitrogen in assimilation. Although usually overlooked, the nitrogen existing in the bacterial cells is an appreciable quantity and often makes up a good fraction of the whole. It has recently been shown that in normal feces nearly one third of the dry weight may often be made up of the bacteria.

The amount of nitrogen excreted increases with the food consumed in general, and especially if this food contains a large amount of indigestible substance. The nitrogen of a meat diet is always more completely utilized than is the nitrogen of beans, for example, where there is considerable cellulose to disintegrate. The nitrogen in this case is largely in the form of protein residues, and may be detected as will be pointed out below. In pathological conditions of the digestive tract there may be a great increase of unutilized nitrogen. This is more especially true of failures in the pancreatic digestion than it is of failure in the work of the stomach.

Proteins. The most important question to consider here is that of proteins themselves in the feces. Nitrogen in other forms has a far different meaning since it may represent bodies which have been already digested and absorbed, and then thrown into the intestine again. But nitrogen as protein represents practically waste in most cases. Among the protein substances which may be found sometimes in feces these may be mentioned: albumins proper, casein, nucleo-proteids, albumoses, peptones and naturally more or less of certain albuminoid bodies which are digested with difficulty. The certain detection of all these bodies under all conditions is not always possible with our present knowledge. Some of the simplest of the tests employed will be briefly mentioned. The soluble substances only are considered here.

Albumins. Acidify the fresh feces with dilute acetic acid and extract with distilled water. The acid prevents casein and mucin from going into solution at the same time. Filter through good Swedish paper and apply tests to the filtrate. Albumose and peptone go into solution with this treatment.

Albumins proper are coagulated by heating the filtrate, while the derived proteins do not respond to this test. The albumins give also the biuret test and are precipitated by solution of potassium ferrocyanide in presence of acetic acid. But albumose, not peptone, responds to the same test.

Albumoses and Peptones. By extracting as above, coagulating any albumin possibly present and filtering, the filtrate may be used for albumose and peptone tests. In the filtrate free from albumin, zinc sulphate or ammonium sulphate may be used to precipitate albumose. In the filtrate from this peptone may be recognized by the biuret test.

Casein. This is sometimes found in the feces of children on a milk diet. To recognize it these tests may be made. The fresh feces may be extracted first with rather weak sodium chloride solution to take out soluble albumins, then with weak acid to complete the removal of such bodies. The casein may next be brought into solution with sodium hydroxide, not too strong, and obtained in the filtrate. In such a filtrate acetic acid produces a voluminous precipitate if casein is present, but mucin is also precipitated. Casein, however, redissolves in an excess of acetic acid while mucin does not. After filtering the casein may usually be thrown out again by cautious addition of alkali to the neutral point, but the precipitate is not as characteristic as in the first instance.

Mucin. This was formerly supposed to be a common and abundant product in normal feces, but this is not the case. Pathologically mucin may be present so as to be recognizable by the eye. A good chemical test for small amounts is still lacking.

Nucleo-proteid. By extracting normal feces with lime water and acidifying with acetic acid, a bulky precipitate is obtained usually, which was supposed to be mucin. It, however, contains phosphorus and belongs to the proteid group. The substance is a normal product in traces and can be found in feces following a diet free from nucleins. It is therefore likely that traces of this protein are brought into the intestines from the breaking down of the intestinal walls. Pathologically much more may be found, but without having a distinct diagnostic indication.

Of all the protein substances mentioned, the casein, if it occurs in large quantities in infants' feces, has perhaps the greatest importance as pointing to imperfect digestive power. There can be no question, of course, as to its origin. Serum or egg albumin as such could rarely be present because such proteins are ordinarily consumed in the coagulated condition. When the tests point to the presence of a true soluble albumin the result shows probably the entrance of albumin from the blood by a reversal of the normal osmotic process. It must be remembered, however, that true albumin is very rarely found in the feces. Occasionally a reaction due to presence of pus or blood may be obtained, but the albumose or peptone reactions are much more frequent. In diarrhœa stools, for example, where insufficient time is given for absorption, these bodies may be found.

Insoluble Proteins. The detection of coagulated proteins and of partly disintegrated albuminoids is practically impossible. Remains of muscle fibers or other complex substances essentially protein may sometimes be recognized by the microscope but they are beyond chemical identification.

OTHER NORMAL AND ABNORMAL SUBSTANCES.

It will not be necessary to discuss the occurrence of the various putrefactive bodies of bacterial origin which are always found in the feces. We have here indol, skatol, various phenols and aromatic acids. Leucine and tyrosine are occasionally found, but their presence is generally pathological, if in quantity more than traces.

Among products of distinctly pathological origin blood and pus may be mentioned; both yield albumin and the corpuscles of each may frequently be recognized by the microscope. It is also possible to recognize the coloring matter of the blood by the spectroscope. It has been mentioned that cholalic acid, a derivative of the two characteristic acids of the bile, may be found with the fats of the feces. The bile acids themselves, glycocholic and taurocholic, are also found; likewise the bile pigments or their disintegration products. Most of the bile coloring matters fail to be reabsorbed from the intestine into which they are discharged, and must be excreted therefore by the feces, and only in small part by the urine.

SECTION III.

THE CHEMISTRY OF THE BLOOD, THE TISSUES AND SECRETIONS OF THE BODY.

CHAPTER XI.

THE BLOOD.

How Supplied. The conversion of food-stuffs into absorbable products has been discussed in the chapters of the last section. It must be shown now how these products are utilized. Sooner or later, by absorption from the stomach or the intestines, mainly from the latter organs, they enter the blood stream through two principal channels, the portal vein and the lacteal lymph vessels leading to the thoracic duct. Ordinarily the amount of absorption from the walls of the stomach is not great; only when a very large quantity of easily digested food is present in this organ or under the influence of special stimuli is the passage of digested substances into the circulation here appreciable. The small intestine with its very considerable surface gives up the bulk of the absorbable products to the blood or lymph stream.

The digested fats pass essentially into the minute lymphatic vessels known as the lacteals. At the time of digestion the contents of these vessels consists of a milky fluid termed *chyle,* but at other times the lymph flowing here is nearly clear. Minute capillary vessels leading to the portal vein take up the larger portions of the carbohydrates and protein bodies from the small intestine and thus convey them to the liver, where a number of important changes take place, the most pronounced being the conversion of the sugar more or less perfectly into glycogen. These reactions will receive attention later. Beyond the liver the hepatic veins lead to the general circulation. In this general way the nutriments reach the blood which is the main channel of distribution, but this fluid is far from being a simple solution or mixture of these nutriments in the condition in which they leave the alimentary canal. The most important of the blood constituents are, in fact, chemically quite distinct from anything produced in the course of digestion. Certain organs of the body have the important function of working over these digestive products and converting them into the things required in the blood. To do this several synthetic reactions are necessary; how these are carried out we do not know, and in some cases we are ignorant also of where they take place. In what follows some of the main facts in this connection will be given.

COMPOSITION OF THE BLOOD.

Quantitative Variations. It is evident that only an average composition can be in general considered since the fluid is in a state of constant change. Soon after a meal certain constituents would naturally be found increased, and after a period of fasting a deficiency in the same would follow. From what has been said it is further apparent that the blood of the portal vein would be found much richer in some substances than that of the hepatic vein or the arteries. It must also be remembered that the blood is not a homogeneous fluid but consists of a true solution in which are suspended certain cell structures. We may therefore consider the average composition of the blood as a whole, or of the corpuscles on the one hand and the fluid portion or plasma on the other. The specific gravity of normal blood varies between 1.05 and 1.07; the average specific gravity of the serum is about 1.03.

Approximately the blood makes up 7 to 8 per cent of the body weight; therefore in an individual weighing 70 kilograms the blood weight would be 4.9 to 5.6 kilograms. Of this blood weight about 60 per cent belongs to the plasma and 40 per cent to the corpuscles. Among the various recorded analyses of human blood as a whole the following may be taken as best illustrating the mean composition, in 1000 parts.

BLOOD ANALYSES.

	Men. Mean of 11 Analyses.	Women. Mean of 8 Analyses.
Water	779	791
Solids	221	209
Fibrin	2.2	2.2
Hemoglobin	134.5	121.7
Albumin and globulin	76.0	76.0
Cholesterol, fat, lecithin	1.6	1.6
Salts and extractives	6.8	7.4

The individual analyses from which these means are taken show rather wide variations. Some more recent analyses made in Bunge's laboratory show the distribution of the mineral matters and may be quoted, the figures here given referring to the blood as a whole.

THE BLOOD.

	Man of 25 years.	Woman of 30 years.
Water	789	824
Solids	211	176
Fibrin	3.9	1.9
Hemoglobin and albumins	199.5	164.8
Salts	7.9	8.6

The salt content was made up in each case as follows:

	Man.	Woman.
Sodium chloride	2.701	3.417
Sodium oxide	.921	1.862 + potassa
Sodium phosphate	.457	.267
Potassium chloride	2.062	1.623
Potassium sulphate	.205	.193
Potassium phosphate	1.202	.835
Calcium phosphate	.193	.418
Magnesium phosphate	.137	
	7.878	8.615

These figures do not show the distribution of the salts between the plasma and corpuscles. In the original analyses from which they are calculated by far the larger part of the sodium salts was found in the plasma, while the potassium salts were found largely in the corpuscles. The calcium and magnesium salts occur mainly in the plasma. In the blood the excess of alkali shown exists probably mainly as carbonate. All analyses seem to indicate a difference between the blood of men and women. The male blood is richer in solids. The female blood on the other hand appears to be slightly richer in the mineral salts.

Blood is characterized particularly by the peculiar compound containing iron present, known as hemoglobin. Many of the tests for the recognition or identification of blood depend on this substance, which is found nowhere else. As a whole blood is distinguished by the phenomenon of coagulation which is connected with the fibrin present. Because of the great importance of this phenomenon it will be briefly discussed here; the details of the subject belong to physiology rather than to chemistry and are not yet sufficiently worked out for clear elementary presentation.

FIBRIN AND THE COAGULATION OF BLOOD.

As has been already pointed out fibrin is the product resulting from a certain reaction in which a forerunner or parent substance called fibrinogen is concerned. As it exists in the blood vessels normally this fibrinogen is soluble and stable, but when the vessel is pierced and the contents allowed to come in contact with the air the soluble fibrinogen becomes the insoluble fibrin, which is the well-known stringy substance described in an earlier chapter. A great deal has been written on the subject of this spontaneous coagulation, which is now generally believed to be brought about by the action of a peculiar ferment formed by the breaking down of the white blood corpuscles. From these cells it appears that a special zymogen which has been called *prothrombin* is first formed; this in the presence of calcium salts yields the true fibrin ferment, or enzyme, called *thrombin*. It may be easily shown that the addition of ammonium oxalate or some other precipitant of calcium salts to freshly drawn blood will prevent its coagulation. It was formerly held that the calcium compounds enter into a chemical combination as part of the fibrin molecule, but Hammarsten's researches seem to show clearly that the part of the calcium is in the formation of the ferment.

In this coagulation it appears that a portion of the original fibrinogen is split off, yielding a product known as fibrin-globulin, which remains in solution; that is, the whole of the fibrinogen does not coagulate as such. The coagulation may be prevented or greatly retarded by addition of oxalates as just referred to, and also by addition of several other foreign substances, as acids, alkalies, strong solutions of alkali salts, sugar, gum, albumose solutions, glycerol, etc. An excess of carbon dioxide delays coagulation, as shown by the slower coagulation of venous blood. Blood collected from a vein in a polished vessel of porcelain or in a vessel whose sides have been covered with oil or vaseline coagulates slowly. On the other hand collecting in a vessel with a rough surface hastens coagulation, as does any mechanical agitation. It has been shown that a polished platinum wire may be passed through a vein without inducing coagulation, while a thread in the same position will collect a layer of fibrin.

The various observations which have been made, while not affording a full answer to the question why the blood does not coagulate spontaneously in the living veins or arteries, suggest several important reasons to account for this absence of the reaction. One of the factors evidently present in all ordinary coagulations is contact with a rough foreign substance. The foreign substance need not be larger than the specks of dust which blood can gather from the air. In leaving a vein or artery blood naturally comes in contact with such particles, and these serve as nuclei for the beginning of coagulation; much as a minute dust particle may be sufficient to start crystallization in a strong solution of alum. In the body the blood is normally in contact with vessels with very smooth walls. If such a vessel be ligatured at two points and the sac thus formed be cut out it will be found that the contained blood will remain fluid some hours or days even. This shows that contact with *living* walls is not the element preventing coagulation.

Apparently blood exists normally in a very peculiar condition of equilibrium, in which not one but several factors are concerned. The same may be said of the equilibrium of many salt solutions. Changes of temperature, the addition of foreign bodies in traces even, stirring, pouring from one vessel into another, or contact with the dust particles of the air in the one case as in the other may induce a change. In the living vessels of the body as well as after leaving the body the equilibrium may be destroyed and a coagulation take place. This is illustrated in the intravascular clotting after wounds in which the vessels as a whole may not be impaired; injury to the lining endothelium results in throwing foreign particles into the blood stream sufficient to induce clotting or coagulation.

EXPERIMENTAL ILLUSTRATIONS.

Some of the simpler phenomena connected with the coagulation of blood may be readily shown by experiment.

Experiment. Have ready two test-tubes. Pour into the first one cc. of a cold saturated solution of sodium sulphate, the other is left clean and dry. Decapitate a rat and allow two cc. of the escaping blood to flow into the tube containing the sodium sulphate. The rest of the blood is collected in the dry tube. In a very few minutes coagulation takes place in the latter tube, while it is prevented by the sodium sulphate in the former.

Allow both tubes to stand at rest a day or two. In the salted tube it will be noticed that most of the corpuscles have settled to the bottom, leaving a clear and lighter colored liquid, while in the other tube the coagulum has begun to shrink into a smaller mass, from which droplets of yellowish serum ooze. The corpuscles in this remain with the fibrin.

Experiment. Collect a quantity of slaughter-house blood by running two volumes of the latter into one volume of saturated solution of sodium sulphate. Shake the mixture and allow it to stand at a low temperature several days. Coagulation does not occur, but a gradual precipitation of the corpuscles is observed, leaving a yellowish liquid known as salted plasma, which may be poured off and used for various experiments.

Experiment. Pour a few cc. of the salted plasma into a test-tube and dilute it with several times its volume of water. On slight warming of the mixture, coagulation follows. The effect of the sodium sulphate is to prevent coagulation. In this case dilution favors it.

Experiment. Pour some fresh blood into a clean vessel and stir it thoroughly with a glass rod, if a small quantity in a beaker is taken, or with a stick if a larger volume, as of slaughter-house blood, is used. The fibrin gradually separates, and entangles most of the corpuscles. Save the serum for tests to be explained and wash the crude fibrin thoroughly under running water to remove the corpuscles and coloring matter. The well-washed fibrin is white and stringy. Fibrin so prepared is employed in many experiments, especially in illustrating digestion phenomena. On the large scale it is used in the manufacture of peptone.

Experiment. To illustrate the ready digestion of fresh fibrin use about half a gram with 10 cc. of 0.25 per cent hydrochloric acid. Keep the mixture some hours at 40° C. The fibrin gradually dissolves to form acid albumin, which may be obtained in solution by filtering from any undigested residue. The careful addition of a little sodium carbonate solution produces a precipitation of the acid albumin.

TIME OF COAGULATION. It has long been observed clinically that the time required for the coagulation of a drop of blood withdrawn by a needle is not constant but varies, and in a marked manner in certain diseases. Based on this observation several forms of apparatus have been devised in which the rapidity of coagulation may be followed and measured. One of the best known forms it that of Wright, which consists of a number of small glass tubes of uniform bore, and open at both ends, into which definite volumes of the blood in question may be drawn. After being filled with blood the tubes are immersed in warm water of body temperature, or in some cases at a lower definite temperature. From time to time a tube is removed from the bath and tested by blowing. As soon as coagulation begins the blood can no longer be blown out easily, and the time required for this is noted. In health this time may be four or five minutes usually, but in jaundice and some other diseases it may be much longer. The time varies somewhat with the form of apparatus used. In the Boggs coagulometer the time required for the clotting of a drop of blood of definite size and shape is followed under the microscope.

BLOOD TESTS.

The serum left after separation of the fibrin by stirring, contains much of the blood coloring matter and may be used as well as the fresh blood for many tests, some of which will be illustrated here.

Experiment. GUAIACUM TEST. To a little blood solution in a test-tube add some fresh tincture of guaiacum and then a few drops of an ethereal solution of hydrogen peroxide. Shake the mixture and observe that the precipitated resin has assumed a blue color, more or less marked. In this test turpentine oil, which has been shaken with air in a bottle, or which has been exposed to the air, can be used instead of the solution of peroxide. Hydrogen peroxide is developed by the action of oxygen on turpentine. In this test the hemoglobin seems to act as a carrier of oxygen to the resin. The oxidation product of the resin is blue.

Experiment. HYDROGEN PEROXIDE TEST. A reaction somewhat similar to the above in principle is observed on mixing 2 cc. of the blood with 10 cc. of the commercial hydrogen peroxide solution. The hemoglobin brings about the decomposition of the peroxide with liberation of oxygen, which escapes, producing froth.

Reaction of Blood. The normal reaction of blood is alkaline, which cannot be observed, however, in the usual way because of the marked color of the pigment. It may be readily seen by working in the following manner:

Experiment. Prepare some small plaster of Paris surfaces by pouring the well-known plastic mixture of plaster of Paris and water on glass plates and allow it to harden several hours at least. The prepared plates are removed from the glass and soaked in a neutral solution of litmus and are then allowed to dry. The test proper can now be made by putting a few drops of the blood on the smooth plaster surface and allowing it to remain there five minutes. It is then washed off with pure water, when it will be found that the part of the plate which has been covered by the blood has become blue from the action of the alkali of the blood on the neutral litmus.

Experiment. HEAT TEST. Heat the solution of blood until it is near the boiling temperature and note that the red color is largely destroyed and that a brownish precipitate forms which contains albumin and decomposed coloring matter. Add now a small amount of sodium hydroxide solution and observe that the precipitate disappears while the blood solution becomes red again by reflected light, but greenish by transmitted light.

Experiment. HEMIN CRYSTALS. When acted on by acids or strong alkalies hemoglobin of blood is broken up into globin and a characteristic compound called hematin. Hematin in turn is acted upon by hydrochloric acid yielding the hydrochloride, hemin, which appears in crystalline form. From the name of their discoverer, these crystals are called "Teichmann's crystals." Their appearance constitutes one of the best tests we have for blood, and can be illustrated by the following: Evaporate a drop of blood on a slide, add two or three drops of glacial acetic acid, and boil. Put on a cover glass and allow to cool. Minute (microscopic) plates or prisms separate out. If old blood, a stain, for instance, is examined, it is necessary to add a small crystal of sodium chloride to the acetic acid, by which means sufficient hydrochloric acid is liberated for the test. The crystals have a dark brown color and are very characteristic. The usual forms as found in human and other blood are shown below.

FIG. 9. Hemin crystals. 1 is from human blood; 2 from a seal; 3 from a calf; 4 from a pig; 5 from a lamb; 6 from a pike; 7 from a rabbit. (LANDOIS.)

FIG. 10. Hemin crystals from stains of human blood. (LANDOIS.)

The most certain means of identifying blood, however, depends on the peculiar behavior of hemoglobin toward light, which will be shortly explained.

HEMOGLOBIN.

Composition. In the systematic classification of the protein bodies hemoglobin is grouped among the proteids or compound substances, inasmuch as it may readily be broken up into a fraction containing iron called hematin, and a histone substance called globin. This cleavage is very easily effected by the action of weak acids and in the mean the hematin fraction is found to amount to about 4.3 per cent. In some experiments as much as 94 per cent of globin has been recovered. It is therefore likely that only the two substances are present. The properties of hemoglobin are not quite constant, inasmuch as from different bloods products of slightly different composition have been obtained. It is possible to secure the hemoglobin in crystalline condition suitable for analysis. A number of such determinations have been made and from them formulas have been calculated. These formulas can be at best only more or less close approximations, but they are interesting as illustrating the great molecular weights here concerned.

Hemoglobin is dextro-rotatory. By an ingenious method Gamgee and Hill have found $a_C = 10.4°$. The globin from it is levorotatory.

Analyses of Hemoglobin. Several results obtained by different observers are here given. The variations must be partly due to differences in methods of preparation and analysis.

	C	H	N	S	Fe	O	Author.
Horse.	51.15	6.76	17.94	0.39	0.335	23.42	Zinnofsky.
"	54.40	7.20	17.61	0.65	0.47	19.67	Huefner.
Dog.	54.57	7.22	16.38	0.57	0.336	20.43	Jaquet.
"	53.85	7.32	16.17	0.39	0.43	21.84	Hoppe-Seyler.
Hen.	52.47	7.19	16.45	0.86	0.335	22.5	Jaquet.

In the first analysis the ratio of the sulphur atoms to the iron atoms is $2:1$; in the third analysis it is $3:1$. On the assumption that the molecule contains but one atom of iron the minimum molecular weight which may be calculated from this analysis is:

$$C_{758}H_{1203}N_{195}O_{218}FeS_2$$

It is interesting to note that the molecular weights found in this way are practically confirmed by the determinations made on the combining power of hemoglobin for carbon monoxide. Assuming that one molecule of carbon monoxide is held by one molecule of hemoglobin, observations of the volume of the gas absorbed by a given weight of the blood pigment lead to practically the same result as was obtained by the iron method.

Combinations of Hemoglobin. The great importance of hemoglobin depends on its power of forming several more or less stable combinations with certain gases. Of these combinations that with oxygen is by far the most important; we distinguish therefore between hemoglobin and oxyhemoglobin. The common form of the substance is really the latter, although it is usually referred to by the simple term—hemoglobin. The oxygen of oxyhemoglobin is very loosely held and may be driven out from its union by the aid of a current of other gases, or by the pump. The amount so held corresponds to two atoms of oxygen for each molecule of hemoglobin. This oxygen combining power in some way depends on the presence of the iron of the hematin.

Oxyhemoglobin. By various methods this substance may be obtained in crystalline form, the crystals being often 2 mm. or more in length. From blood of different animals different crystalline forms have been observed. In all cases the crystals are red and soluble in water; they are more easily soluble in water containing a little sodium carbonate. They are insoluble in ether, benzene and chloroform, and the water solubility varies with the nature of the blood from which they were made, that from the blood of man and the ox, for example, being easily soluble, while the oxyhemoglobin from the blood of the horse or dog is rather slowly soluble. Because of their solubility it is very difficult to secure crystals from human blood, but from dog's blood they may be made as follows:

FIG. 11. Crystals of hemoglobin. *a* and *b* from human blood; *c* from the cat; *d* from the guinea-pig; *e* from the marmot; *f* from the squirrel.

Experiment. Beat up 100 cc. of the blood thoroughly, cool to a low temperature and add 10 cc. of ether and a little water. Shake this mixture thoroughly and allow it to stand on ice over night. Filter on porous paper, squeeze out the mother-liquor as far as possible, dissolve in a little water, filter again and to the new filtrate add one-fourth its volume of alcohol, meanwhile stirring constantly. Allow the mixture to stand to crystallize. An illustration is given of the usual forms.

A simple test may be also made by mixing a drop of dog's blood with a drop of water on a slide and allowing it to partly evaporate. A cover glass is then put on and the crystals are looked for with the microscope.

The conditions of combination between hemoglobin and oxygen have been studied by several authors. It has been found that by exhausting blood under the air pump the greater part of the oxygen becomes free. It has been found further that 1 gm. of hemoglobin may be made to take up or give off something over 1.3 cc. of oxygen. This reduces to 2 atoms of oxygen for 1 atom of iron present in the hemoglobin. Various chemical agents have the same effect. In the case of certain solutions the action is a chemical one, while with several inert gases the action is physical.

These reactions are accompanied by a change of color in the oxyhemoglobin or blood solution experimented upon. Oxyhemoglobin solutions show a brighter red color than do those containing the reduced hemoglobin. This difference is well illustrated in the contrasting shades of arterial and venous blood, the former containing plenty of oxygen in combination while the latter is deficient. The loss of oxygen is illustrated by some simple experiments:

Experiment. Shake about 10 cc. of diluted defibrinated blood with a few drops of ammonium sulphide solution or with Stokes' reagent. (Stokes' reagent is a solution of ferrous sulphate, to which a small amount of tartaric acid has been added, and then ammonia enough to give an alkaline reaction.) Warm gently, and observe that the bright color of arterial blood gives place to the darker purple of venous. On shaking the mixture now with air the bright red color returns. For the success of this experiment where Stokes' reagent is employed it should be freshly prepared before use. Various other substances behave in similar manner.

Experiment. Generate some hydrogen gas in the usual manner, and allow it to bubble through diluted defibrinated blood. A change of color follows after a time, due to the mechanical loss of oxygen. The same result may be accomplished by exhausting the oxygen of the blood by means of an air pump. Exposure to the air restores the color in a short time, as before.

The color change may be noticed readily with the unaided eye, but is much more marked when observed in the spectroscope, as will be pointed out below.

The maximum amount of oxygen which may be held by the hemoglobin was given above as 1 molecule for 1 molecule of the pigment. This holds only for strong oxygen pressure, however. Under lower atmospheric pressure a part of the oxygen becomes dissociated, as illustrated by these figures given by Huefner for a 14 per cent hemoglobin solution:

Atmospheric Pressure in Mm.	Per cent of Oxygen Free.	Atmospheric Pressure in Mm.	Per cent of Oxygen Free.
760.0 mm.	1.49	238.5 mm.	4.60
715.6	1.58	119.3	8.79
620.8	1.81	47.7	19.36
524.8	2.14	23.8	32.51
477.1	2.15	4.8	70.67
357.8	3.11		

The loss of oxygen does not become marked until comparatively low pressures are reached.

Carbon Monoxide Hemoglobin. When a current of carbon monoxide is led into a blood solution it displaces the oxygen and forms

THE BLOOD.

a very stable compound. One molecule of the monoxide takes the place of the molecule of oxygen combined as oxyhemoglobin. This reaction is accompanied by a change of color, not as marked, however, as the change from reduced to oxyhemoglobin. The combination with carbon monoxide is the reaction which takes place in cases of poisoning with illuminating gas, which contains 10 to 25 per cent of the monoxide. The addition of pure air does not displace the combined gas except where a great excess is used.

Experiment. Lead a current of illuminating gas, best the so-called "water gas," into 50 cc. of blood in a flask. Continue the passage of the gas until a distinct cherry red color is produced. When the combination appears to be complete treat a few cc. of the liquid with Stokes' solution, which fails to effect a reduction. With portions of the mixture further tests should be made to illustrate methods of differentiating between normal blood and blood containing much monoxide. The differentiation in each case depends on the greater stability of the monoxide hemoglobin with the reagent in question.

Experiment. Add some strong solution of sodium hydroxide to ordinary blood. This gives a brownish green precipitate at first and then a red solution. Treat blood saturated with carbon monoxide in the same manner. This gives a red precipitate and finally a red solution.

Experiment. Dilute some of the monoxide blood with four volumes of water and to the mixture add an equal volume of a 3 per cent tannic acid solution. The red color persists much longer than it would in the case of a simple oxyhemoglobin solution, which should be tried for comparison.

Experiment. To about half a cubic centimeter of the monoxide blood add 20 cc. of water and 10 drops of strong yellow ammonium sulphide solution. Shake thoroughly and then add enough dilute acetic acid to give a faint acid reaction. A rose red color appears, while with normal blood decomposition products are formed which have a dirty gray color.

Experiment. Mix 1 cc. of the monoxide blood with 5 cc. of basic lead acetate solution and shake well. The mixture remains red, while with normal blood under the same conditions a brown color results.

Experiments have been carried out to determine what portion of the hemoglobin must be combined with monoxide to have death follow. It appears that if about half the pigment in the blood is still unchanged recovery may be expected by free use of air. The mere action of a great excess of air may gradually displace the combined monoxide. The spectroscopic appearance of the monoxide hemoglobin will be referred to below.

Nitric Oxide Hemoglobin. Under certain conditions nitric oxide, NO, may be combined with hemoglobin to form a very stable compound. The union takes place molecule with molecule and may be obtained by treating carbon monoxide hemoglobin with the nitric oxide, which has the power of breaking up the monoxide combination. The direct action of nitric oxide on oxyhemoglobin does not lead to the desired result, as oxidation of the NO follows and the acid formed destroys the hemoglobin. In presence of urea, however, the direct union is possible. The substance forms crystals like those of oxyhemoglobin and gives a very similar spectrum.

Sulphohemoglobin. When hydrogen sulphide is led into a solution of oxyhemoglobin in presence of air a compound is formed which, however, is not permanent. Decomposition soon follows and a greenish brown mixture results. With reduced hemoglobin away from the air a true compound is formed which gives a characteristic spectrum, and which is much more stable.

Other Combinations. Several authors have described other compounds formed by the union of hemoglobin and gases. Of these, so-called carbohemoglobins, acetylenehemoglobin and cyanhemoglobin are the best known. These combinations are but slightly stable and have no special importance. As acetylene was formerly prepared on the laboratory scale it was poisonous, and this property was assumed to be due to its action on hemoglobin, which was thought to resemble that of carbon monoxide. Since the manufacture of acetylene from calcium carbide was begun this notion has been dispelled. The action of acetylene on the blood is very weak. In the early laboratory product impurities present were probably responsible for the observed effects.

DERIVATIVES OF HEMOGLOBIN.

Some of these are practically identical with hemoglobin, while others are products of complete decomposition. The first to be considered is:

Methemoglobin. Oxyhemoglobin in solution or in crystal form, alone or in presence of certain reagents, shows a great tendency to pass over into this modification which contains just as much oxygen as the original, held, however, in stable combination. Under the air pump methemoglobin does not give up any oxygen, while from oxyhemoglobin nearly the whole of the extra molecule may be abstracted. While this formation of methemoglobin takes place spontaneously it is greatly hastened by the action of several substances, some of which are oxidizing agents, while others are reducers. Of the oxidizing substances ozone, potassium permanganate, potassium chlorate, iodine, potassium ferricyanide and nitrates have been used, while such reducing agents as pyrogallol, pyrocatechol, hydroquinol and hydrogen even, acting on the blood have brought about a formation of the stable methemoglobin. Certain substances given as remedies have the power of converting the oxyhemoglobin into methemoglobin. Amyl nitrite, acetanilid and nitrobenzene may be mentioned here. The poisonous action of large doses of potassium chlorate has long been supposed to be due in part to the same reaction.

Solutions of methemoglobin are not bright red but reddish brown, and the crystalline substance is also brown. The color of an alkaline solution is red, but this is not due to a reconversion into oxyhemoglobin. Certain reducing agents have the power of gradually changing the methemoglobin back into oxyhemoglobin and then into reduced hemoglobin. Ammonium sulphide and Stokes' reagent work in this way. The conversion may be followed by aid of the spectroscope.

A product known as acid hemoglobin is formed by the action of weak acids on hemoglobin. This appears to be a step in the formation of methemoglobin, the spectrum of which it resembles. With strong acids decomposition takes place and hematin results.

Hematin. It has been already explained that hemoglobin breaks up readily into globin, about 96 per cent, and hematin, about 4 per cent. This decomposition follows, as just mentioned, by the treatment with strong acids, and also by various other reactions. The product from reduced hemoglobin is known as hemochromogen, while from oxyhemoglobin oxyhematin or common hematin is obtained. The relations may be thus illustrated:

$$\text{Hemoglobin} \begin{cases} \text{globin} \\ \text{hemochromogen} \end{cases} \begin{cases} +\text{O} = \text{hematin} \\ -\text{Fe} = \text{hematolin} \end{cases}$$

$$\text{Oxyhemoglobin} \begin{cases} \text{globin} \\ \text{hematin} \end{cases} \begin{cases} +\text{HCl} = \text{hemin} \\ -\text{O} = \text{hemochromogen} \\ -\text{Fe} = \text{hematoporphyrin} \end{cases}$$

Hematin is usually obtained as an acid combination or ester. In one process frequently followed blood is warmed with an excess of glacial acetic acid. Crystals containing acetic acid separate on cooling. In another process the hydrochloride is obtained; in either case the free hematin is secured by saponification with weak sodium hydroxide solution. Several formulas have been given for hematin; the one most commonly accepted is

$$C_{32}H_{32}N_4FeO_4,$$

while for hemin crystals the formula

$$C_{32}H_{30}N_4FeO_3HCl$$

has been given. Küster has recently given the formula $C_{34}H_{34}N_4FeO_5$. It is possible that different analysts have obtained, not the same, but closely related products. Hemin is secured in minute brownish crystalline plates, hematin as an amorphous bluish black insoluble powder. The spectrum, which is important, will be referred to later.

Hematoporphyrin. This is a derivative obtained by the action of acids on hematin. In this treatment the iron is removed, as illustrated by the following reaction, when hydrobromic acid is employed as the decomposing agent:

$$C_{32}H_{32}N_4FeO_4 + 2H_2O + 2HBr = 2C_{16}H_{18}N_2O_3 + FeBr_2 + H_2.$$

The substance appears to be related to and isomeric with bilirubin. The alkaline solutions of hematoporphyrin are deep red, the acid solutions incline to deep violet or purple. The acid and alkali spectra are very different and characteristic.

The relation of the blood coloring matter to the bile pigments is illustrated by these formulas:

$C_{32}H_{32}N_4O_4Fe$ .... hematin
$C_{32}H_{32}N_4O_5$ .... hematoporphyrin
$C_{32}H_{36}N_4O_6$ .... bilirubin
$C_{32}H_{36}N_4O_8$ .... biliverdin

Hematolin is the name given to an iron-free compound obtained by decomposing hematin in absence of air.

Hemochromogen. This is obtained by reducing a hematin solution with ammonium sulphide or with zinc dust and alkali. It forms a dark red powder insoluble in water but soluble in alkalies. The solution exposed to the air absorbs oxygen and appears to regenerate hematin.

Hematoidin is a red-colored pigment which has been found in old blood extravasations. It seems to be identical with bilirubin.

THE OTHER PROTEINS OF THE BLOOD.

In addition to fibrin and hemoglobin blood contains serum albumin and serum globulin, which have been described already in a previous chapter. The combined substances make up about 7 per cent. They may be approximately separated by precipitating the globulin from blood serum by addition of a large volume of water and also by salt precipitation, which may be illustrated in this way:

Experiment. Prepare blood serum as free as possible from corpuscles as already shown and mix about 100 cc. with an equal volume of cold saturated ammonium sulphate solution. A separation of the globulin follows. Filter; the filtrate contains practically all the serum albumin which may be coagulated by boiling. The albumin may be purified by long dialysis. To recognize the globulin in the precipitate, first wash the latter with more half-saturated ammonium sulphate and then dissolve in slight excess of water. It may be necessary to add a little common salt to assist in the solution. On heating a portion of this solution coagulation follows. On diluting some of it very largely with water precipitation of the globulin takes place. From the first water solution most of the salts may be separated by long continued dialysis.

Magnesium sulphate, added in powder form to saturation, is sometimes used in the place of ammonium sulphate to effect the separation of the albumin and globulin. The reaction in both cases depends on the fact that serum albumin may be salted out only with great difficulty.

It is an interesting fact that other proteins do not appear to be present in the blood. The various proteins consumed as food suffer peculiar changes somewhere in the body and are converted into these two. These in turn serve for the preparation of the various other related bodies found in the several tissues of the organism. Gelatin may be formed in this way from the proteins of the blood, but it does not appear that gelatin can replace other proteins as a food since it is deficient in one of the essential protein component groups, viz.: the tyrosine group.

It is not yet known how constant the relation of serum albumin to serum globulin is or whether this relation is the same in all kinds of blood. Egg albumin is not equivalent to serum albumin physiologically, since if injected into the blood it appears soon unchanged in the urine. The albumins of related animal species seem to be nearly alike, but this does not hold absolutely true for animals of widely different species.

THE SUGAR OF THE BLOOD.

This is found in the plasma and has generally been assumed to be glucose, $C_6H_{12}O_6$, although good reasons may be assigned for the assumption of other sugars as well. Ordinarily the simple sugar finally formed in the digestive process is glucose and the possible passage of other sugars into the blood has commonly been overlooked. As the amount of sugar in the blood is small, about 0.15 per cent in the mean, its certain identification is a matter of extreme difficulty; it must be remembered that separation from the large amounts of proteins present must be complete before any accurate identification of the remaining trace of sugar may be thought of. The older observers depended almost solely on the common reduction tests which are not very sensitive in dealing with traces. Recent investigators have shown that a left-rotating sugar is present and apparently, also, pentoses in traces. As glucose and fructose yield the same osozone this simple reaction cannot be applied to detect a fructose content. Occasionally small amounts of disaccharides appear to be present. Of these maltose passes into dextrose by inversion, while saccharose and lactose would be eliminated as such by the kidney. Glucoronic acid in combination is also present and this may be confounded with a sugar in some of the tests. More will be found on this point in a following chapter.

SALTS OF THE BLOOD.

The total mineral matters of the blood, exclusive of the iron of the hemoglobin, amount to a fraction of one per cent only, but still are of very considerable importance. These salts are largely the chlorides, phosphates and carbonates of the alkali metals, the potassium salts being most abundant in the corpuscles, while the sodium salts are most characteristic of the plasma. It is believed that the variations in this salt content are very small normally. The nearly constant osmotic pressure of the blood points to this. Slight changes are speedily corrected by the kidneys.

Experiment. The presence of reducing carbohydrate and salts in the blood may be demonstrated in this way. Mix about 50 cc. of fresh blood with 300 cc. of water and boil vigorously a few minutes. A drop or two of acetic acid may be added during the boiling to maintain a nearly neutral reaction. Filter and divide the filtrate, which should be perfectly clear, into two parts. Concentrate one-half to a volume of about 10 cc. and apply the Fehling test for sugar. Concentrate the other half likewise and use portions for tests for phosphates and chlorides. The sulphate test usually fails with the volume of blood taken. Evaporate a small portion of this concentrate nearly to dryness on a glass slide, allow what is left to cool and crystallize. Sodium chloride crystals may be recognized by the microscope. To some of the evaporated residue apply the flame test (with spectroscope) for potassium.

GASES OF THE BLOOD.

The blood holds several gases in loose combination. These are principally *oxygen, nitrogen* and *carbon dioxide*. Minute traces of *argon* seem to be present also, which like the more abundant nitrogen must exist in a condition of simple solution. The methods of accurate gas analysis as applied to blood were developed by Lothar Meyer, who made a number of determinations in blood from different sources. These methods have been further improved by others who have placed many results on record.

The mean amount of nitrogen is about 2 volume per cent. The oxygen and carbon dioxide are variable. In arterial blood the oxygen, which is held mainly through the agency of hemoglobin, amounts to about 22 per cent by volume; that is, from 100 cc. of arterial blood 22 cc. of oxygen in the mean may be obtained by aid of the vacuum pump. The venous blood always contains less oxygen, probably not over 15 per cent by volume. These amounts are far larger than could be absorbed from the air through the partial pressure of the oxygen present. In fact it may be shown that only a fraction of 1 volume per cent may be held by the blood plasma perfectly free from corpuscles.

The loosely combined carbon dioxide may vary from 30 to 40 volume per cent in the arterial blood, while in venous blood it is much higher, reaching nearly 50 volume per cent in the mean. This gas is held partly in the form of bicarbonate and partly through the agency of the proteins, especially the hemoglobin. Most of the carbon dioxide is, however, held by the serum and may be largely drawn out by aid of the vacuum pump. On withdrawal of the gas other weak acid bodies are able to take its place in alkali combination. It is held by some observers that the globulins have this power. In acid intoxications where mineral or organic acids increase in the blood the carbon dioxide rapidly decreases and may fall to a tenth or twentieth even of its usual value. These stronger acids take the alkali and there is therefore nothing left to hold the carbon dioxide.

Some of these points will be taken up in a later chapter in discussing respiration phenomena.

Other Substances. Besides the substances mentioned above the blood always contains a number of other bodies of more or less importance. Among these may be mentioned the fats, soaps, cholesterol, lecithin and jecorin. The total *fats* amount ordinarily to about 0.2 per cent, but after a meal may be temporarily much increased. Minute traces of fatty acids as soaps may be also present. *Cholesterol* appears to be present in free form in traces and also as an acid combination or ester. *Lecithins* are present in very small amount in both corpuscles and plasma, but anything like a quantitative determination does not appear to be possible. The name *jecorin* is applied to a peculiar substance containing phosphorus, described by several observers as occurring in blood. It is soluble in ether, like lecithin, and seems to exist in combination with a carbohydrate group or similar reducing residue. The substance has never been obtained in form pure enough for analysis, and it is even possible that it may be a mixture of several compounds, one of which is a combination of glucuronic acid.

Variations in Disease. In disease the normal proportions of the various substances may suffer marked changes. A decrease in the normal number of corpuscles (about 5 millions to the cubic millimeter for men, 4 to 4.5 millions for women) may follow to the extent of 10 per cent or more in certain anemic conditions. There may also be a change in the proportion of hemoglobin without a change in the number of corpuscles. The methods of estimating the amount of hemoglobin will be given later. The salts in the blood suffer a percentage decrease after consumption of large quantities of water, but only temporarily. An actual decrease may occur in cholera and inflammatory diseases. The normal minute amount of sugar is increased in diabetes, but not greatly, because of the eliminating power of the kidneys. It may be temporarily increased by the use of certain drugs such as curare, amyl nitrite, chloral, or by inhalation of chloroform vapor. After meals rich in fats there is a temporary increase of fat in the blood, but a persistent increase is noticed in the blood of drunkards and of corpulent individuals. In diseases where there is rapid breaking down of proteins there is usually observed an increase of fat.

A loss of blood to the extent of one-third is not necessarily dangerous if it be withdrawn slowly. If one-half the blood is lost there is great danger of death. Blood may be added by transfusion, but for safety should be from an animal of the same species. The serum of one animal has usually a destructive action on the corpuscles of another. Transfused blood then may be a source of danger rather than a means of saving life. This peculiar action of serum will be referred to later in some detail.

CHAPTER XII.

THE OPTICAL PROPERTIES OF BLOOD. USE OF THE SPECTROSCOPE AND OTHER INSTRUMENTS.

Solutions of hemoglobin and the various modifications and derivatives described in the last chapter absorb light from certain regions of the spectrum. The character and extent of this absorption are such as to afford a ready means of identifying blood or its pigments through the aid of the spectroscope.

THE SPECTRUM FIELD.

The absorption spectra with which we are here concerned are all found in the middle portions of the spectrum between the Fraunhofer lines C and F, that is in a region easily observed. For practical purposes an elaborate instrument is not necessary. Excellent service is rendered by many of the smaller direct vision spectroscopes. For quantitative tests, however, much more complete apparatus is required. The blood spectrum differs from that of all other red solutions and is very easily recognized with a little practice. As the absorptive power of hemoglobin is very great dilute solutions only are used and these must be observed in a rather shallow cell, preferably in one with parallel sides about 1 centimeter apart. For illumination a good oil lamp flame is excellent; a steady gas flame may also be employed.

The general arrangement of the essential parts of the spectroscope is shown by the following figures. Fig. 12 illustrates, diagrammatically, the path of the light rays through the instrument. From the source F the light enters the collimater tube through a narrow slit and reaches the prism P, where it suffers refraction and dispersion. Beyond the prism the rays are received by the double convex lens of the ocular tube and thrown to the eyepiece at E. A magnified virtual image is formed as shown by the dotted lines. The third tube carries a scale, the image of which is reflected into the ocular and shows with the spectrum. In absorption spectrum analysis, with which we are concerned here, the light at F must be white and between this and the collimator slit a cell must be placed to hold the colored solution or diluted blood. This is shown in the next figure, where B is an ordinary kerosene lamp with flat wick. The edge of the flame is turned toward the absorption cell and slit. The apparatus here figured is arranged for absorption analysis and, with parts to be described later, may be used for quantitative work.

For most simple blood examinations the small direct vision spectroscope shown below may be used. With proper combination of crown and flint glass prisms it is possible to practically correct the refraction and leave a field with satisfactory dispersion.

Variation in Spectra by Dilution. In all dilutions the positions of the absorption bands remain the same, but their density and width

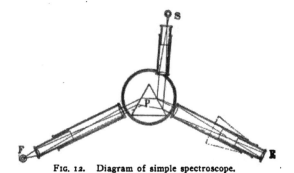

FIG. 12. Diagram of simple spectroscope.

vary with the concentration. In a relatively strong blood solution, for example, there appears to be but one broad oxyhemoglobin band be-

FIG. 13. Spectroscope arranged for absorption analysis.

tween D and E, but on proper dilution this breaks up into two characteristic bands. The question of dilution is therefore important and

FIG. 14. Direct-vision spectroscope.

for any given instrument and light the observer should settle this by a few preliminary experiments.

Spectrum of Oxyhemoglobin. This consists of two bands in the yellowish green portion of the spectrum between D and E. The bands have not the same width, the one near E being slightly broader than the other. The preparation of proper dilutions may be illustrated in this way:

Experiment. Measure out 5 cc. of defibrinated blood and dilute it accurately with 120 cc. of water. Filter into a clean flask and use the clear filtrate for tests to follow. Mark this mixture Solution No. I.

Dilute 50 cc. of No. I with 50 cc. of water and mark the mixture Solution No. II. and continue this until seven dilutions in all are secured, the first one being 1 in 25, as above, and the last 1 in 1600. This is almost colorless.

Take seven test-tubes of thin, colorless glass, and as uniform as possible in diameter. Number them 1 to 7 and two-thirds fill each one with the dilute blood solution corresponding to its number. Place each tube before the narrow slit of the spectroscope and adjust the flame of an oil or gas lamp so that its light may pass through the solution into the slit. Pull out the draw tube until the light is properly focused and observe that the bright field is traversed by two black bands which cut out portions of the yellow and green. With strong blood solutions all light except red is shut out, but with solutions of the dilutions 2 to 7 the field is obscured only by the two bands. In solution No. 2 they are very dark and well defined. With increasing dilution they grow fainter and are scarcely visible in solution No. 7. In all the solutions examined note the position of these bands with reference to the characteristic colors. Note also that the bands grow narrower with increasing dilution, and that it becomes more and more difficult to locate the edges of the bands sharply. This fact has some bearing on questions of quantitative determinations to be referred to later. Some of the common absorption spectra are illustrated.

With instruments furnished with a simple scale it soon becomes an easy matter to fix approximately the limits between which each band is found and also the point of deepest absorption in each band. These data may be expressed in arbitrary scale divisions, in fractions of the distance from D to E, or most definitely, in wave lengths of light, if the instrument has been graduated in that way. In all accurate comparisons of spectra some such method of recording the observations must be adopted.

Spectrum of Reduced Hemoglobin. It was pointed out in the last chapter that the spectrum of reduced hemoglobin is very different from that of the ordinary oxyhemoglobin. In place of two bands we have after reduction a single broad band filling three fourths of the space between D and E. This is the simple effect observed with Stokes' solution. If ammonium sulphide is employed in place of Stokes' solution the same broad band appears and in addition a single narrow band, the center of which is in the red to the left of D. This narrow band may be due to some sulphohemoglobin formed at the same time. It will be recalled that the reduced hemoglobin solution is purplish red in place of deep bright red.

Experiment. To a dilute solution of blood, about 1 part to 50 of water, add a few drops of strong ammonium sulphide solution and warm gently in a test-tube until the change of color noted above is reached. Now place the tube before the

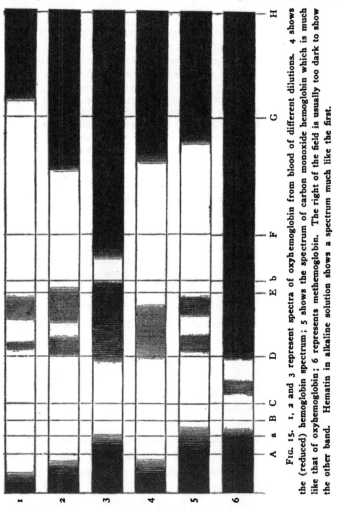

FIG. 15. 1, 2 and 3 represent spectra of oxyhemoglobin from blood of different dilutions. 4 shows the (reduced) hemoglobin spectrum; 5 shows the spectrum of carbon monoxide hemoglobin which is much like that of oxyhemoglobin; 6 represents methemoglobin. The right of the field is usually too dark to show the other band. Hematin in alkaline solution shows a spectrum much like the first.

slit of the spectroscope and observe the bands referred to, especially the narrow one in the red. Hydrogen sulphide gives practically the same result.

Experiment. Repeat the above experiment, using Stokes' solution instead of the sulphide. A single broad band appears now; if the liquid is shaken briskly the air acts on the reduced coloring matter with oxidizing effect, as shown by a division of the band, but only temporarily. On standing a short time the single broad band, not very sharply defined, returns. For these tests it would be well to employ several dilutions, beginning with No. II of the series given above.

Spectrum of Carbon Monoxide Hemoglobin. It has been already mentioned that the spectrum of carbon monoxide hemoglobin is very similar to that of oxyhemoglobin. This may be found by a simple test.

Experiment. Into diluted blood, as before, pass a stream of common illuminating gas until the liquid is saturated, which requires but a few minutes. On placing the tube in front of the spectroscope the two dark bands described will be seen, and slightly farther from the yellow than is the case with oxyhemoglobin.

These bands do not change in extent or position by agitation of the liquid with air, as follows with reduced hemoglobin, and when the liquid is treated with ammonium sulphide or with Stokes' reagent no reduction takes place. The two bands persist.

Spectrum of Methemoglobin. This is characterized by a band in the yellowish red and by a broad band in the blue. To some moderately dilute blood add a few drops of fresh solution of potassium ferricyanide. The mixture becomes brown. It has been already explained that by careful reduction with a small amount of ammonium sulphide hemoglobin is regenerated. This may be followed by the spectroscope. Add a few drops of the sulphide solution, allow the mixture to stand a short time and then shake vigorously in the air. Oxyhemoglobin bands now appear.

The Hematin Spectra. Of these the spectrum of the pigment in weak acid mixture is the most characteristic. This may be obtained by first coagulating 10 cc. of blood and 50 cc. of water by vigorous boiling, enough weak sulphuric acid being added to maintain an acid reaction. The coagulum is separated, pressed dry and rubbed up in a mortar with 25 cc. of absolute alcohol and 1 cc. of strong sulphuric acid gradually added. The mixture is then transferred to a flask and heated half an hour on the water-bath. After cooling the filtered liquid may be examined. A strong dark band in the red is plainly seen.

Different spectra are found after alkaline treatment. Warm some diluted blood with a few drops of sodium hydroxide solution until a brownish-green color results. The absorption spectrum of this liquid is not characteristic. The whole of the field from red to violet is dark. On careful reduction with a little ammonium sulphide or Stokes' solution the spectrum of hemochromogen or reduced hematin may be obtained. This consists of two sharp bands between D and E, somewhat like those of oxyhemoglobin, but nearer E.

QUANTITATIVE SPECTRUM ANALYSIS.

The amount of hemoglobin in a solution may be very accurately estimated by the aid of the spectroscope, but an instrument with special attachments for the purpose is required. Several distinct methods have been applied for the purpose, but the methods now followed involve a simple direct comparison between light which has been weakened by passing through a blood solution and light passing into the spectroscope directly. Such a comparison is easily made by means of instruments with double collimator slit, as first introduced by Vierordt. The arrangement of the apparatus made by Kruess is shown in Fig. 13, while the double collimator slit and ocular and reading scale are shown in Figs. 16 and 17.

The method of measurement depends on the principle that there is a simple relation between the amount of light absorbed by a solution and the concentration, that is the number of absorbing molecules in the same. By finding therefore the fraction of the original light absorbed we can arrive at the amount of absorbing substance in solution. The loss of light in passing through solutions of increasing concentration follows the law worked out by Lambert for the loss in passing a series of glass plates of same thickness and color. Each new layer absorbs the same fraction of the light reaching it, and in the same way each unit of added concentration absorbs the same fraction absorbed by the first unit.

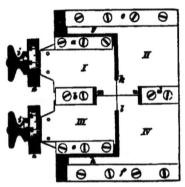

FIG. 16. Symmetrical double slit for the absorption spectroscope.

Supposing the increased absorption of light to follow through the addition of new layers of absorbing substance, the relation between the original and residual intensities may be reached in this manner. Calling the original intensity I and the intensity after passing the first layer (or first unit of concentration) I' we have

$$I' = I\frac{1}{n},$$

the original intensity being reduced to $1/n$ by the first layer. By a second, third and following layers we have

$$I\frac{1}{n}\cdot\frac{1}{n}, \quad I\frac{1}{n}\cdot\frac{1}{n}\cdot\frac{1}{n}\cdots, \quad I\frac{1}{n^m}.$$

The last expression shows the intensity after passing m layers. For purposes of calculation this can be put in another form, taking the original intensity as unity:

$$I' = \frac{1}{n^m} \text{ gives } \log I' = -m \log n. \quad \log n = -\frac{\log I'}{m}.$$

FIG. 17. Ocular and reading scale of the Kruess spectrophotometer.

In comparing the light-absorbing powers of solutions some arbitrary basis must be taken. Practically the thickness of layer which will reduce the original intensity to $\frac{1}{10}$ its value is so taken. The light-extinguishing power of a substance or its *coefficient of extinction*, has been defined as *the reciprocal value of the thickness of a layer of the substance necessary to reduce the intensity of the transmitted light to $\frac{1}{10}$ its original value.*

Representing the extinction coefficient by E and the reduced intensity by I' we have from the above formulas:

$$E = \frac{1}{m} \text{ and } I' = \frac{1}{10}, \quad \log n = -\frac{\log \frac{1}{10}}{\frac{1}{E}} = E.$$

Therefore

$$E = -\frac{\log I'}{m}.$$

In practice m may be given a constant value and called 1 (the thickness of cell, for example). The formula becomes

$$E = -\log I'.$$

It was said above that increasing the thickness of a layer of absorbing substance has the same effect as increasing its concentration in the same degree. From this it follows that the extinction coefficient must be directly proportional to the concentration. Let E and E' represent two extinction coefficients and C and C' the corresponding concentrations, then

$$E : C :: E' : C'.$$

The relations

$$\frac{E}{C}, \frac{E'}{C'}, \frac{E''}{C''},$$

etc., must be all equal and constant for the same substance. This constant ratio is a characteristic which connects the light-absorbing power of a solution with its strength; it may be represented by the letter A and be termed the *absorption ratio*. Hence, for a given color

$$A = \frac{C}{E}.$$

The value of the constant A must be found for a given spectrum region by employing a series of suitable concentrations. The determination of E consists in finding the value of the reduced light as compared with the original; from the formula given above the extinction coefficient is equal to the negative logarithm of this diminished intensity. In the various forms of photometers employed in this kind of work the peculiar measuring mechanism permits the direct and simple estimation of the intensity of the light after absorption as compared with the light before absorption. We find I' then as a fraction, and E is the negative logarithm of this. For example, suppose we have in a liter 0.25 gm. of an absorbing substance. The concentration, or C, is 0.00025, which represents the value per cubic centimeter, taken as the unit of volume. Next, suppose we find with our special measuring instrument that the value of the light after absorption is only 0.0436 of the original, that is about one twenty-third.

Substituting in our formula we have

$$E = -\log I' = -\log 0.0436 = 1.36051$$

and finally

$$A = \frac{C}{E} = \frac{0.00025}{1.36051} = 0.000184.$$

In this way, by repeating the observations with a number of different strengths of solution of the substances, we find the value of the constant. As the individual observations may differ a little the mean must be taken. A becomes thus fixed once for all for a given spectral region, and its value may be employed to determine the concentration of *unknown* solutions since

$$C = EA.$$

Quantitative spectrum analysis by absorption is based on these very simple principles. Blood or other substance to be examined is placed in a cell with plane parallel sides, preferably exactly 1 cm. apart. The cell should be half filled and is brought into proper position in front of a spectroscope with a double slit, the level of the liquid just reaching to the top of the lower slit. The light from the illuminating lamp enters the upper slit directly and the lower one through the

Fig. 18.

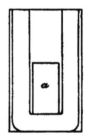

Fig. 19.

Absorption cell and Schulz glass prism as used in quantitative analysis by absorption. The position of the prism is shown in Fig. 19.

colored liquid. If the two slit openings were the same to begin with the upper one must now be narrowed until the light passing through it is reduced to the intensity of the light through the blood and the lower slit. The width of each slit may be measured on a micrometer screw head or in some other convenient way. In these observations but a small portion of the spectrum is brought into the field of view. The eyepiece in the spectroscope is therefore furnished with a screen which can be opened or narrowed at will and symmetrically, that is from both sides, so as to expose some definite small portion of the spectrum. The instrument should be so constructed as to permit any desired portion of the spectrum to be quickly and accurately brought into the field.

In place of using a simple cell it is much better in practice to employ a cell with so-called Schulz glass prism. With the simple cell the meniscus formed at the top of the liquid projects a broad dark band across the field horizontally, which makes the comparison of the upper and lower spectra very difficult. With the cell furnished with a Schulz prism this difficulty is largely overcome, but the details of the arrangement can not be explained here. They will be easily understood by use of the instrument. When the Schulz prism is employed light enters the *upper* slit through 1.1 cm. of solution and the lower slit through 0.1 cm. of solution and the 1.0 cm. of the clear glass prism.

Spectral Region. λ =	Name of Substance.					
	Oxyhemo-globin.	Hemo-globin.	Methemo-globin.	CO-Hemo-globin.	Bilirubin in Chloroform.	Bilirubin in Alcohol.
569.3–555.5	0.00133	0.00122	0.00260	0.00131		
549.9–540.0	0.00100	0.00150	0.00199	0.00115		
558.1–534.3					0.00113	0.000215
501.2–494.3					0.0000598	0.000142
494.3–486.1					0.0000356	0.000116
486.1–480.6					0.0000209	0.000102
480.6–474.4					0.0000148	0.0000842
474.4–468.4					0.0000126	0.0000700
468.4–461.7					0.0000118	0.0000667

The absorption ratios for a number of physiologically important substances are shown in the above table. The spectral regions are given in the usual wave lengths.

CLINICAL METHODS OF ESTIMATING OXYHEMOGLOBIN.

The spectrophotometric estimation of oxyhemoglobin as described above is not simple enough for quick clinical determinations, which have to be made in the course of daily practice by medical men. Other forms of apparatus have been devised for this purpose and are in common use. In all of these, comparison is made between the blood under examination, properly diluted, and a standard color assumed to represent normal blood correspondingly diluted. Some of these appliances give pretty good results, but others are very faulty and the values they furnish quite untrustworthy. In the following pages several of the commoner forms will be briefly described.

Fig. 20. Fleischl hemometer, showing divided cell for blood and water and reflecting mirror to secure uniform illumination.

Fleischl's Hemometer. This instrument consists essentially of a circular cell with glass bottom divided by a vertical partition into two equal compartments as shown below. In one of these the accurately diluted blood is placed in given volume. The other compartment is filled with pure water to the same level. The cell rests on a stage below which there is mounted a white reflecting mirror by means of which light may be thrown upward to illuminate the two compartments of the cell uniformly. Immediately under the water compartment a long colored glass wedge is placed in such a manner that light must pass through it into the water. By a rack and pinion mechanism this wedge may be moved to the right or left under the water cell so as to bring a thinner or thicker portion of the glass below the water. The glass is colored by means of purple of Cassius to resemble diluted blood as nearly as possible, and the light shining through it into the water imparts a more or less perfect blood color to it. The wedge is moved

FIG. 21. Dare's hemoglobinometer. R, milled wheel acting by a friction bearing on the rim of the color disc; S, case inclosing color disc, and provided with a stage to which the blood chamber is fitted; T, movable wing which is swung outward during the observation, to serve as a screen for the observer's eyes, and which acts as a cover to inclose the color disc when the instrument is not in use; U, telescoping camera tube, in position for examination; V, aperture admitting light for illumination of the color disc; X, capillary blood chamber adjusted to stage of instrument, the slip of opaque glass, W, being nearest to the source of light; Y, detachable candle-holder; Z, rectangular slot through which the hemoglobin scale indicated on the rim of the color disc is read.

FIG. 22. Horizontal Section of Dare's Hemoglobinometer, in which the arrangement of parts is clearly shown. L is the standard wedge-shaped color disc. The blood is inclosed between the glass plates O and P and is illuminated by the flame J. The eye observes the blood and the color scale through the apertures M and M'.

until the liquids in the two compartments of the cell appear to have the same color. The color in a certain portion of the wedge is intended to correspond to normal blood, or blood with 100 per cent of the normal oxyhemoglobin, and degrees placed at proper intervals along the wedge represent corresponding higher or lower percentages. For example, if when the colors in the two compartments are matched the reading on the wedge scale is 97, it means that the blood contains color due to 97 per cent of the normal or average oxyhemoglobin content.

This Fleischl instrument is one of the best in principle but it must be carefully used to furnish good results. The color of the wedge does not correspond to blood color unless a certain kind of white light is employed. A wedge made for candle light cannot be used with sunlight. In addition to this difficulty the wedges themselves are often at fault. They sometimes fail to produce a blood red with any kind of light.

Miescher has suggested several improvements in the Fleischl instrument which render it much more accurate. Readings may be made from several different dilutions from which a mean value may be taken.

The Hemoglobinometer of Gowers. This instrument has been made in several forms. The construction is essentially this. Two narrow glass tubes of the same diameter are used; one receives as a standard a 1 per cent solution of normal blood, while the other is graduated from below from 0 to 100 degrees and is intended to receive the blood under examination. A measured portion of this blood, usually 20 cubic millimeters, is poured into the tube and diluted with distilled water, a little at a time. After each addition of water the tube is shaken thoroughly to mix and a comparison made with the standard tube. When the colors are finally the same, as read horizontally, that is across, not down through the tubes, the degree of dilution reached in the graduated tube is noted. This indicates the percentage of color present as compared with the standard. A blood which can be diluted to 100 degrees (100 times the original small volume taken) contains 100 per cent of the normal hemoglobin content or is normal, while if it can be diluted to 75 degrees only, the comparison shows that this blood contains but 75 per cent of the average hemoglobin.

In place of using blood as a standard a gelatin solution stained with picrocarmine or other stain is frequently employed. But in time such color standards always fade, and an abnormally high result is recorded as a consequence.

Dare's Hemoglobinometer. In this instrument the principle employed in the Fleischl apparatus is used, but the comparison is made between the colored glass standard and undiluted blood. The possible error due to dilution is thus avoided. Some idea of the apparatus is given by the illustrations above. A drop of perfectly fresh blood is placed over the opening in a capillary flat cell into which it is immediately drawn, much as a drop of water is drawn in between a slide and cover glass not in absolute close contact, when the water is put on the edge of the cover. The capillary observation cell is mounted at the end of an eyepiece through which it may be clearly seen. A small portion of the colored glass standard may be seen at the same time. The blood cell and red glass are evenly illuminated by a candle flame placed in fixed position in front of the apparatus. The colored glass standard is given the form of a circular disk, which may be rotated by a screw motion. This disk is beveled from one side to the other, giving a wedge effect as in the Fleischl apparatus. The rotation of the disk brings therefore thicker or thinner portions of the edge in the field of the eyepiece along with the cell holding the blood. When the colors are matched the corresponding hemoglobin value is read off on a scale. In practice this instrument is somewhat more convenient than the Fleischl. The accuracy is about the same.

Tallquist's Chart. This consists of a small book containing blank sheets of a special fine grained filter paper and a colored chart with a number of shades corresponding to blood stains with different hemoglobin content. To make the test a small drop of blood is drawn and placed on one of the sheets of filter paper, into which it soaks and spreads. The color produced is compared with one of the ten shades in the color scale.

The test is extremely simple, but its accuracy is dependent on the accuracy with which the colors in the chart are printed, and their permanence. Unfortunately the colors change as the chart, in use, is exposed to light and air.

CHAPTER XIII.

FURTHER PHYSICAL METHODS IN BLOOD EXAMINATION. FREEZING POINT AND ELECTRICAL CONDUCTIVITY. THE HEMATOCRIT.

OSMOTIC PRESSURE.

In many of the phenomena of the body the osmotic pressure of dissolved substances plays an extremely important part. This is especially true in the study of the blood as a whole, and it is therefore proper at this point to enter upon a short explanation of what is meant by osmotic pressure and what its relations are.

Nature of Osmotic Pressure. Solids in solution exert a pressure in all directions quite analogous to that observed with gases, and in general the laws connecting increase in pressure with concentration and temperature are the same as for gases. With many solids, however, dissociation in solution or separation into ions takes place and each separate ion behaves as a whole molecule as far as pressure is concerned.

Some of the simple effects of this pressure are easily observed. When a drop of a strong solution of blue vitriol is placed carefully on the surface of a weak solution of potassium ferrocyanide a precipitate of copper ferrocyanide forms as a sheath or membrane around the vitriol drop and holds it in nearly spherical form. If the drop is properly deposited, which requires some care, it will gradually enlarge by the entrance of water, which dilutes the enclosed copper sulphate. None of the latter passes out and the ferrocyanide solution evidently does not enter since no more precipitate forms within the drop. The copper ferrocyanide membrane must possess therefore some interesting properties; it is permeable for water, it is not permeable for either of the salt solutions. Similar membranes may be made with a number of substances and their impermeability for many salt or other solid molecules may be shown. Some membranes are permeable for certain salts but not for others. The fact of the existence of *pressure* within such a membrane may be shown by the following well-known experiment, in which a copper ferrocyanide sheath or membrane is made in a different manner.

Experiment. Procure a small fine grained porous battery cell, about 3 to 4 inches long and 1 inch in outside diameter, and clean it thoroughly. This may be done by washing the cell with water, then with weak hydrochloric acid and finally with water very thoroughly. Close the cell with a perforated rubber stopper, pass a glass tube through the perforation and connect the outer end of this with a suction pump. On dipping the cell in water or the acid it may be drawn through the pores of the cell to effect the cleaning. When the cell is clean it is placed in a potassium ferrocyanide solution containing about 150 gm. per liter and solution drawn through by means of the pump until the pores are thoroughly filled. Then the cell is washed, inside and out, with distilled water and immersed in a blue vitriol solution containing about 250 gm. per liter. A precipitate is thus formed within the pores of the cell, which is allowed to remain some hours in the solution. The cell is then removed, washed with water and is ready for use. Fill it with a 5 per cent cane sugar solution, close with the rubber stopper and long narrow glass tube and immerse the cell in a beaker of distilled water the temperature of which should be the same as that of the sugar solution. After a short time liquid begins to rise in the glass tube which serves as a kind of manometer. This is in consequence of the entrance of water to the sugar solution. Sugar can not pass out in the other direction as the precipitate membrane is not permeable for it, but it is readily permeable for the water. The sugar in its effort to pass out to the water exerts a pressure on the retaining membrane, and it is because of this pressure that the water is able to enter the cell. The flow of the water continues until its hydrostatic pressure exactly balances the sugar or osmotic pressure. In some cases mercury manometers attached to such cells register pressure, of several atmospheres.

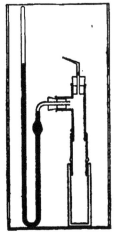

FIG. 23. Apparatus for observing and measuring osmotic pressure.

The pressure actually observed in such an apparatus is just short of that required to press the solvent, water, through the membrane in the opposite direction. Theoretically it should amount to 22.4 atmospheres for a solution containing a gram molecular weight dissolved per liter, since it has been found that the osmotic pressure of a body is the same it would possess if it existed in the condition of a gas at the same temperature and in the same volume. A gram molecular weight of hydrogen (2.014 gms.), of oxygen (32 gms.), of nitrogen or other gas occupies a volume of 22.4 liters under normal temperature and pressure conditions. If condensed into 1 liter their pressure would be 22.4 atmospheres. Experiment has shown that a gram molecular weight of sugar or similar solid in water to make a liter volume exerts a pressure of 22.4 atmospheres. In the case of salts which break up into component parts or ions the pressure becomes correspondingly greater. In very dilute solutions a molecule of sodium chloride, for example, exerts practically double the pressure observed for a molecule of sugar. In this dilute condition the component parts, or ions, of sodium and chlorine seem to exert a pressure corresponding to whole molecules.

The above experiment is a somewhat crude one and is intended merely as an illustration of the development of pressure. For accurate measurements much more elaborate apparatus must be employed and numerous precautions observed. Practically, however, osmotic pressure is always measured by indirect methods to be explained later. A familiar illustration of a semi-permeable sheath or membrane is found in the red blood corpuscle. Normally this holds its hemoglobin and certain salts because it is suspended in a liquid which has the same osmotic pressure. But if the corpuscles be placed in pure water they are seen to swell and finally break because of the passage of water through the cell sheath which is not permeable for the solid contents. By means of the hematocrit, as will be explained, it is possible to find the average volume occupied by the corpuscles in a given sample of blood. When mixed with water or solutions with lower osmotic pressure the corpuscle volume increases; in stronger salt solutions, on the other hand, the individual corpuscles shrink in size and their total volume becomes less. The hematocrit may therefore be used to measure or compare osmotic pressures in certain cases.

INDIRECT METHODS. CRYOSCOPY.

Although the blood contains about 20 per cent of organic substances and about 1 per cent of mineral matters its osmotic pressure depends largely on the latter. This is because of the simple fact that the large gross weight of organic matter represents relatively but a small number of molecules, and the actual pressure is measured by the total number of molecules or ions present. This osmotic pressure in health remains practically constant; even after great loss of blood when the total volume is restored by drawing on the lymph serum, although the relative number of corpuscles may be much reduced, the osmotic pressure of the new blood is practically unchanged. This is due to the fact that the blood and lymph serum are isosmotic.

The Freezing Point Method. The measurement of the osmotic pressure of blood or any other solution by the direct method suggested by the experiment given above is extremely difficult. Several indirect methods may be followed, two of which are in common application. One of these only is suitable for the examination of blood. In the first of these methods the elevation of the boiling point of the solution is observed; in the second the depression of its freezing point. Comparatively simple relations obtain between the three phenomena. In a solution the tension of the vapor is decreased in proportion as the osmotic pressure of the dissolved substance increases and more heat must therefore be applied to actually lift the atmosphere or boil off the solvent. For each gram-molecule per liter dissolved this elevation of the boiling point of water is about $0.52°$. This method, by noting the elevation of the boiling point, cannot be applied to blood, because of its coagulation, but there is no drawback to the method depending on the separation of the solvent by freezing. With increase in amount of salt or foreign substance dissolved in the water, the lower must its temperature be brought to effect a partial separation by freezing out a portion of the solvent. The lowering of the freezing point is accurately proportional to the number of molecules (or ions) present. The molecular freezing point depression for water is 1.85°; that is, the freezing point of a solution containing one molecular weight in grams of a substance, such as sugar or urea, dissolved in a liter, is 1.85° below the freezing point of water. The osmotic pressure of a substance of which 1 gram molecule per liter is dissolved in water, is 22.4 atmospheres. Therefore a freezing point depression of 1° C. corresponds to an osmotic pressure of 12.1 atmospheres.

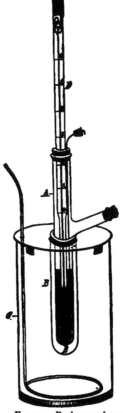

Fig. 24. Beckmann freezing point apparatus. D is a fine thermometer, C the containing jar, B the outside or air mantle tube and A the tube in which the mixture to be observed is placed. Two stirrers are shown; one for the cooling mixture in the jar and one for the experimental mixture.

Apparatus. Various forms of apparatus have been devised for the experimental determination of freezing point. The Beckmann apparatus is most commonly employed. It consists essentially of a strong test-tube to contain the substance to be examined. This is suspended in a somewhat larger tube which serves as an air bath. The large tube finally is supported in a strong beaker or battery jar which receives the freezing mixture to reduce the temperature of the substance under experiment. The freezing mixture may consist of ice, water and salt, which must be stirred up frequently to maintain a uniform degree of cold. A very delicate thermometer passes down into the substance in the inner tube, which is also furnished with a stirrer of platinum wire. The blood or other liquid is stirred until coagulation begins, the thermometer being meanwhile carefully watched. The temperature goes down at first a little below the normal freezing point, because of overcooling, but soon rises and remains stationary. In experimenting with aqueous solutions a known weight of pure water is taken and its freezing point with the thermometer used is accurately found. Then the salt or other body, which has been accurately weighed, is added and a new determination made. While the principle is simple the details call for some skill in manipulation. Full descriptions of the method may be found in works on physical or organic chemistry, as it finds applications in many directions, and especially in the determination of molecular weight. The general appearance of the simple apparatus is here shown.

The freezing point of normal human blood is about —0.56°. As a reduction of 1° corresponds to an atmospheric pressure of 12.1 atmospheres, the normal osmotic pressure of the blood is about 6.8 atmospheres. It makes but little difference here whether we consider the whole blood or the plasma free from corpuscles and fibrin. The result is mainly due to the small molecules present, and these are inorganic. A solution of 20 gms. of serum albumin in water to make 100 cc. would have a freezing point of about —0.03°; the effect of the other proteins would be practically the same. A solution of urea containing 10 gm. in 100 cc. has a freezing point of — 3.08°, one of glucose with 10 gm. in 100 cc. a freezing point of — 1.03°, while a solution of common salt with the same weight dissolved would show a depression of about 5°.

Variations. This observed freezing point depression is normally constant and nearly the same for the blood of all the common animals. But temporary variations may occur. After consumption of large quantities of water it may sink to — 0.51°, while following a meal rich in salty food a further depression to — 0.62°, or even lower, may be observed. But these changes are very speedily rectified through the elimination of proper quantities of salts and water by the kidneys. If an examination of the blood shows a greater depression than that which may be accounted for by absorption of food constituents a failure of some kind in the functions of the kidneys is indicated. Through injury to the mechanism of these organs the osmotic pressure of the blood may rise to over 12 atmospheres, corresponding to a depression of the freezing point of a whole degree or more.

Because of these observed facts the determination of the freezing point of the blood has become a test of practical importance in the diagnosis of disorders of the kidney. With proper facilities the experiment may be quickly made and will serve to detect an abnormality in the blood more readily than this may be accomplished by chemical analysis. It is customary at the present time to designate this freezing point depression by Δ. Thus, normally, for human blood

$$\Delta = -0.56°.$$

ISOTONIC COEFFICIENT.

When a few drops of blood are mixed with an excess of salt solution of a certain strength and the mixture allowed to stand at rest the corpuscles gradually settle and leave a colorless liquid above. If the same volume of a certain weaker salt solution is taken with the blood the mixture after shaking is found to leave no longer a colorless liquid above the settled corpuscles, but a somewhat reddish liquid. This color shows that the corpuscles have been broken and that a little of the hemoglobin has escaped. An experiment will illustrate the fact; it is due to Hamburger.

Experiment. Prepare a series of common salt solutions of the following strengths: 0.7 per cent, 0.65 per cent, 0.60 per cent, 0.55 per cent, 0.50 per cent and 0.45 per cent. Measure out accurately 20 cc. of each into test-tubes and add to each 5 drops of defibrinated bullock's blood. Shake and allow to stand. Notice that in some of the tubes the corpuscles have settled, leaving the salt solution practically clear and colorless; in others there is color, which is greatest in the tube with the least salt. In the tube with 0.60 per cent of salt there should be no color, while in the tube with the next weaker solution some appears. There must be therefore some solution between these two in which the corpuscles just fail to give up color. Hamburger found this to be one with 0.58 per cent of salt.

Osmotic Tension. Hamburger made a large number of experiments of this description and found the limiting value of the strength of solutions for which no loss of color follows. He spoke of these solutions as being *isotonic,* or as having the same osmotic tension as the content of the corpuscles. The numerical values found bear a close relation to the molecular weights of the salts used. Thus, the following values were noted:

	Molecular Weight.	Isotonic Value.
NaCl	58.5	0.585 per cent.
NaBr	103.0	1.02 "
NaI	149.9	1.55 "
KNO$_3$	101.2	1.01 "
KBr	119.1	1.17 "
KI	166.0	1.64 "

These values, while isotonic and isosmotic with each other, are not, however, quite isosmotic with the blood. A common salt solution having a percentage strength of 0.9 per cent has practically the same freezing point as the blood. Blood corpuscles (human) in such a solution do not swell or shrink, consequently lose no hemoglobin. But if placed in weaker salt solutions water is gradually absorbed to make the outside and inside osmotic pressure the same. After a time, however, with decreasing strength of the salt solution, so much water is absorbed that the limit of strength of the corpuscle sheath is reached and a break follows. The escape of hemoglobin shows this point. With salt solutions this break takes place with practically corresponding osmotic pressures, but there are many substances which do not follow the rule at all. This is particularly true of solutions of urea, glycerol, ammonium carbonate, sodium carbonate and ammonium chloride. Even with rather strong solutions of these bodies the corpuscles fail to hold their hemoglobin. A satisfactory explanation of the abnormal behavior of these bodies is not known. Blood so changed is said to be *lake-colored.* Following the Hamburger designations normal blood was said to be hyperisotonic, since it contains more than enough salts to hold the corpuscle intact.

HEMATOCRIT METHODS.

It has been shown above that the red blood corpuscles maintain their normal volume in liquids which have the same osmotic pressure as the blood. In liquids with a lower pressure they swell, while in solutions possessing a higher osmotic pressure than the blood they contract. The corpuscles are extremely sensitive to such influences, and changes in volume follow with even very trifling changes in the osmotic pressure of a liquid with which the blood may be mixed. The blood corpuscle may be used then as a kind of indicator to disclose variations in osmotic pressure, and substances may be compared as to the osmotic pressure they exert by noting their behavior with the corpuscles.

It would of course be very difficult to prove anything by measurements on a single corpuscle, but it is possible to make the observation on a large volume. If blood is drawn up into a narrow tube of capillary dimensions, placed in a centrifuge and rapidly rotated the corpuscles are thrown to the outer end of the tube, which must be closed of course. The volume occupied by the corpuscles compared with the original blood volume may be easily seen.

Köppe's Hematocrit. An instrument in which such an observation may be accurately and easily made was devised by Hedin and called the hematocrit. A special form of this apparatus was constructed by Köppe and is used for the purpose of comparison of corpuscle volumes. The essential part of the apparatus, as shown in the figure, is a graduated capillary pipette about 7 cm. in length, which may be closed at both ends by small metal plates. At the upper end the capillary bore is widened out so as to form a small mixing vessel. The pipette proper has a graduation of 100 divisions. By means of a syringe attached to the pipette by a bit of rubber tubing blood may be drawn up into the capillary and its volume accurately noted. The pipette may be closed and rotated now rapidly in the centrifugal machine, which throws the corpuscles to the outer end. To prevent coagulation it is best to moisten the pipette first with a layer of cedar oil which does not interfere with the reading of the blood volume. The relation between blood volume and corpuscle volume may thus be read off on the graduation. A drop of similar fresh blood is next drawn into the capillary and its volume noted; following this some solution is drawn in also, and then with the blood up into the wider mixing part, where by means of a bright, fine wire the two liquids may be stirred together. The plates are then put on the ends of the pipette, where they are held by springs. The pipette may be rotated as before in the centrifugal machine, the rotation being continued until the volume occupied by the corpuscles becomes constant. By using a number of pipette tubes it is possible to employ different mixtures and soon find one in which the corpuscle volume remains normal. If a series of sugar or salt solutions of known osmotic pressure are employed, that of the blood must be taken as equivalent to the pressure in the solution for which no change in the volume of the corpuscles occurs.

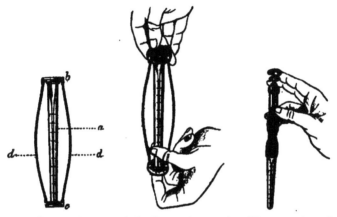

FIG. 25. The essential part of the Koeppe hematocrit. The measuring tube a is closed by two plates, b and c, which are held fast by the springs d. The tube is filled by means of a peculiar syringe shown at the right.

Conversely the apparatus may be, and is frequently employed to find osmotic pressures of solutions. The volume of the corpuscles is found in some solution, of cane sugar for example, which has about the same osmotic pressure as the blood, but which must be accurately known. Then other solutions of a new substance are tested until two are found which give volumes, one greater and the other less than that with the sugar. A simple calculation will then give the concentration of the solution of the substance under comparison which has the same osmotic pressure as the standard sugar solution. The method would naturally fail for any substance which acts chemically on the blood or which destroys the corpuscles, such as urea or glycerol.

CLINICAL USES OF THE HEMATOCRIT.

On the assumption that the volume occupied by the corpuscles varies with the number of cells, attempts have been made to use the hematocrit in place of the cell counter. With normal blood cells the relation is practically constant and a volume of 50 per cent in the hematocrit corresponds very closely to the average 5,000,000 cells per cubic millimeter. But unfortunately where such a simple method of making a blood cell count is the most desirable it is at the same time the least reliable, since in disease the corpuscles do not always retain their normal size. A factor of perhaps greater importance, however, is obtained by taking the ratio of the volume as found by the hematocrit to the corpuscle count as made by a hemacytometer. With undiluted blood the hematocrit may be used to determine whether or not *pigmentation* has taken place. If the corpuscles are intact a nearly colorless serum is secured; a more or less reddish serum points to disintegration of the corpuscles.

THE ELECTRICAL CONDUCTIVITY OF BLOOD.

Electrolytes. It has been found by experiment that certain solutions conduct the electric current while others do not. Pure liquids do not conduct at all, as a rule. Thus absolutely pure water, glycerol, alcohol, anhydrous sulphuric acid and similar substances are practically non-conductors. Solutions of many organic substances are likewise non-conductors, practically. The sugars, for example, belong to this class. But organic acids and salts and many so-called basic bodies are, like the corresponding inorganic substances, conductors. In general, liquid conductors or electrolytes are compounds which in solution separate or dissociate into component parts or ions more or less perfectly.

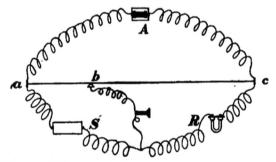

FIG. 26. Diagram of Wheatstone bridge connections. A represents a cell or induction coil, ac the bridge wire, S the standard resistance with which comparison is made, R the conductivity cell containing the substance under examination. In most conductivity experiments A is a small induction coil, a telephone, as shown, being employed as the current indicator.

The mineral salts and inorganic acids and alkalies are in general good conductors, as they "ionize" to a considerable degree.

Blood serum has the power of conducting the current and mainly because of its content of salts. The proteins in absolutely pure condition, salt free, are probably non-conductors. Some of them, however, because of their acid character exist in combinations resembling salts and these have a weak conducting power. But, because of their high molecular weight, the part which this conductivity plays in the total conductivity of the blood is very small. As applied to the blood, therefore, conductivity measurements give us an idea of the number of salt molecules present, or inorganic concentration.

In practice conductivity is the reciprocal of resistance, which is the factor actually measured. Resistance is expressed in terms of some standard arbitrarily chosen, and comparisons are usually with the "ohm" as the final standard. The legal ohm is the resistance of a column of pure mercury 106.3 cm. long and 1 mm. in section at

FIG. 27. Simple Wheatstone bridge. The bridge wire of exact length is stretched over the graduated scale. The several wire connections are made at the binding posts lettered.

a temperature of 0°, but for practical use resistance standards of wire are employed. A series of standard wire resistances running from a tenth, or hundredth of an ohm even, to 1000 ohms or more is generally employed in the form of a resistance "set" or "box."

In dealing with solutions the unit of conductivity is taken as the reciprocal of the resistance of a substance which, in the form of a column 1 cm. square and 1 cm. long (a symmetrical cubic centimeter), has a resistance of 1 ohm. That is, the conductivity, κ, is measured in terms of that of an ideal liquid, one symmetrical cubic centimeter of which has a conductivity of 1 between opposite faces, or which offers between the same faces a resistance of 1 ohm. The resistance of liquids is always found in small vessels of glass made in different shapes and sizes according to the character of the liquid. Small platinum plates are mounted in the vessels and it is the resistance of the column between these which is measured. Before use the *resistance capacity* of the vessel must be found. This is done by measuring in it, with the plates in fixed position, the resistance of some liquid the conductivity of which has been previously determined by some standard method. The data for several solutions have been very accurately determined and are everywhere used for purposes of graduation of conductivity vessels. With such a standard liquid with conductivity κ we find in our cell the resistance, R. The resistance capacity C, is given by the relation:

$$C = R\kappa.$$

That is, C is the resistance which would be found in the vessel if it were filled with a liquid of unit conductivity, and is used as a constant in all following calculations with the same vessel when we wish to find κ. R we always find by direct measurement in ohms and with C known we have now:

$$\kappa = \frac{C}{R}$$

The resistance of liquids cannot be found as is that of a solid by means of the Wheatstone bridge combination and a galvanometer, since under such circumstances liquids suffer hydrolysis with rapid change of resistance. In place of the direct current and galvanometer Kohlrausch suggested the use of a weak induction current, with a telephone as current indicator. With this arrangement, which is illustrated by the annexed diagram, it is possible to measure the conductivity of the serum or other liquid, very readily; a simple form of Wheatstone bridge is shown also.

ac represents the graduated wire of the Wheatstone bridge, S the standard resistance with which comparison is made, R the cell containing the serum or other liquid under investigation, T the telephone which ceases to buzz when no current passes through it to or from b. This gives the "null" point in the combination and when this is found the following proportion holds:

$$ab : bc :: S : R.$$

ab and *bc* are read off directly as bridge wire lengths, S is the known comparison resistance. Hence the unknown cell resistance is given by

$$R = S \frac{bc}{ab}$$

As S in practice is always taken as 10, 100 or 1000 ohms and *ac* is always divided decimally, tables are constructed giving directly the value of R for any value of *ab* read off. In practice the cell R is always kept at a constant temperature, as the conductivity of liquids varies greatly with temperature changes. To maintain this

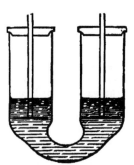

Fig. 28. Simple form of Kohlrausch conductivity cell.

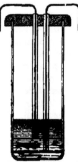

Fig. 29. Conductivity cell for poor conductors or small quantities.

constant temperature the cell is usually immersed in a large water thermostat, so constructed that it may be readily controlled. Forms of cells are illustrated.

The electrical conductivity of urine is also an important factor which may be found by the same kind of apparatus, and which will be discussed later.

Value of the Conductivity for Blood. Expressed in the terms just explained, the value of the conductivity of blood serum is about $\kappa = 0.012$, or expressed in another form very convenient for calculation, 120×10^{-4}. A good part of this conductivity is due to the sodium chloride present. If the chlorides be accurately determined by one of the usual methods of quantitative analysis and the proper conductivity corresponding to this salt content be calculated, which is possible with a considerable degree of accuracy, and subtracted from the total or observed conductivity a remainder is obtained which measures the "achloridic" conductivity, that is the conductivity due to the sulphates, phosphates and carbonates present.

The conductivity of the salts in the serum is somewhat less than in pure water, but it is possible to make a correction for this interference of the proteins and obtain satisfactory values. The conductivity determination coupled with a few simple chemical tests gives probably a better view of the inorganic combinations in the serum than would be found by an examination of the ash of the blood, since the ash must contain sulphur and phosphorus salts resulting from the oxidation of the organic compounds of these elements. The general method of calculating conductivities in a mixed fluid like the blood will be discussed under the head of conductivity of the urine. The information furnished by conductivity measurements is, it will be seen, an extension of that furnished by the osmotic pressure determinations. By a combination of the two processes it is possible to distinguish approximately between the concentrations of several classes of molecules present, and to follow variations in these concentrations rapidly. As yet the clinical value of the method is somewhat uncertain, however.

CHAPTER XIV.

SOME SPECIAL PROPERTIES OF BLOOD SERUM. BACTERICIDAL ACTION. PRECIPITINS, AGGLUTININS, BACTERIOLYSINS, HEMOLYSINS.

SELF PRESERVATION OF THE BLOOD.

In earlier experiments on transfusion of blood to supply a loss brought about by excessive bleeding it was recognized that the added blood sometimes seemed to act as a poison to the individual to whom it was given. It was found later that this toxic action followed the passage of blood from one species of animal to another, but that the transfusion from man to man, from dog to dog or from rabbit to rabbit was not accompanied by the same danger. Such observations were frequently made and gradually led to the conclusion that the plasma or serum of the blood of each animal contains a something which has a destructive action on the corpuscles of other bloods and which may be designed to protect the blood from the action of any foreign substance. Various theories have been put forward to explain this recognized property of the serum. As yet our knowledge in this field is largely of the empirical order, and scarcely suitable for clear elementary presentation. But the importance of the subject as thus far developed is so great that a short chapter on what seems most satisfactorily established may not be out of place. The phenomena in question are certainly chemical and from this side must receive their final explanation. Numerous related phenomena are found to call for the same kind of consideration.

It has long been known that the large white cells of the blood, the so-called leucocytes, have the peculiar power of destroying bacteria or other foreign cells which find their way into the blood stream. Hence these corpuscles have been called *phagocytes* or devouring cells. This destruction of small invading organisms they seem to accomplish by a kind of digestive process which in a general way may be followed under the microscope. To some extent the destruction of one kind of blood by another may possibly be accounted for in this way. But the chief action is certainly of a different character. While the white cells are in a measure protective agents, active in destroying elements which would be harmful if left in the blood, it seems altogether likely that the most important conserving forces in the blood are soluble compounds, possibly of the nature of enzymes. This view has been gradually developed and rests on a basis of experiment and observation.

Harmful foreign bodies entering the blood may be in the nature of cells, as of bacteria, or they may be the poisons called toxins produced by bacteria. Anything in the blood which resists or overcomes the force of this invasion is called an *anti* body. Normal serum seems to contain a number of anti substances, which have received different names, depending on how or against what they act. Some are called precipitins, others agglutinins, cytotoxins, etc., which terms will be explained later. In addition to the anti bodies normally present in sera in variable amounts, and which confer a certain degree of immunity, there may be produced artificially a greatly increased specific immunity against some particular invasion. It was the discovery of this fact which in reality led to the systematic study of the whole phenomenon.

The castor oil bean contains a peculiar poisonous principle known as ricin, which if given in relatively large doses is fatal, but against which an animal may be immunized by treatment with gradually increasing small doses. An experiment made by Ehrlich, to whom much is due in this field of investigation, showed that the serum of the treated animal must contain now a specific anti body capable of neutralizing the physiological action of ricin. He found that if the ricin poison and the serum of the immunized animal are mixed *in vitro* in certain proportion, and then injected into a fresh non-immunized animal no toxic action follows. The serum of the first or immunized animal has acquired the property of chemically combining with or in some manner neutralizing the action of the poison. That something akin to a chemical action is here in question is shown by the fact brought out by further experiments that certain proportions must be observed in the mixing of the serum and toxic substance just as in the complete neutralization of an acid by an alkali. It was further found that this combination may be hastened by heat and retarded by cold, which is true of most chemical reactions. The behavior is also *specific;* that is, the animal immunized against the castor bean poison is not immunized thereby against other vegetable toxic substances, as abrin, for example, and the serum of the animal will not, *in vitro,* neutralize the toxic abrin solution.

The same general condition has been recognized in connection with other immunizations and the characteristically specific nature of the anti body produced in the serum has been shown beyond question. The anti bodies protective against diphtheria have no effect against the toxins of tetanus or other disease and *vice versa.* With such facts established, inquiry was naturally directed toward the question of the chemical nature of these substances and to the question of their mode of action. In the voluminous discussions which have been carried on over these points it is not always easy to distinguish between observed facts and stoutly maintained theories.

GENERAL CHARACTER OF THE ANTI BODIES.

Antitoxins Proper. Foreign harmful agents gaining access to the blood may be of several kinds. Some of these are soluble toxic compounds, products of cell action, which in their behavior bear some relation to strong alkaloidal poisons. Many of these toxins are produced by bacteria in the animal body during the progress of disease and the symptoms observed are often due to the action of these poisons rather than to mechanical disturbances brought about by the bacteria directly. The toxins as soluble products have the power of wandering with the blood stream and thus reaching particularly vulnerable or susceptible organs. The soluble serum constituent which is normally present in small amount or which may be developed there to neutralize the toxin in some manner is called an *antitoxin* in the restricted sense. There is reason to believe that the two things combine with each other in a true chemical union and leave a soluble inert product.

Precipitins. The serum of normal blood contains constituents antagonistic not only to toxic substances but to other sera as well. The serum of one animal tends to precipitate or render cloudy the serum of another. This effect may be greatly increased by a kind of cultivation, which may be illustrated in this way. Rabbits' blood has normally some antagonism for ox blood, but if sterile ox serum be injected into the rabbit, intraperitoneally or intravenously, beginning with small doses and increasing through a number of days, a condition is finally reached in which the rabbit blood serum shows a very strong precipitating power for ox serum. The small amount of anti body in the rabbit's blood has evidently increased enormously through this treatment. The organism through the attack of the foreign serum gradually develops a protective agent which acts through exclusion or precipitation. This serum constituent is called a *precipitin.*

A vast number of experiments have been made in this field and the subject has importance in different directions. We recognize not only the normal effort of the blood to protect itself in this way, but also the remarkable power of development in the peculiar anti body here concerned. From another standpoint, however, the phenomenon has assumed even greater importance and that is in the identification of blood. This precipitin reaction like the others is specific and the serum of the rabbit immunized with ox serum will react only with the ox serum. But precipitins do not seem to be formed in the blood of animals which are closely related. The serum of a rabbit which has been treated with pigeon serum will not react with chicken serum; an anti rabbit serum cannot be secured by treating a guinea-pig with the serum of rabbit's blood. These general facts have been confirmed by many observations.

Blood Tests with Serum. The method of utilizing these generalizations is essentially this. Rabbits are the animals commonly used for experiments, since they bear the treatment in general well and yield a fairly large quantity of immunized blood later. Each rabbit is treated by injection with the blood serum of one animal, these injections being repeated a number of times, through several weeks. Then the rabbit is killed, bled and the blood allowed to stand for separation of clot. The clear serum is preserved in sealed tubes for future use, sometimes with, sometimes without addition of an antiseptic, as putrefaction does not appear to impair the reaction. By immunizing rabbits separately with the blood of man, the ox, horse, pig, dog, sheep, goat, chicken, etc., a whole series of *test anti sera* will be obtained, with which it is possible to identify most of the common bloods. Not much blood is required in the tests. A drop or two of blood, from a dried clot for example, is soaked in water or normal salt solution, the liquid obtained filtered to clarify it and treated in a small test-tube with two or three drops of the test-serum. The liquid to be tested need not be strong. In practice it should be divided into a number of small portions in test-tubes and each portion should receive a few drops of an immunized rabbit serum. Precipitation or clouding will occur in the tube to which the corresponding anti serum is added. For example, if the original clot of blood was human blood the extracted dilute serum in all the tubes, except the one to which rabbit blood immunized with human blood was added, will remain clear; other tubes with portions of the extracted clot show no reaction with the few drops of rabbit sera immunized with the blood of other animals.

The medico-legal importance of this reaction has already been recognized and tested in many ways. The blood of certain monkeys seems to react as does human blood, but those who have practiced the test most testify as to its certainty and wide applicability in distinguishing between human blood and the blood of the common domestic animals.

The Cytotoxins. This name is given to certain anti compounds in blood which are destructive of form elements. The anti bodies before considered deal with soluble substances, but here we have to consider something whose power extends to the breaking down of cell structures, whether of the blood corpuscle or of bacteria. In the one case the term *hemolysin* is used to describe the anti body; in the other case the term *bacteriolysin* is employed. In their mode of action these agents appear to be much alike and both are found in normal bloods. Both also may be greatly increased artificially.

The hemolytic action of one blood on another was first observed in experiments on blood transfusion which have been referred to already.

A foreign blood introduced into the circulation of an animal of a different species brings about a variety of changes; clots are sometimes formed and from resultant changes in pressure serous exudations may follow. Hemoglobinuria is a general consequence and this of course results from a breaking down of blood corpuscles in quantity. It has been assumed by some writers that this hemolytic effect is possibly due to altered osmotic pressure in the blood, as similar phenomena are brought about by the admixture of blood with weak solutions. But the peculiar specificity of artificial hemolysis shows that this explanation is not satisfactory.

If the blood of man, for example, receives an injection of human blood under proper condition no harm results, but if ox blood be used the case is different. A large transfusion of the ox blood might be at once fatal, the hemolysins of that destroying the human corpuscles. On the other hand transfusion of small amounts of ox blood would have different effects varying with the manner of transfusion. With but little ox blood added the human hemolysins would be greatly in excess and by their chemical mass action would bring about a relatively great destruction of the ox blood corpuscles, while the corpuscles of the human blood would suffer but little change. But more than this would probably take place as illustrated by what has been observed with certain lower animals. The serum of the eel is especially destructive of the corpuscles of rabbit's blood and a large injection of eel serum into the rabbit would produce death. With repeated small doses, however, the rabbit's blood is stimulated to develop the antitoxin or antihemolysin which protects against the eel serum poison. A condition of immunity is thus reached, and what has been shown for the rabbit has been shown for other animals. An explanation of the origin of the hemolysins will be offered below, but now we are concerned only with the fact.

In the above cited experiment the eel serum develops gradually an antihemolysin which works to prevent further destruction of the rabbit corpuscles. But the action does not stop here. By injection of blood, hemolysins for the corpuscles of this blood are also formed. Numerous experiments of this kind have been made with animals. For example, when rabbit's blood is gradually injected into the dog the production of hemolysins is stimulated and the serum of the dog so treated is found to be far more toxic for the rabbit than was the original serum. This toxic action, by test-tube experiments, has been found to be parallel to the hemolytic action of the dog serum on the rabbit corpuscles, thus showing that the toxicity *may* depend on the destruction of the corpuscles. The hemolysins produced as just explained are also in general specific in their character, which can be followed by experiments *in vitro* as well as *in corpore*.

In general the bactericidal action of serum resembles its hemolytic action, although control experiments *in vitro* cannot be as readily performed. We have therefore the *bacteriolysins* to consider along with the other cell destroyers. These bodies exist to some extent in normal blood and other body fluids and serve to protect the organism against the attack of bacteria which in any way gain admission to the body. Milk is relatively rich in bacteriolysins and hence the well-known germicidal action which has been long recognized. In this respect the behavior of mother's milk is more marked than that of cow's milk. Besides the cytotoxins of this class normally present in blood, specific bodies may be developed by the general methods followed in other cases, that is by the gradual introduction of cultures of specific bacteria, beginning with cultures of relatively little virulence. In this way the blood of the treated animal becomes immune for some one bacterium species and develops the power of destroying that bacterium only for which it was specially immunized. The same animal may be immunized against several kinds of bacteria at the same time and the different specific bacteriolysins do not appear to have any destructive action on each other. They exist together in the blood just as the different proteins may exist side by side.

Through the process of immunization the blood of the animal acquires not only the power of attacking the specific bacterium, but also the toxins of this bacterium. At least two kinds of anti bodies are therefore produced and there are conditions in which only one of these may be active. A serum may be active in the breaking down of bacterial cells, but inert as against the poisons produced by such cells. The complexities of the phenomena, however, cannot be detailed here. It should be said further that bacteria produce hemolysins, which are probably part of the toxins secreted. At any rate some of the toxins found in cultures are strongly hemolytic.

Agglutinins. Among the several modes of defense observed in sera of various animals that of *agglutination* of invading cells must next be briefly considered. We have seen that blood cells and bacterial cells may suffer a kind of dissolution through the action of *hemolysins* or *bacteriolysins,* and that a foreign serum is attacked by the *precipitins.* In addition to these defensive anti bodies there are present others which work by agglutinating or precipitating cells. A certain similarity exists between these bodies and the precipitins, but investigations appear to show that they are distinct. The agglutinating power is found in normal serum, and like the other anti agencies it may be greatly increased artificially and by the same general means. Agglutinins as precipitating agents enter into a loose kind of combination with the cells which they throw down. There is here a suggestion of combination in some kind of chemical proportions.

Bacteria agglutinins and blood cell agglutinins are to some extent specific, but apparently less so than are the precipitins. Because of this specificity the phenomenon has been applied in a method of diagnosis. Following observations of Gruber and others, Widal suggested a test which is now commonly employed in diagnosis of typhoid fever. It is essentially this:

Widal Test. A small amount of the blood or serum of the suspected typhoid fever patient is mixed on a slide with a bouillon culture of typhoid bacilli. After a time the mixture is examined with the microscope. If the suspected blood contains the agglutinins developed in the disease, "clumping" or precipitation of the bacteria from the culture must follow. (It is held as characteristic that loss of motility must also be observed.) The intensity of the agglutinin reaction may be estimated by noting the degree of dilution in which the blood serum will still agglutinate the bacteria from the bouillon culture.

Opsonins. There has been much discussion as to the manner in which the phagocytic power of the leucocytes, already referred to, may be increased. By one school of observers it is held that increased phagocytosis is due to a modification in the leucocytes themselves, which modification is produced by a variable element in the serum in question. The name *stimulin* has been given to this agent which is able to increase the activity of the white cells in the destruction of bacteria. Another and more generally accepted view is that increased phagocytic action is due, not so much to these cells themselves, but to a peculiar change in the bacterial organisms, which are to be destroyed. Wright and others following him hold that the serum contains a specific ferment-like body whose function it is to prepare or so modify invading bacterial cells that they may be readily engulfed and destroyed by the phagocytes. This specific agent is called an *opsonin*, and it has been shown that the opsonic power of a serum is subject to great fluctuation. In disease it may be greatly diminished; on the other hand, it may be artificially stimulated in certain directions so as to develop a marked bactericidal action. From this point of view bloods may be compared through the so-called *opsonic index*, which, briefly, is the ratio of the bacteria-engulfing power of 100 washed white cells in contact with the serum of a certain blood, to the power possessed by 100 similar cells in contact with the serum of a normal blood, taken as a standard, the cells and sera in each case being brought in contact with the same number of bacteria. The observation is made by the aid of a high-power microscope. The opsonic index observation has become of such importance that it is frequently used in diagnosis, and the *opsonic treatment* of disease is directed toward increasing this function or power of the blood of a patient, by properly graduated injections of tuberculin, for example, that a higher index is gradually developed. It is but fair to state that the doctrine of opsonins has its active critics as well as adherents.

Other Anti Bodies. In the above brief survey of the subject of anti bodies present or developed in the blood only those which have been the object of most frequent investigations have been mentioned. Bacteriologists have called attention to numerous other varieties or subdivisions, but it is not the purpose of this chapter to take up the discussion of details. What has been given is sufficient to call attention to the broader principles concerned.

CHEMICAL NATURE OF THE ANTI BODIES.

On this topic much has been written, but as yet no satisfactory answer can be given to the question: What are they chemically? As formed in the serum of blood or in milk it may reasonably be assumed that they must bear some relation in composition to the protein bodies. On this basis attempts have been made to separate them by fractional precipitation reactions such as were developed by Hofmeister and others for the proteins and which have been detailed in former chapters.

It appears from the evidence thus far offered that some of these bodies, at least, must be classed among the globulins. Pick and others have recently been able to separate the active substances from several kinds of immune sera and establish pretty accurately the limits of precipitation. The active fractions separated contained the real anti bodies in minute amount only, probably. In some cases they were found in the euglobulin fraction, and in other cases in the pseudoglobulin fraction of the precipitate.

There has been much speculation as to the part of the blood which gives rise to these various anti bodies. They are soluble and may not be separated by filtration but on dialysis they behave as other substances of very high molecular weight. In many respects they resemble highly active proteolytic ferments or enzymes, as the characteristic phenomena are exhibited even in dilutions of the active serum of

1 : 20,000, and heat and chemical reagents interfere with the active properties much as in the case of the enzymes. But there are apparently some exceptions which have led certain authors to deny their enzyme-like character. From the sum of the facts observed in the occurrence and action of the anti bodies several writers have been led to think of them as derived from the breaking down of the highly complex polynuclear white corpuscles of the blood. The behavior of these in the "living" condition has been already referred to; in their disintegration it is possible they may give off more and more of the groups on which their activity depends. But there are other possibilities and these will be referred to in the following section.

ORIGIN AND MODE OF ACTION OF THE ANTI BODIES. EHRLICH'S THEORY.

In the early days of observations on blood serum immunity the doctrine of the phagocytes received considerable attention. With increase of knowledge this theory was seen to be inadequate to account for accumulating facts, and the assumption of the soluble proteolytic ferments, the *alexins,* was next to attract attention. These may be formed from the polynuclear leucocytes or more remotely in the organs where these cells may have their origin, in the spleen, for example, and in the marrow of bones. As the name indicates the alexins are protective substances, but the simple assumption of these bodies acting alone, as a chemical reagent would for example, in the annihilation of intruding bacteria was soon seen to be too narrow to accord with experience. The phenomenon of immunity through the alexins or other bodies is a complex one, but has received much elucidation through numerous observations of recent years. One of the most important of these is concerned with the so-called Pfeiffer experiment.

Pfeiffer's Phenomenon. In his well-known experiments on the behavior of cholera bacteria Pfeiffer found that the serum of an animal which had been immunized against cholera, when tested *in vitro* against the vibrios, seemed deficient in bacteriolytic power, possessing no greater activity, evidently, than that due to the proteolytic ferment of the normal serum. Activity may be given to the immunized serûm, however, by injecting it back into the peritoneal cavity of the animal; cholera vibrios injected at the same time are quickly destroyed. The action is strictly specific, since typhoid bacilli injected in the same way remain active. It was found also that the same result can be reached with the vibrios by adding, *in vitro,* fresh normal serum to the latent immunized serum. The vibrios succumb as they did *in corpore,* showing that a vital action is not in play. Numerous similar observations were made by Bordet and others. From all these experiments it is evident that at least two things are concerned in the bacteriolytic action and the analogous hemolytic phenomena. The immune body developed in the course of immunization does its work as a cell destroyer, whether a blood corpuscle or a bacterium, only by the aid of something normally present in fresh serum. The specific immune body was found to be thermostable, that is it withstands a temperature of $55°-70°$ without loss of its special properties. If after being warmed to this extent the immune serum is cooled to below $55°$ and mixed with fresh normal serum the full cytolytic activity appears. On the other hand the ferment in the normal serum is thermolabile; a temperature of $55°$ or above destroys it permanently. For this thermolabile normal ferment Ehrlich proposed the name *addiment*. This is the same as the alexin body. The term *complement* is now more commonly used to describe the same thing which seems necessary to make the specific immune body really active.

The Side Chain Theory. Thus far we have been concerned with the results of experiments, with facts about which there cannot be much question. But a comprehensive theory to correlate all these generalizations became necessary. Many attempts have indeed been made to establish a theoretical basis for the doctrines of immunity, but it remained for Ehrlich to suggest something which is really tangible from the chemical standpoint. To follow the Ehrlich notions some other matters must be explained first.

Years ago, in attempting to explain some of the properties of large organic molecules, Pasteur introduced the notion of molecular asymmetry into chemical science. He showed the value of the notion of configuration in dealing with certain classes of chemical problems. This general idea was greatly advanced by Emil Fischer in his papers on the chemistry of the sugars. Certain ferments were found to decompose one sugar of an isomeric group, but to leave the other almost identical sugars untouched. In other words the ferments were found to observe a specific selection, and to work as ferments, the enzymes in question must possess a certain stereochemical structure bearing some relation to the stereochemical structure of the sugar. Without this relation fermentation cannot take place. In order to make his meaning plain Fischer employed a figure which has since become famous. He said, in speaking of certain glucosides, "Enzyme and glucoside must fit into each other as a key into a lock in order that the one may be able to exert a chemical action on the other." In one of his papers Fischer suggested that the idea of related molecular configuration of ferment and fermentable body may prove of value in physiological investigations as well as in chemistry, and in the development of the theory of Ehrlich this prediction has been verified. Toxins and anti bodies combine with each other only when they possess corresponding atom groups, and specificity is regarded as dependent on this relative configuration.

Without going into minute details the chemical part of the Ehrlich theory is briefly this: Bacteria, animal cells and toxins are all complex aggregations of more or less complex molecules. The latter have certain configurations dependent on the presence of side chains or side groups, to borrow an expression from organic chemistry. These side chains are directly or indirectly the points of attack or defense in the action of the several bodies on each other. In order that a substance may act as an enzymic poison or toxin to cells of the body both cells and toxins must therefore possess certain reciprocal configurations. It has been suggested further that these side groups are concerned in all the actions of the cells and that it is through them, for example, that the latter absorb their necessary nutriment and elaborate new structures from it. Some of the side chains may be constructed to combine with fats, some with carbohydrates and some with proteins, but in the presence of toxins or bacteria with the right kind of side chains combination with these may take place instead. Many of these combinations, perhaps all, take place not directly, but indirectly, through the presence of an intermediary body or group which itself must possess two linking complexes or groups with proper configurations.

To describe these various groups certain special names have been suggested. *Immune body* is the specific substance formed in the immunizing process against cells and is known also by several other names, among which *amboceptor* and *intermediary body* are the most commonly used. The *complement, addiment* or *alexin* is the ferment-like body found in normal fresh serum, and which added to the immune body makes up the real cytotoxin. It is not specific and, as intimated above, is sensitive to heat, and also to light and air (oxygen). The various groups of the large cell complex, whether of a body cell or of a bacterium, which have the power of uniting with other groups are called *receptors*. The part of the receptor which is free to combine with a food molecule or analogous substance is called its *haptophorous* group. Ehrlich pictures the part played by the immune body and the complement in this way. Assuming that a bacterial cell enters a medium where these complexes are present the immune body attaches itself to the haptophorous group of the cell. In this condition no action follows immediately; but the immune complex has itself two haptophorous groups or side chains, through one of which the union with the bacterial cell is effected, while with the other it joins on to the addiment or complement through its haptophorous group. In this way the complement, which alone is inactive or unable to attack the cell, is brought into the immediate neighborhood of the latter, where its proteolytic efforts are more effective. Every fresh normal serum seems to have present enough of the complement groups; the question of destroying the invading cell depends then on the number of immune or intermediary groups in the field. Another part of the Ehrlich theory attempts to account for these.

The Immune Group. Ehrlich traces the development of the immune body to the spontaneous effort on the part of the cell to protect and regenerate itself in case of partial destruction. The various cells of the body exist in a kind of equilibrium with each other. An injury to one, that is the loss of some of its side chains, immediately leads to an effort at compensation. The hyperplasia observed in an organ may extend to the single cells and in consequence of this we have over-compensation. The cell's efforts at regeneration lead to the production of more side chains than are actually necessary and some of these combine with the aid of their receptor groups with toxin or with complement. Many are formed in excess and are thrown off into the circulation. These free receptors constitute the various anti or immune bodies. Combining with complement groups they form the true cytotoxins. The larger the number of free receptors thrown off into the blood by the over-compensating efforts of the attacked cells the stronger is its cytotoxic or antitoxic character since these receptors hold either the toxin or foreign cell and thus protect the parent native cell from attack or destruction.

The fundamental point then in the Ehrlich theory of serum immunity is this formation of side chains in excess by over-compensation, and is founded on the somewhat earlier Weigert doctrine of cell regeneration and over-compensation in general. Ehrlich has added the chemical conceptions of side chain groups and has drawn numerous illustrations from organic chemistry to show how they may act. Large molecules holding amino, sulphonic acid or halogen addition groups, for example, may lose these or take them on again or take others like them without losing their identity. Reagents acting on such a large molecule attack, not the nucleus, but these side chains in general. The simple organic molecule has not the power of *self regeneration*, but the cell, which is a collection of many such molecules, has the power of forming new materials from the nutritive substances furnished to it. If whole new cells are formed why not parts of cells or the outlying side groups as well, and this is the Ehrlich assumption, which is not unreasonable.

As explained above these side chains or receptors are of various kinds. Three distinct types or orders are easily recognized. Receptors of the First Order have one haptophorous group and form *antitoxins*. That is, they combine chemically with the soluble toxins in the serum and in a sense neutralize them. Receptors of the Second Order have one haptophorous group with which a foreign molecule or group may be held and one special group which performs the function of an *agglutinin* or *precipitin*. Receptors of the Third Order or *amboceptors* have two haptophorous groups with which two things may be united. One of these is the foreign cell (through its corresponding haptophorous group) and the other the complement. In this way the complement or alexin is able to work on the invading cell and attack it through its "zymotoxic" group. These amboceptors are in themselves inactive and can behave as cytotoxins (hemolysins or bacteriolysins) only when joined to the complement or ferment group. They are formed in the serum by immunization with foreign cells, and in turn combine with cells.

Another product of immunization with cells is the agglutinin receptor, while immunization with toxins leads to the formation of receptors of the First Order. Cytotoxins produced by one animal species A, brought into the serum of another animal species B, lead to the formation of anticytotoxins which may be either anti complements or anti amboceptors.

Toxin molecules on standing or by heating seem to lose some of their activity or toxic power, while their power of combining chemically with or neutralizing antitoxins is not diminished. In this condition they are called by Ehrlich *toxoids*, and he explains the behavior by assuming that the toxins have two characteristic groups, one of which, a haptophorous group, persists and combines with the antitoxin while the other is less stable and may be lost; this he calls the *toxophorous* group. By warming to $55°-60°$ the complement bodies of serum become converted into active *complementoids*, which retain the haptophorous group but lose the zymotoxic group. Amboceptors may lose in the same way one of their combining groups and become *amboceptoids*.

In the further development of the Ehrlich nomenclature the term *toxon* was introduced to describe another form of modified toxin.

The toxons are bodies of relatively slight toxicity, and exist with the toxin from the start, in place of being developed on standing. They have the power of combining with antitoxins.

It would not be proper in this place to go more fully into the details of the Ehrlich theory; enough has been given to furnish the student an outline of the most important points in the theory. It is of course true that much of the present view is artificial and tentative and, with closer fixation of facts, must be modified. This has been the history of the development of all chemical theories. In its main features the Ehrlich doctrine gives us a tangible picture of how the serum may act toward foreign bodies. For the ultimate reasons for the formation of immune side chains by stimulation we have no more explanation than we have for many of the manifestations of chemical affinity. In its outlines this theory of the action of immune serum appears wholly fanciful, but in reality it makes no greater claim on the imagination than do some of the oldest accepted theories of general chemistry.

CHAPTER XV.

TRANSUDATIONS RELATED TO THE BLOOD.

THE LYMPH.

The capillary vessels convey the arterial blood with its store of nutriment to the various tissues, which by transudation receive the required amount of nourishing matter. The communication between the blood vessels and the tissue cells is not a direct one; on the contrary, this transudation takes place into the multitudes of star-shaped spaces which break the continuity of the cells, and which communicate with each other by means of fine canals. A liquid passes from the blood into this network of spaces which form the beginning of a new vascular system. The transuded liquid is the lymph which serves the double purpose of nourishing the tissues and draining them also, since this liquid not only gives up large molecules of absorbable matter, but takes up at the same time various products of metabolism. What is left over after this contact with the tissues collects in the minute lymph capillaries and then into the larger lymphatic circulation proper.

Composition. Being thus related to the blood the lymph must have a composition not greatly different from that of the plasma. The normal lymph is a nearly clear fluid with a specific gravity somewhat less than that of the serum as a rule. It contains salts and organic substances as does the serum of the blood, but is naturally poorer in protein elements since a portion of these has been taken up to nourish the neighboring cells. The lymph contains a small amount of fibrinogen. Very few red corpuscles are present, but as the formation of leucocytes or white corpuscles takes place in the so-called lymphatic glands these form elements are abundant in the final flow.

It has been already shown that potassium salts are common to the corpuscles of the blood, while the salts of sodium are abundant in the plasma. We naturally find the same thing in the lymph which contains, in addition to proteins, fat, sugar, cholesterol, etc., inorganic salts, having about the following composition, according to some old analyses by Schmidt:

Sodium chloride	5.67 per 1000
Sodium oxide	1.27
Potassium oxide	0.16

Sulphuric acid, SO_3 .... 0.09 per 1000
P_2O_5 as combined with alkalies 0.02
Calcium and magnesium phosphates 0.26

This result is approximately what one would expect from an analysis of blood serum and is in fact about what has been found. The two fluids have the same osmotic pressure, due largely in both cases to salt content.

Function of the Lymph. The amount of metabolic substances returned finally to the venous circulation through the lymphatics does not appear to be great. The chief product of oxidation, CO_2, seems to be thrown back directly from the lymph spaces to the smaller vessels leading to the venous system. The lymph spaces into which the transuded serum flows have apparently two ways of discharge. The bulk of the liquid with some of the absorbed metabolic products passes, as intimated, into the gradually enlarging lymphatic system, but certain other complexes, and among them possibly the most abundant oxidation products, evidently find their way immediately into the capillary beginnings of the venous circulation. It is not possible to give exact figures as to the relative amounts of these products going the two courses, but it is accurately known that the carbon dioxide pressure in the lymph is much less than in the venous circulation, and the urea appears also to be less.

It appears to be pretty well established that the leucocytes are active in hastening the destruction of complex products of tissue waste. These cells seem to possess a marked chemical activity which is manifested in a kind of digestion of the grosser complexes separated in the tissue metabolism, but, unfortunately, our knowledge here is meager. The normal end products of this breaking down process are not formed at once. Possibly the leucocyte is one of the assisting agents. It has been therefore held by many writers that the formation of these lymph cells is probably the most important part of the work in the lymphatic system.

The amount of lymph which is formed daily is relatively large and possibly equal to the volume of blood. A large part of this comes from the flow through the lacteals. Certain substances have the power of stimulating the flow of lymph. Various salts, sugars and urea have this property; they are called lymphagogues. Muscular exertion also increases the production of lymph, because it is needed to supply tissue waste, and a more rapid flow is called for to carry off the products of disintegration. Of the manner in which the lymph gives up its content of nutrients to the tissues absolutely nothing is known; but the object of this intermediary system is readily seen. It serves as a regulating mechanism to prevent too rapid changes in the blood composition which would follow if it should come in direct contact with the tissues.

CHYLE.

During the digestion of fatty foods the lymph absorbed from the intestinal walls contains numerous minute fat globules in the form of an emulsion. This portion of the lymph is known as chyle and is carried along by the lacteals and finally discharged into the lower part of the thoracic duct.

In composition chyle differs from the lymph from other sources mainly in its fat content. In the periods when digestion is not in progress the lacteal lymph is also clear. These vessels are then partly collapsed and hard to see.

TRANSUDATIONS PROPER.

The lymph has sometimes been considered a transudation of the blood, but the term is now more commonly used to describe the flow of liquid from the blood into certain cavities of the body under pathological conditions. A transudation proper is then a modified lymph and results often from an imperfect elimination of water by the kidneys, or from some disturbance in the circulation. Inflammatory transudations are sometimes distinguished as *exudations*, and in these the cell elements are much increased. If they are excessive the discharge is known as *pus*.

For example, the pleural and peritoneal cavities contain but little fluid. The serous surfaces are moist, but it would not be possible to collect enough fluid substances for satisfactory analysis under normal conditions. In the advanced stage of pleurisy a considerable quantity of fluid collects in the pleural cavity and its composition resembles that of the lymph, but it is poorer in solids ordinarily. In some forms of acute peritonitis a collection of similar fluid may take place in the peritoneal cavity and this may amount in bad cases to several liters.

The various forms of dropsy described by physicians are essentially characterized by analogous transudations of serous fluid without inflammation. Ascites, or dropsy of the abdomen, hydrocele, or dropsy of the testicle, and hydrothorax, dropsy of the pleura, are illustrations. Some analyses are given below, showing the general nature of the fluids collected in such cases. It must not be supposed, however, that exactly similar results would always be obtained by analyses of fluids from the same organs. The composition of pus serum is somewhat similar; it contains, however, more products of protein disintegration.

	Hydrocele Fluid (Hammarsten).		Pus Serum (Hoppe-Seyler).
Water	938.85	Water	909.63
Serum albumin	35.94	Proteins	70.22
Globulin	13.25	Lecithin	1.03
Fibrin	0.59	Fat	0.27
Ether extractives	4.02	Cholesterol	0.70
Soluble salts	8.60	Alcohol extractives	1.13
Insoluble salts	0.66	Water extractives	9.22
		Salts	7.75

	Pleural Transudate (Scherer).		Peritoneal Transudate (Hoppe-Seyler).
Water	935.52	Water	969.64
Albumin	49.77	Albumin	19.29
Fibrin	0.62	Urea	0.31
Ether extract	2.14	Ether extract	0.43
Alcohol extract	1.84	Alcohol extract	1.37
Water extract	1.62	Water extract	0.98
Salts	7.93	Salts	7.98

Amniotic Fluid. This may be considered as a kind of transudate. A number of analyses have been made which show about 98.5 per cent of water, 1 per cent of salts and 0.5 per cent of organic solids, largely proteins.

THE LYMPH CELLS.

These large cells or leucocytes have already been referred to as formed in the lymph glands. They are also formed in large numbers in the spleen and the thymus gland. From whatever source produced they are supposed to have the same general composition and chemical function. The following analysis by Lilienfeld gives an idea of their general composition. The dry substance of the cells amounted to 114.9 parts per 1000 and was made up of the following constituents, the figures referring to per cent amounts of the dry matter. The cells analyzed were from the thymus of the calf.

Leuco-nuclein	68.79
Albumins	1.76
Histone	8.67
Lecithin	7.51
Fat	4.02
Cholesterol	4.40
Glycogen	0.80

In addition to these organic substances mineral matters are present, with salts of potassium characteristic. The substance described as leuco-nuclein is apparently the nucleic acid complex which in the original cell is combined with the histone as a nucleate. The lecithin is an important fraction in these cells, and may exist in part as a protein combination. The methods of separating lecithin are far from exact.

Pus Cells. In their origin and characteristics these may be considered as very similar to the leucocytes, if not indeed identical with them. The few analyses made show a general agreement when reduced to the same terms. It must be remembered that such analyses are far from simple operations, especially in the separation of the several protein constituents. The following figures by Hoppe-Seyler should be compared in that light with the above. The numbers refer as before to the dry matter:

Nuclein and albumin	67.40
Lecithin	7.56
Fat	7.50
Cholesterol	7.28
Cerebrin and extractives	10.28

Cerebrin is the name given to a body containing nitrogen in small amount, but which is not a protein. It is usually found in products derived from the brain. As to its exact nature but little is known.

The pus cells float in a fluid known as the pus serum, which closely resembles the other transudates, as shown by the analysis above. The cells may be separated from the serum by the centrifuge, and if mixed with a little strong alkali yield a gelatinous slime, which is characteristic. This test is sometimes applied for the detection of pus in urine. On bacterial decomposition pus yields a number of products easily recognized as derived from the nuclein fraction of the protein.

The Spleen. While our knowledge of the functions of the spleen is very imperfect, a few words may be said in this connection, since as far as is known the lymph glands in producing leucocytes do about the same kind of work. The one thing most apparent about the spleen is that in this organ large numbers of white cells are formed and given to the blood; it is also known that these cells suffer destruction there, as the spleen pulp contains considerable quantities of the xanthine bases, which are among the common products of cell nuclei destruction. The cells so destroyed may possibly be those which have already served their purpose in the blood as the disintegrating agents concerned in the breaking down of other bodies. Uric acid, as derived from the xanthine bases, is known to result when blood is rubbed up with the spleen substance. The spleen is enlarged in many cases of infectious diseases. This is possibly from the abnormally great production of leucocytes needed in the blood in overcoming the effects of the toxic agents or invading bacteria.

Of the chemical nature of the spleen substance little is known, as it is practically impossible to free it from blood for analysis. In addition to the xanthine bodies and other decomposition products there seems to be present an albuminous substance containing iron, which is considered as an albuminate; but of its uses nothing definite is known, beyond the possibility that it may be concerned in the production of red corpuscles.

The chemical work performed by the spleen may evidently be done by other organs, as it may be completely extirpated without leading to fatal results. In its absence the production of great numbers of leucocytes falls to the lymph glands, and the red marrow of bones may take over the work of generating red blood corpuscles.

CHAPTER XVI.

MILK.

The qualitative composition of milk as produced by the mammary glands of different animals is nearly the same whatever the species of animal. But in quantitative composition very great differences obtain. Cow's milk has always been taken as the type with which comparisons are made, as it is the kind everywhere in general use. The essential differences between it and mother's milk will be pointed out in what follows.

COW'S MILK.

In an earlier chapter an analysis of cow's milk is given which represents a general average of composition of good market milk. But the normal milk of individual cows may be very different from that there described. The qualitative composition is always the same, but in the amounts of fat, sugar and protein present the greatest divergences are noticed. These variations depend on the race of the animal, the period of lactation and especially on the feed. It is also a well-known fact that the richness of milk varies during the time of milking, the first portions of milk withdrawn from the udder being poorer in fat than the last part or "strippings." In speaking of normal milk, then, these facts must be kept in mind; a milk may be normal but not necessarily rich or good, from the standpoint of food value. The following table illustrates the variations found in the analyses of milk from a large number of cows. The mean specific gravity is from 1.029 to 1.033.

	Mean.	Maximum.	Minimum.
Water	87.4	91.5	84.0
Fat	3.5	6.2	2.0
Sugar	4.5	6.1	2.0
Proteins	3.9	6.6	2.0
Salts	0.7	1.0	0.3

When the water of milk is found as high as 91.5 per cent of the whole the sum of the fat, protein, sugar and salts can be only 8.5 per cent in place of 12.5 per cent, which should be expected in the mixed market milk.

Market Milk. Experience has shown that the mixed milk from a herd of well-kept cows should have a composition not far from that given in the table above under "mean." Laws have been passed in most of the large cities of the United States and Europe requiring that milk sold as pure must be of a quality not inferior to this mean value. Indeed, in some places a market milk of still higher standard is required.

PHYSICAL COMPOSITION OF MILK.

The exact nature of the mixture of the component parts of milk has long been a debated question. When taken from the udder the fats, proteins, sugar and salts are mixed homogeneously, and no immediate tendency is observed toward a separation of the light fat from the other and heavier solids. In time, however, such a separation takes place and the fat rises in the form of cream. Milk cannot be looked upon as a transudation from the blood because it contains substances not found in that fluid; the casein of milk and the lactose are different from the proteins and sugar normally existent in the blood, and the fat of milk is more complex probably than the blood fat. It is necessary to admit, then, that some of the milk components are produced in or from the substance of the mammary glands themselves. It is held by some authorities that the nucleo-proteids of the gland cells are similar to or identical with the casein, which therefore has its origin in the gradual breaking down of those cells. In regard to the milk fat it is known that certain fats can pass but little changed from the food through the blood and appear finally in the milk, imparting peculiar properties. But, on the other hand, milk fat is produced when the food of the animal contains no fat whatever, and certainly no fats resembling the characteristic volatile fats of the milk. In the carnivora, confined to an essentially protein diet, milk fat is formed, and in the herbivora on a diet containing largely pentoses and other carbohydrates milk fat is likewise produced normally and in quantity.

These facts, then, seem to be clear, that while under some circumstances fats as such pass from the blood into the milk, and this is further evident by the experience of feeding cows with certain foods rich in fats, the milk glands have the power of producing the several individual fats as occurring in milk from compounds which are not fat to begin with. In discussing the chemistry of proteins in an earlier chapter it was shown that in the breaking down of these bodies under the influence of various agents fatty acids are found among the decomposition products. The complex protein molecule may be all that is necessary to give rise to the milk fats if other things are not available.

The origin of milk sugar is not at all clear. Lactose is not a constituent of our ordinary foods and at best the blood contains probably only inverted sugars or monosaccharides. In some cases the formation of milk sugar may be traced indirectly to the carbohydrates of the food; but this will not explain the production of sugar in the carnivora. Here as before we are probably obliged to fall back on the behavior of the complex proteins. Among the groups they contain, or at any rate yield in decomposition, the presence of sugar groups has been certainly shown. This was explained in a former chapter. The milk lactose probably results from a synthesis of these simpler sugars.

From what is in general known of the nature of complex protein matter such as exists in the milk glands it seems therefore possible to trace the origin of the milk proteins, sugar and fats to the disintegration of this original protein substance. But of the agents of disintegration, and following necessary syntheses, we know absolutely nothing. The presence of certain enzymes has been assumed, but as they have not been isolated or identified, their part in the reactions remains speculative.

CHEMISTRY OF THE MILK COMPONENTS.

Fats. In the older literature milk fat was given a comparatively simple composition. It was assumed to consist of stearin, palmitin, olein and butyrin essentially, the last named volatile fat imparting the flavor to the separated butter. At the present time we must admit that our knowledge is far from exact on the subject, but we know that the composition of milk fat is by no means as simple as once assumed. In the chapter on the fats an analysis is given which conforms better to our modern notions. We find, then, besides butyrin several glycerol esters of the same series of comparatively volatile acids. Among the heavier fatty acids myristic acid seems to have some importance, as disclosed by a number of analyses. A small amount of lecithin appears to be also present.

Although produced from a variety of materials in feeding experiments, milk fat, as butter, maintains a rather constant composition as disclosed by both chemical and physical tests. The melting point of the fat is usually between $31°$ and $32.5°$ C. and its specific gravity at $38°$ is about 0.912. Butter fat is easily saponified and from the saponified mass the fatty acids which are non-volatile and practically insoluble in hot water may be separated. These insoluble acids amount in the mean to about 87.5 per cent of the weight of the original butter fat. That is, 10 grams of average butter fat should yield 8.75 gm. of insoluble acids, the difference representing the lighter soluble fatty acids and glycerol. In the fat from very rich milk the insoluble acids may be somewhat lower than this, while in poor milk they would be higher. These facts are important in distinguishing between butter and its substitutes.

Fat Globules. In milk the fat exists in the form of minute globules of different sizes. The diameters of these globules vary between about 0.0016 mm. and 0.01 mm. A cubic centimeter of normal milk containing 3.5 per cent of fat contains 100 millions or more of these globules. In the milk they are described as existing in the form of an emulsion, but of the exact nature of this emulsion our knowledge is imperfect. It has been held also that a membrane of casein encloses the fat globules and that this prevents the ready extraction of fat when ether or similar solvent has been added to milk. If the milk is previously shaken with a little acid which is supposed to break or destroy this membrane the ether added will now dissolve it. But this membrane cannot be directly detected with the microscope, and experiments on the formation of fat emulsions by the aid of casein and weak alkalies have shown that the presence of a membrane is not necessary to account for the round form or the failure to dissolve readily in ether. The surface of the globule is not the same as the interior portion, as it appears to take a stain by certain agents which does not penetrate. But in the conflict of views advanced it is not yet known what the surface actually is.

Casein and Lactalbumin. These compounds have been mentioned in the chapter on proteins and their place in the general scheme of classification pointed out. In the free, pure state the casein is a distinctly acid body which neutralizes alkali and forms salts with rather sharply defined properties. Casein may be easily separated from milk in this way:

Experiment. Dilute 500 cc. of skimmed milk with about 2 liters of water in a large jar; add enough dilute acetic acid to make not over 0.1 per cent of the whole. This causes a precipitation of the casein in fine white flakes which soon settle, leaving a nearly clear whey. After some hours decant this whey and add a greater volume of distilled water and stir up well. Allow this mixture to settle and pour off the water. Add a liter of water and enough weak sodium or ammonium hydroxide to dissolve all the casein and produce an opalescent solution. This in turn is reprecipitated with dilute acetic acid after adding considerable water, and these operations are repeated several times. In this way a casein nearly free from calcium salts is obtained. It is washed well with water by decantation, then poured on a Buchner funnel, drained, washed with alcohol, until the water is removed and finally several times with ether to take out the fat. On drying a fine white powder is obtained with which the important properties of casein may be shown.

Experiment. Weigh out 5 to 10 gms. of casein into a beaker or flask and add distilled water. Note that it appears to be quite insoluble (This might be shown by filtering and testing the filtrate by evaporation.) Add a few drops of phenol-phthalein reagent and run in standard sodium hydroxide solution until a permanent pink appears. In this way the equivalent or combining weight of the casein may be found very closely. It is over 1000.

The alkali salt of the casein forms a somewhat viscid solution. If exposed to the air it dries down to a gummy mass which is very adhesive and acts like a mucilage. In the arts similar, but crude, solutions are used as sizing material and as a constituent of certain paints. These products are made from the cheap whey from the creameries.

Experiment. While the alkali salts of casein are readily soluble in water the heavier metal combinations are not. This may be shown by adding to the alkali solution as obtained above solutions of salts of other metals. Precipitates are formed readily with most of them. The calcium salt is moderately soluble, as may be shown by rubbing up some casein with calcium carbonate and water. On filtering and adding a drop of acetic acid to the filtrate a casein precipitate comes down.

Experiment. The *lactalbumin* may be shown by boiling the decanted liquid from the first acetic acid precipitation given above. A coagulum forms as in a dilute white of egg solution to which a little acid had been added.

The phosphorus in casein appears to be combined in at least two forms. On digesting casein with pepsin and hydrochloric acid a product known as pseudo-nuclein is separated because of its failure to digest. In long continued digestion some phosphorus seems to pass into the form of orthophosphoric acid, while another portion remains in the albumoses formed, in the organic condition. The digestion residue, however, is not nucleic acid, which distinguishes the casein from the true nucleo-proteids. In the precipitation of casein from milk by the treatment given above the combined phosphorus, whether in acid or organic combination, does not appear to be touched. The mineral phosphates are separated, however, but not completely, as the finally washed and dried casein always contains a trace of ash, a part of which is calcium phosphate. This ash probably has nothing to do with the true casein, but is present because of imperfect separation.

Milk Sugar. This crystallizes with one molecule of water, $C_{12}H_{22}O_{11} + H_2O$, and yields glucose and galactose on inversion. It is separated in large quantities from the whey of cheese factories and is employed in the manufacture of invalid and infant foods. The general properties of the sugar have already been given.

The Mineral Substances in Milk. In the analyses quoted at the beginning of the chapter the ash of the milk is given as about 0.7 per cent in the mean. This amount appears small, but still it is of the highest importance, as it makes up between 5 and 6 per cent of the total solids of the milk. The composition of milk ash has been the subject of many investigations. While it cannot represent exactly the condition of the inorganic substances in the original milk, the agreement is an approximate one and is probably near enough for practical purposes. In obtaining ash for analysis, sulphur and phosphorus in organic combination are thrown into oxidized form and combined as salts, sulphates and phosphates, in which form we find them in our subsequent tests. The following figures from König represent the composition of milk ash as the mean of 9 analyses:

	Per Cent.
K_2O	24.06
Na_2O	6.05
CaO	23.17
MgO	2.63
Fe_2O_3	0.44
P_2O_5	27.98
SO_3	1.26
Cl	13.45

Accepting these figures as fairly accurate, and they agree pretty well with the results of all analysts who have dealt with the question, 1 liter of average cow's milk would contain the following amounts of the several constituents:

K_2O	1.74 gm.
Na_2O	0.44 gm.
CaO	1.67 gm.
MgO	0.19 gm.
Fe_2O_3	0.03 gm.
P_2O_5	2.02 gm.
SO_3	0.09 gm.
Cl	0.97 gm.
	7.15 gm.

Noteworthy here are the relatively large amounts of the phosphates of calcium and potassium. These salts represent all the mineral matters needed in nourishing the body. As found in the milk they exist in the combinations from which they are most readily assimilated.

The Colostrum. This is the milk secreted before and for a few days after parturition, and is characterized by higher specific gravity and content of solids. It contains a large amount of coagulable proteins and therefore thickens on boiling. Some idea of the general composition is given by the following figures which represent the means of a number of analyses:

	Per Cent.
Water	74.05
Casein	4.66
Albumin	13.62
Fat	3.43
Sugar	2.66
Salts	1.58

Whey is the fluid left after separation of the larger part of the fat and casein in the cheese industry or by analogous coagulation. The sugar and salts remain practically unchanged, while the fat and casein are reduced to traces. The lactalbumin left averages about 0.5 per cent.

Buttermilk differs from ordinary milk essentially in its lower content of fat. It is usually sour because of being separated from ripened cream, and contains therefore an appreciable amount of lactic acid formed at the expense of some of the sugar.

Skimmed milk is in composition similar to buttermilk but is usually sweet. In the modern methods of separation by centrifugal machines the fat may be reduced to less than half of one per cent; the protein is also somewhat reduced.

SOME EXPERIMENTS WITH MILK.

A few simple tests may be made to illustrate the composition of milk.

THE TEST FOR FAT. Pour about 20 cc. of milk in a porcelain dish, add an equal volume of clean, dry quartz sand and evaporate, with frequent stirring, about an hour on the water-bath. Then loosen the dry mass as well as possible by means of a spatula, or glass rod, and pour over it 25 cc. of light benzine. Stir up well and cover with a sheet of paper and allow to stand 15 minutes. Then pour the liquid through a small, dry filter into a small, dry beaker, and place this in hot water to volatilize the benzine.

A residue of fat will be left. Do not attempt to evaporate the benzine over a flame, or on a water-bath under which a lamp is burning. Heat the water, then extinguish the flame and immerse the vessel containing the benzine in the hot water.

THE TEST FOR SUGAR. Measure out about 10 cc. of milk, and dilute it with water to make 200 cc. Add to this 5 cc. of a copper sulphate solution, such as is used in making the Fehling solution (69.3 gm. per liter), and then enough potassium or sodium hydroxide solution to produce a voluminous precipitate containing copper with all the proteins and fat. For this purpose about 3.5 cc. of a 1 per cent sodium hydroxide solution will be required. Allow the precipitate to subside, pour or filter off some of the supernatant liquid, and boil it with Fehling's solution. The characteristic red precipitate forms, showing presence of sugar.

PROTEIN TEST. The presence of proteins in milk can readily be shown as follows: Mix equal volumes of milk and Millon's reagent in a test-tube, and boil. The bulky red precipitate which forms proves the presence of the body in question.

Action of Rennet on Milk. The mucous membrane of the stomachs of most animals, and especially that of the young calf, contains an enzyme known as the "milk curdling ferment," the "rennet ferment," or rennin, the nature of which has already been explained in the chapter on the ferments.

A crude extract of the mucous membrane of the stomach from the calf is commonly called rennet and has long been in use for the curdling of milk in the production of cheese. This curdling consists essentially in the coagulation or precipitation of the casein, which, it will be recalled, is not readily thrown down by the usual methods.

An active rennet can be readily obtained by digesting the stomach of the calf with glycerol or brine. If a brine extract is precipitated by alcohol in excess a white powder separates, which when collected and dried, has very active properties. Several powders of this description are now in the market. Let the student try the following experiment with such a product:

Experiment. Warm some fresh milk to a temperature of 38° to 40° C. in a testtube or small beaker, then add about half a gram of commercial "rennin," and after stirring it well keep for 15 minutes at a temperature not above 40°. Then as the milk cools it assumes the consistence of a firm jelly. It is essential in this experiment that the temperature be kept within the proper limits, as the enzyme is not active at low temperature and it is, like others, destroyed by high temperature. The casein or cheese which is obtained in this way is not the same as that precipitated by acids as it contains much calcium in combination. This form of casein is usually called para-casein.

Repeat the experiment by adding about 5 drops of a concentrated sodium carbonate solution to the milk and then the rennet. Coagulation now fails or is partial.

The Action of Pancreatic Extract on Milk. The behavior of milk with extract of pancreas is somewhat complicated because of the complex nature of the milk; the sugar, the fat, and the protein bodies all suffer some change under the influence of the several pancreatic enzymes.

The most interesting of these changes, however, is that produced in the proteins, and is commonly called peptonization.

At the present time the digestion, or peptonization of milk, is a very common practice in the preparation of food for the sick room, and can be illustrated by the following experiment:

Experiment. Dilute about 10 cc. of milk with an equal volume of water, and add half a gram of sodium bicarbonate. Next add a few drops of a liquid extract of pancreas, or a very small amount (10 to 20 mg.) of one of the concentrated "pancreatin" powders on the market. Shake the mixture and keep it at a temperature of 40 degrees on the water-bath half an hour. At the end of this time filter and apply the peptone test—potassium hydroxide and dilute copper sulphate—and observe the pink color. As the action of the pancreatic extract is continued the liquid resulting becomes very bitter from the formation of digestion products other than "peptone." The reaction should therefore be checked by cooling before this very bitter stage is reached.

It will be observed that these experiments illustrate the conditions in two kinds of digestion. The pancreatic digestion of proteins in milk is favored by a neutral or slightly alkaline reaction. Alkali interferes with the rennet coagulation. In the stomach the clotting of the milk is favored by the combined action of the acid and ferment. In the normal stomach coagulation the presence of calcium salts seems to be essential. If milk be treated with a small amount of sodium oxalate solution and then rennet, coagulation fails. Calcium chloride solution added later, the proper temperature being meanwhile maintained, brings it about.

THE ANALYSIS OF MILK.

The above experiments suggest some of the steps in the quantitative analysis of milk, a brief outline of which follows:

Water and Total Solids. Weigh out about 5 grams in a small platinum dish and evaporate to dryness over a water-bath which requires some hours. Then transfer the dish to a hot air oven and maintain at a temperature of 105° through half an hour. Cool the dish in a desiccator and weigh. The loss of weight represents the water, as practically nothing else of consequence is volatile.

Ash or Mineral Matter. After weighing the dry residue or total solids above place the dish on a triangle over a clear Bunsen flame and heat until all the organic matter, and finally the excess of carbon, is driven off. The ash left must be perfectly white. Cool and weigh as before. There is some slight loss of volatile salts in this ignition.

Fat. Where many analyses are made as a routine operation, as in the control of market milk, fat is generally now determined, by separating it from the milk in a centrifugal machine and reading off the volume. A definite quantity of milk is measured out, mixed with a little acid to facilitate the breaking up of the fat globules, placed in a special bottle with graduated neck or stem and rapidly rotated. The liberated fat collects in the stem and is read off. With the Babcock machine in common use the method is rapid and very accurate

Fat is very frequently determined by evaporating milk mixed with broken glass or quartz sand to dryness and extracting with a good solvent, preferably light petroleum benzine or perfectly anhydrous ether. A better method is to distribute about 5 to 10 grams of milk from a pipette over the surface of a strip of specially prepared absorbent paper. This is coiled up somewhat. loosely, placed in an air oven, dried thoroughly and then transferred to a Soxhlet extraction apparatus, where it is treated with the solvent by percolation through two or three hours. The solvent carries the fat down into a small weighed flask. On evaporation of the solvent the dry fat is left and may be so weighed.

Sugar. To determine the sugar, the fat and proteins must be first separated, which may be conveniently done by the copper process as illustrated above. 25 cc. of milk is diluted with water to 400 cc. and 10 cc. of the Fehling copper solution added. Then from a corresponding sodium hydroxide solution (containing 10.2 gm. to the liter) alkali is added in amount just sufficient to throw down a bulky precipitate containing all the proteins and fats with the copper. This requires about 7 cc. of the alkali. The mixture is diluted to 500 cc. and a portion is filtered off for tests. If the precipitation was properly made a clear filtrate is secured which contains only a trace of copper, and not enough to appreciably affect the accuracy of titration by the Fehling solution as described in an earlier chapter. The proper factor for lactose must be used in the calculation.

The protein and fat may be precipitated by use of a solution of lead acetate or mercuric nitrate without dilution. On filtering a clear filtrate is obtained which may be tested by the polariscope. 50 cc. of milk should be used, and after precipitation and filtration made up to 100 cc. for the polarization test. The details cannot be given here.

The Proteins. If the sugar, fat and ash are accurately found the proteins may be estimated by difference; that is by subtracting the sum of these from the total solids found. But this plan should not be followed except as a control. A direct determination of casein may be made in this way: 10 cc. of milk is diluted to 50 and mixed with dilute acetic acid to produce complete precipitation. Something less than 1.5 cc. of 10 per cent acid will be needed for this. The precipitate is collected on a Gooch funnel, washed with water, hot alcohol and finally enough ether to remove all the fat. What is left is dried and weighed as casein. From the first filtrate plus the wash water the albumin may be precipitated by boiling. The coagulum is collected on a Gooch funnel, washed with water and alcohol and dried as before.

It is also possible to determine the total nitrogen by the Kjeldahl method, and multiply this by the factor 6.25 to obtain corresponding total protein. This gives a fairly good control.

MILK PRESERVATIVES.

Milk shippers and dealers often attempt to keep milk from spoiling—turning sour usually—by the addition of some anti-ferment substance. The propriety of such an addition has been much discussed. In general the use of food preservatives should be kept within certain defined limits, as the consumer has the right to know what he is using. The chemical substances employed in this way possess different degrees of activity. Boric acid and formaldehyde have been most frequently added for the purpose, but they have been rather generally condemned for this and other foods. In the case of milk it is sometimes a question of the lesser evil; the trace of formaldehyde required to effectually preserve it from acid fermentation is very small. If *no more* than this minimum is used it is not likely that the harm from using it would be very great, if at all noticeable. The use of such milk is probably preferable to that of the sour, unpreserved milk often used by children in the poorer quarters of our cities.

In many of our large cities attempts are now made to pasteurize a good portion of the milk supply. The degree of safety afforded by this operation, as carried out in practice, is, however, very illusory.

MOTHER'S MILK.

We turn now to a short discussion of the chief points of difference between mother's milk and cow's milk, which is a subject of great practical importance. Success in substituting cow's milk for mother's milk in the feeding of small children depends very largely on the extent and accuracy of our knowledge here. It is a singular fact that we know much less about the chemistry of human milk than we know of other milks, and this is in part due to the difficulty in securing a perfectly normal secretion for analysis.

Because of the presence of certain salts milk has a so-called amphoteric reaction, that is, it shows an acid behavior with blue litmus and an alkaline with red. In mother's milk the alkaline reaction is stronger than in cow's milk, but the attempts to determine it by titration with the usual indicators lead to results of relatively little value because of the disturbing action of the proteins present. The salts of mother's milk are lower than in cow's milk.

Analyses. The analysis of human milk seems to present several points of difficulty and the published results do not show very good agreement. The separation of the proteins offers the greatest difficulty, as the simple and accurate methods employed in the analysis of cow's milk fail to give equally satisfactory results when applied to mother's milk. The explanation of this will be given below. There are variations in the composition of human milk as in that from other species, but average values are about as given below:

Water	87.5
Fat	3.8
Casein	1.6
Albumin	0.5
Sugar	6.2
Salts	0.4
	100.0

This analysis must be accepted as representing the facts only in a general way. Indeed, some authors go so far as to assert that no mean value for woman's milk is possible, as the variations from individual to individual are too great to permit an average result to have any legitimate meaning. This much, however, is well established: the fat in woman's milk is not greatly different in amount from that in cow's milk; the sugar is about fifty per cent higher in the mean; the salts are lower, sometimes as little as 0.2 or 0.3 per cent of ash being found; the total proteins are about half as much as in cow's milk. But as to the relation of the casein to the albumin and as to the nature of the casein itself, the greatest divergence of views exists. Some analysts have actually found more albumin than casein as a result of experiments. This is probably due to the employment of a faulty method for the precipitation of the casein; it has been pretty well established that the conditions of precipitation or coagulation are entirely different from those obtaining for cow's milk. It is indeed likely that the protein called casein in woman's milk is quite distinct from that of cow's milk. Under the action of rennet the former coagulates in fine flakes while the curd of cow's milk as at first produced is in very large flakes. The two caseins have apparently different contents of sulphur and phosphorus and give up their nitrogen in digestion experiments in different ways. It has been recently suggested that the product coagulated as casein from human milk may contain other proteins in sufficient amount to give it the peculiar properties noticed.

It must be remembered that the salts put down in the analysis of milk are always obtained as ash from the incineration of an evaporated residue. In the original milk they do not occur in this form, but in part, at least, in organic combination. Most of the sulphur and phosphorus occur in this condition. The lower casein content of mother's milk must be responsible for part of the salt difference.

Modified Milk. From all this it is evident that attempts to modify cow's milk so as to make it resemble mother's milk must be more or less abortive, as we are not able to duplicate the unknown proteins in the human secretion. However, many suggestions have been made in this direction and the line followed is essentially this: Cow's milk is first diluted with an equal volume of water or whey to reduce the proteins to the proper percentage amount. Then a certain volume of cream is added to restore the fat, and enough milk sugar or cane sugar to bring that constituent up to about 6 per cent. Unfortunately, the addition of fat is uncertain because of the great variations in market cream. Good cream should contain at least 20 per cent of fat, but is usually much inferior to this standard. The cream sold in cities often contains from 5 to 10 per cent of fat only. Assuming, however, a cream containing 20 per cent of fat, 3.5 per cent of casein and 3.5 per cent of sugar, and taking the gram and cubic centimeter as equivalent for our purpose, the following illustration will serve as an example of such a modification. Starting with average market milk of the composition given some pages back, 500 cc. may be mixed with water, cream and sugar to give a result as follows:

	In 500 cc. of Market Milk.	1000 cc. contains, approximately, after addition of 400 cc. of Water, 100 cc. of Cream, 35 gm. of Milk Sugar.	
Fat..................	17.5 gm.	37.5 gm.	3.8 per cent.
Sugar................	22.5	61.0	6.1
Proteins.............	19.5	23.0	2.3
Salts................	3.5	4.0	0.4

This mixture has a percentage composition pretty close to that of mother's milk. Sometimes the dilution is made with whey in place of water; the final result in this case is a product containing a little more protein because of the content of albumin in the whey.

Another important distinction, however, must not be lost sight of. While the milk of the cow is sterile when it leaves the udder it takes up from the hands of the milker or from the air a large number of bacteria which speedily increase to give a content of millions to the cubic centimeter. Most of these are doubtless harmless and have no bad effect on the milk; of others this cannot be said, as their presence soon leads to changes in the milk which may render it absolutely unfit for use as an infant food. Mother's milk is and remains sterile and is therefore free from this danger. It must be recalled further that the bactericidal behavior of human milk is relatively very strong. While all milks seem to have a certain content of bacteriolysins, these anti bodies in mother's milk are most potent as far as the destruction of the ordinary bacteria is concerned. It is quite likely that no small portion of the superiority of human milk as infant food is due to this observed fact.

All kinds of milk are affected to some extent by peculiar flavoring or other accidental substances in the food of the parent animal. It is a well-known fact that cows having access to certain weeds yield a milk with characteristic taste and odor. In the same way many substances given as remedies pass to some extent into the milk of the mother and may have an effect on the nursing child. Bay rum used for bathing the breasts of a nursing mother has been known to pass, in part at least, into the milk and give to it a very strong odor and taste.

THE MILK OF OTHER ANIMALS.

In some countries the milk of the goat and the ass have economic importance, and mare's milk is used by certain Asiatic peoples in producing a fermented beverage. Analyses of several kinds of milk are on record; some of these are given in the following table, taken mainly from the König compilation:

	Goat.	Ass.	Mare.	Sow.	Bitch.	Cat.	Sheep.	Elephant.
Water	86.9	90.0	90.0	82.4	75.4	81.6	81.3	67.0
Fat	4.1	1.3	1.1	6.4	9.6	3.4	6.8	22.0
Sugar	4.4	6.3	6.7	4.0	3.1	4.9	4.7	7.4
Proteins	3.7	2.1	1.9	6.1	11.2	9.4	6.4	3.0
Salts	0.9	0.3	0.3	1.1	0.7	0.7	0.8	0.6

Bunge has called attention to a relation which exists between the composition of a milk and the rapidity of growth of the animal feeding on it. In the case of the young of the dog, cat and sheep, for example, the rate of growth immediately after birth is very rapid and the milk of the mothers correspondingly rich in proteins and calcium phosphate. The young of the horse and ass are slow growers, that is, a relatively long period is required for them to double in weight. These mothers' milks are low in proteins and salts, but relatively high in sugar. In the human species these relations are even more pronounced.

CHAPTER XVII.

THE CHEMISTRY OF THE LIVER. BILE. CELLS IN GENERAL.

From the earliest days of physiological chemical investigation the composition of the liver cells and the nature of the processes taking place there have been the subject of many studies. It is well known that the liver has a certain definite work to do in the animal organism and of some of the functions we have fairly accurate ideas. Of other functions there is much yet in dispute, but it may be said that a number of synthetic reactions are unquestionably carried out through the activity of cell enzymes there formed. Before taking up the special work of the liver cells something should be said of the composition of animal cells in general.

COMPOSITION OF CELLS.

In structure all animal cells agree in consisting of two essential parts, a nucleus and surrounding protoplasm. In young cells these two parts are usually easily recognized, but in the old cells of complex structures they assume various forms, bearing apparently little resemblance to the original type. Cells are in general the center of the various chemical reactions taking place in the body. Some of these reactions take place in the fluids outside the cell, but by the aid of ferments of cell origin; most reactions, however, seem to be carried on within the cell, which may be illustrated by the familiar conversion of sugar into alcohol and carbon dioxide by the yeast cell. The enzyme which does this work may, however, be extracted, as has been already shown.

Of the chemical nature of nucleus and protoplasm not a great deal is known. It is extremely difficult to isolate original cells from their modified products or tissues in general, and a sharp chemical differentiation between the two component parts of the cells is not yet possible, however simple the microscopic differentiation may be. But some points have been worked out and these will be briefly referred to. Our information here comes mainly from analyses of the simplest cells, as in cells of the more complex tissues the true cell characteristics are obscured.

The Nucleus. The most important chemical constituent of the nucleus is the complex protein substance known as *nuclein*, already referred to in an earlier chapter. Nucleins of different character are obtained from different sources. These nucleins exist in combination as nucleo-proteids, and in turn break up into nucleic acids and a protein fraction, which was explained in the chapter referred to. The cell nucleus appears to consist very largely if not wholly of the nucleo-proteid. On digestion with pepsin and hydrochloric acid the nuclein is separated and may be purified by washing with water, dissolving in very weak alkali and reprecipitating with acid. By digestion with pancreatic extract the nucleic acid is left. The pure nuclein is a white amorphous substance which gives the Millon test and the biuret test. The various nuclein substances in cells are characterized by a strong affinity for dye stuffs, especially for some of the coal tar dyes; this property is utilized in the microscopic examination of tissues. Nuclein fused with sodium carbonate and nitrate yields phosphate, but heated without the alkali an acid residue (metaphosphoric acid) is left. Of other constituents of the nucleus but little is known. Lecithin may be present.

By various decompositions nuclein substances yield a number of peculiar basic bodies known as the xanthine or purine bases, which will be considered in a following chapter. The cell nucleus contains in combination a number of metallic elements among which iron is perhaps the most important. Potassium salts are present, while those of sodium are present only in traces, if at all.

The Protoplasm. This soft spongy portion of the cell consists largely of water. The solid part, making up 10 to 20 per cent usually, contains several albumins proper and nucleo-proteids. Lecithin is an important and relatively abundant constituent of the protoplasm. Its presence seems to be intimately associated with phenomena of reproduction and building up of new tissues. The chemistry of lecithin, as a phosphoric acid fat, has been explained already, and it must be recalled that this term includes a number of phosphatides. In the cell protoplasm they doubtless exist in part as a complex lecitho-protein. This may account for the fact that the separation of the lecithin is sometimes difficult.

In all cases the protoplasm seems to contain the complex alcohol substance cholesterol, glycogen and ordinary fats. How these various complexes exist, to what extent they are necessary or essential in the cell structure, we can not say. As cells have various functions to perform, they have the power of producing different ferments for the purpose and such products can not be distinguished by our present methods of analysis from the material of the cell itself.

FUNCTIONS OF THE LIVER CELLS.

The anatomical location of the liver gives it a most important relation to the other organs of the body. With the exception of the fats most of the products of the digestion of foods pass through the portal vein into the liver and there undergo certain preparatory changes. Substances not true foods take the same course and many toxic bodies, metallic and alkaloidal, find a resting place in the liver. In toxicological examinations the liver, after the stomach, is the most important organ for analysis.

Not only are the fundamental food stuffs, the proteins and the carbohydrates, worked over and more or less altered in the liver, but partly metabolized products seem to be further changed in passing through this organ and are there brought into a condition for final excretion. The evidences that such reactions take place have been worked out in a number of cases experimentally and will be referred to below.

Composition of the Liver. We have here the materials found in cells in general and also others having special functions. The *protein* substances separated belong to several groups; albumin, globulin and a nucleo-proteid have been recognized. Iron exists in combination with several of these protein bodies. One of these is known as ferratin and contains the iron in complex combination; others appear to be albuminates in which the iron is more readily recognized.

Next to the proteins the *fats* are relatively abundant in the liver and may amount to 3 or 4 per cent by weight normally. Pathologically, by fatty degeneration, or by filtration from other tissues, the fat may be greatly increased, even to 30 or 35 per cent of the weight of the whole organ. The liver fat is usually comparatively soft, but that formed in some degenerations is harder.

The proportion of *lecithins* in the liver is variable and is ordinarily below the true fats. The average amount is from 2 to 3 per cent. It has a more important function to perform than have the fats proper, since it is found by experiment that in starvation the lecithin fat is the last to disappear. The ether extract of the organ in this case is largely lecithin.

Much has been written of the functions of the lecithin bodies, and part of their behavior is possibly physical. Recent investigations have suggested that in certain ferment phenomena they may play the part of *activators* for the pro-ferments. In the liver, where a multiplicity of such reactions occur, the presence of such large quantities of these phosphatides may have a special meaning.

The most important substance found in the liver is probably *glyco-*

gen, which is a transformation product and variable in quantity. The amount present at any one moment depends on the carbohydrate consumption and the time which has elapsed since a meal. It may be as high as 15 per cent of the whole weight of the organ or may run down to a fraction of 1 per cent, after fasting or after the performance of hard work. The formation of glycogen will be discussed below.

The liver, consisting largely of cells in rapid state of change, furnishes a relatively large amount of the so-called nitrogenous extractives. These include the *xanthine* and related bodies, *urea, uric acid, leucine, cystin* and other substances representing certain stages in metabolism. The total amount of these compounds present at any one time is very small, and probably not over 0.5 per cent of the dried organ; but even this small amount is important, as will appear below.

We have finally several *mineral substances* present. These include essentially the chlorides and phosphates of the alkali and alkali earth metals with some iron compounds. Of the latter those with proteins have been mentioned above, but other iron salts are present and the quantity may be increased by the administration of inorganic substances as remedies. In normal conditions the iron content is extremely variable. The amount may be accurately determined only after washing out the blood (containing hemoglobin) by aid of salt solution of proper strength, about 0.9 per cent. Recent investigations have shown that in the livers of women the iron varies from 0.05 per cent to 0.09 per cent of the dry substance, while in men the content is more irregular, running from 0.05 per cent to 0.37 per cent. The amount seems to increase with age, but no explanation for the variations can be given. In children and very young animals the content is also high. It sinks, and rises again, later in life. In addition to the iron a trace of copper is said to be always present and may have some physiological function. Other metals occasionally found are probably of accidental occurrence, as the liver retains such foreign substances through a long period.

CHEMICAL CHANGES IN THE LIVER.

In recent years much has been written on this obscure but highly important topic. Many of the changes taking place in the liver come under the head of fermentations, enzymic reactions. Hofmeister pointed out, a number of years ago, that there are at least eleven of these in play. He mentioned a proteolytic and a nuclein-splitting ferment, one which splits off ammonia from amino acids, a rennet ferment, a fibrin ferment, an autolyzing ferment, a bactericidal ferment, an oxydase, a lipase, a maltase and a glucase. But since then our views have been much broadened. We have, in addition to these reactions, which result in general in the breaking down of molecules, a number of others which are synthetic in their nature. A brief study of what is known of all these changes is sufficient to indicate the immense importance of the liver in the metabolic phenomena of the body.

CARBOHYDRATE CHANGES.

These reactions will be considered first because they have been the most thoroughly studied and also because of their intrinsic importance.

Glycogen Formation. It was long ago established that the food carbohydrates after digestion reach the circulation almost exclusively by way of the portal vein and the liver. In the normal food of man and the herbivora the carbohydrate food is usually starch and this becomes dextrin, maltose and finally glucose before absorption. As no marked accumulation of the sugar takes place in the blood after a meal it must follow that it or some derived reserve product must be temporarily retained somewhere. The place of this retention is the liver and the form in which the sugar is held is glycogen. The chemical reactions of glycogen have been discussed in the chapter on the carbohydrates, but in this place other relations must be considered. No simple answer can be given to the question as to the method of formation of glycogen from sugar. Although the formula is commonly written $C_6H_{10}O_5$, it is, like common starch, certainly a multiple of this. Hence a simple equation connecting glucose and glycogen of the form

$$C_6H_{12}O_6 - H_2O = C_6H_{10}O_5$$

is not strictly correct. Besides, several other facts appear which complicate the problem. While glucose is ordinarily the sugar which passes through the portal vein, other sugars are also consumed and in the digestive process do not become changed to glucose. From cane sugar we have some fructose and from milk sugar some galactose, and with these in the food it appears that glycogen is still formed. Moreover, it has been shown that substances not carbohydrate at all may give rise to glycogen. Animals have been starved until the liver was practically free from glycogen (as known by previous trials with other animals) and then fed on fibrin or washed out lean meat. On killing the animals a short time later a store of glycogen was found in the liver, indicating its formation from something in the protein. With such facts in mind it is not possible to form any simple theory of the production of the reserve substance. From the sugars it is likely that some such reaction takes place as occurs in the formation of starch in plants. The carbohydrate built up in the plant from water and carbonic acid is a sugar and this is transformed by some enzymic reaction into starch as a reserve material. The mechanism of this change, however, is quite obscure.

Attempts have been made to connect the formation from proteins with the sugar group of the gluco-proteids, but casein and gelatin fed to animals lead also to production of glycogen, and these bodies in pure condition do not furnish a sugar complex by laboratory treatment. In addition to this it is impossible that the sugar group could be abundant enough in the other common proteins to account for the large amount of glycogen which may be formed by protein diet. These facts lead to the view that a synthesis must be concerned in the reaction. Such protein derivatives as leucine, the hexone bases and other bodies have been thought of as leading possibly to the end, but direct experiments with animals have given no satisfactory proof of such a hypothesis. For the present, therefore, the method of production from proteins must be left without explanation.

A diet of fat leads also to glycogen accumulation or formation in small amount, according to some recent observations. This latter reaction requires some kind of an oxidation and is more difficult of explanation than the other. It must be remembered that an *accumulation* of glycogen may follow from diminished destruction as well as from increased production, and where the amount in question is small, an apparent increase may be traced to errors of observation or experiment. In a mixed diet it is practically impossible to trace the effect of any one substance. The behavior of pentoses is an illustration; according to the statements of some authors these carbohydrates increase glycogen. It may be, however, that they simply behave as sparers of glycogen by undergoing oxidation, which otherwise the glycogen would have to undergo.

Not all the carbohydrate reaching the portal vein is transformed in the liver; apparently only a certain portion is so changed, while the excess is stored up temporarily in other organs. This is evident from the fact frequently observed in animal experiments that the amount of glycogen in the liver is below what should be expected from the food when this is excessive in carbohydrates. With ordinary or deficient feeding the liver doubtless is able to store as glycogen all the sugar conveyed to it, but an excess must find lodgment elsewhere. The muscles undoubtedly receive the greater share of this excess. In extreme cases the liver may hold 200 grams of glycogen, which would correspond to the same weight of starch.

Glycogen Destruction. This stored up glycogen disappears in normal conditions gradually after its accumulation; the disappearance is hastened by work or by lowering of temperature, showing that it may be called upon for supply of heat as well as for direct mechanical work. Before being utilized, however, for these purposes the glycogen must be thrown back into the form of sugar; how this is done is still a matter of discussion. According to one view the action is a "vital" one, depending on the life of the cells of the liver themselves; by another view this conversion is wholly enzymic, a peculiar ferment bringing about the change during life as well as post-mortem. This question has lost much of its importance since the work of Buchner on the zymase, or enzyme of yeast active in alcohol formation, as we now know that enzymes are present where, by earlier methods of experiment, they were supposed to be absent.

The more recent careful experiments seem to show beyond question that a true glycogen-splitting ferment is present. By proper manipulation the cell effect may be excluded, while that of the ferment is left intact. This may be accomplished in the following way: The fresh organ is washed free from blood by forcing water through the portal vein until that escaping by the hepatic veins is clear and colorless. The liver is then chopped fine and allowed to stand a day in a large excess of alcohol for dehydration. The alcohol is poured off, the residue pressed, dried at a low temperature and ground to a powder. In this form it is suitable for extraction with something which does not interfere with enzymic power, but which prevents bacterial or other cell activity. For this purpose chloroform water, or solutions of sodium fluoride have been used. A good extracting mixture may contain in 100 cc. of water 0.2 gm. of sodium fluoride and 0.9 gm. of sodium chloride. The liver powder is exhausted with such a solution at a temperature of 38° and the filtered liquid obtained may be used in two ways. On standing, the sugar in the solution increases while the glycogen decreases. In addition, if pure glycogen be added to such an extract it is found also to diminish with corresponding increase of sugar. It is further found that boiling the fluoride extract destroys all converting power, which fact speaks likewise for enzyme action.

The glycogen-converting power of solutions made as above is considerable and sufficient to fully account for the post-mortem increase of sugar always found in the liver. By extracting not the whole liver but portions it is possible to compare the distribution of the ferment. Experiments made with this end in view have shown that this is pretty uniform. By following the same general method Pick has compared the ferment activity of the liver with that of other organs where glycogen may be stored. In such experiments the diastatic action of the liver has been found to be in excess as should be expected, since this is the organ where the greatest accumulation normally takes place. This normal conversion of glycogen by the liver ferment is interfered with by various substances which may be taken as remedies; quinine salts seem to be especially active.

AUTOLYTIC FERMENTATION.

The liver, or other organ, removed from the body and left to itself speedily undergoes a change. Unless precautions are taken to prevent it the bacterial decomposition may become pronounced and obscure other reactions. Some years ago Salkowski gave the name *auto-digestion* to the fermentations taking place in the liver, in which a change in the nitrogenous constituents is mainly involved. Other chemists followed the subject further, taking precautions to exclude all bacterial influences, and have brought to light a number of very peculiar reactions which follow from the presence of ferments in the organs. The name *autolysis* has been given to these self-digestion reactions in general. They are not confined to the liver, but are observed in all organs. An enormous literature has accumulated already on this topic, because it has great practical as well as scientific importance. In these spontaneous digestions various products are formed, some of which are volatile; a general softening of the tissues concerned may also take place and the sum of these changes is important in bringing about the difference between fresh meat, and stored, "ripe" or "hung" meat, for example. While bacteria play an important part in curing meat it is well known that changes go on within the tissues which cannot be due to bacterial action. These are the autolytic changes which were first clearly followed in the liver, and which will be here briefly discussed.

The Production of Organic Acids. This is one of the simplest phenomena observed. If the livers of dogs or other animals are carefully removed and kept under chloroform or toluene a gradual gain in acidity is observed. The liver must be minced before being covered with the protecting liquid. The autolysis in this case is slow, weeks or months being required to show any large amount of acid. The best temperature for the experiment is 38–40° C. Instead of employing antiseptics it is possible with care to remove and store the liver in sterile jars *aseptically*. Under these conditions the spontaneous change is very rapid, more acid being formed in one day, ordinarily, than after a month of the antiseptic treatment. By making several pieces of the liver on removal from the animal, putting each in a separate jar and testing one from time to time, it is possible to follow the course of the autolysis. Among the acids produced formic, acetic, fermentation lactic and paralactic, butyric and succinic have been recognized. In experiments described by Magnus-Levy the total acid formed in one day in 100 grams of liver, by the aseptic treatment, may correspond to over 20 cc. of normal alkali. If this were calculated as lactic acid it would amount to 1.8 gm. The relation between the volatile and non-volatile acids varies with the animal, but not regularly.

It is not possible to trace exactly the source of all these acids, but apparently they come in part from a decomposition of the sugar of the liver, since this is found to decrease as the acid increases. Lactic acid may be formed first from sugar and butyric acid from the lactic as in the bacterial fermentations. The appearance of hydrogen and carbon dioxide at the same time favors this view.

The Alteration in the Proteins. When subjected to aseptic auto-digestion, or to the same digestion with chloroform or toluene, the protein substances gradually break down into simpler products. Among these the amino acids may be most readily recognized; there is also an increase in the nitrogen which may be distilled off with magnesia. The behavior here is somewhat similar to that which follows in acid hydrolysis of the proteins, or which occurs in prolonged boiling with water under pressure; in both cases a kind of hydrolysis results and this may be what takes place in auto-digestion.

In prolonged aseptic auto-digestion of the liver very considerable quantities of leucine and tyrosine are formed; on the outer surfaces, where evaporation can take place, the latter may even separate in crystalline bunches easily recognized. The hexone bases and bodies of the xanthine group also result but not always in very great quantities. The greater number of these reactions are those of hydrolytic cleavage, and that they follow spontaneously is one of the best proofs of the character of the ferment agents present. While in a general way similar, it has been found that certain organs yield amino acid and other groups not liberated in the autolysis of other organs. This is an extremely interesting fact, as it points to the specificity of function suggested also by other reactions. It has been pointed out, further, that the corresponding organs in different animals show certain differences in this respect.

Pathological Importance. This possibility of self-digestion in the liver and other organs may help explain some of the phenomena observed in pathological conditions. The acids found sometimes in the urine as well as the leucine and tyrosine have usually been traced to the liver. These experiments show the rapidity with which such products may be formed by a degenerative process. Pathologically the urine sometimes shows a very high reducing power which cannot be associated with sugar or uric acid or creatinine. The liquid formed in the liver autolysis is always strongly reducing in action and this may suggest an explanation for the observation of the urine.

Bactericidal Products. It is worthy of note that in these autolytic decompositions substances are formed which have a marked bactericidal action. This has been shown in many ways and the suggestion appears reasonable that in the continuous breaking down processes going on in various organs we have some of the factors of natural immunity. These autolytic products must not be confounded with the alexins already referred to. In a few experiments on record injections with pressed out juice from autolyzed organs have been sufficient to prevent death in small animals infected with virulent cultures. The bactericidal action of the fresh liver or other organ is comparatively slight.

OTHER FERMENT ACTIONS.

Other ferments present in the liver have not been very thoroughly studied. The presence of a fat-splitting ferment or lipase has been shown, but, as yet, little is definitely known of the extent of its action in the body. The oxidase ferments are better known and the action of liver extracts in bringing about oxidations of various organic substances has been studied with the object of throwing some light on normal oxidations in the body. How many of these oxidizing ferments the liver may contain is of course not known. The reaction thus far the most carefully studied is that between water extracts of the liver and salicylic aldehyde. In the process this becomes salicylic acid.

The action of a liquid obtained by pressing the minced liver ground up with sand has been studied with reference to its power of hydrolyzing certain esters. Ethyl butyrate seems to be readily split by this liver juice. The reaction points to the presence of a lipase-like ferment which doubtless has the power of splitting other bodies of this type. The boiled liquid is without the ester-splitting power. It has been found further that the active element can be completely salted out by addition of ammonium sulphate to saturation, and it may be precipitated by addition of a strong solution of uranium acetate.

THE BEHAVIOR OF THE LIVER WITH POISONS.

The fact has been referred to already that many metallic and some organic substances combine with the liver cells. All this has a practical bearing on toxicological investigations, in which experience has shown the importance of including the liver in the analytical tests. Recent experiments have thrown some light on the question of the manner of combination of poisons. Corrosive sublimate, for example, fed in very small portions to dogs was found later by post-mortem examinations in the globulin fraction of the liver extract. The fixation of arsenic is different; it combines with a nuclein substance and in very stable form, which explains the practical difficulty of separating this substance in forensic investigation.

Experiments have also been published showing the behavior of small doses of morphine sulphate and strychnine sulphate in the liver. It appears that the retaining power of the liver for these poisons is relatively large when they are administered by the mouth or injected into the portal vein. The retention of the alkaloids by the organ has been experimentally shown. Such observations have an important bearing in explaining the fact that many poisons are far more active when injected hypodermically than when given through the stomach. This seems to be true of many substances besides the metallic poisons and the alkaloids. The phenols, for example, are likewise retained to a marked extent by the liver.

SYNTHETIC PROCESSES IN THE LIVER.

It has long been known that the liver is the seat of the formation of a large number of metabolic products, some of which involve syntheses. Several of these reactions may be briefly explained in this place, but nothing like a full discussion will be attempted. A few illustrative cases only will be taken to show in a general way what is best known in this field. The reactions mentioned take place in other organs, as well as in the liver, but as the latter seems to be mainly concerned, this is a good place to discuss them.

The Formation of Urea. Of all the synthetic reactions known to occur wholly or in part in the liver this one has been the most thoroughly studied. The older notion of the formation of urea exclusively from the more complex uric acid is no longer held; the belief that the latter complex represents a portion of the protein residue which in some manner escaped its normal and proper fate, that is, conversion into urea, has long since been abandoned in view of much accumulated evidence to the contrary. Indeed, at the present time it appears more likely that a *part* of the uric acid excretion may be traced to a synthesis from urea.

A great many observations unite in suggesting the liver as the organ in which urea is most abundantly produced, and certain ammonium salts as being largely or mainly concerned in this production. These observations have been made in the laboratory as well as clinically. In diseases in which the liver is involved there has frequently been noticed a marked reduction in the portion of the excreted nitrogen appearing as urea. It is also known that the administration of ammonium salts is not followed by an increase of ammonia in the urine. Parallel with this observation we have the further one made on the blood, which has shown that the fluid of the portal vein is far richer in ammonia than is that from the hepatic vein. Such observations have been followed up by experiments in which fresh blood is forced through a living or a recently removed liver by means of specially constructed apparatus. The same blood may be caused to pass the liver many times. After passing a few times and reaching uniformity in composition various ammonium and related compounds are added to the blood and the circulation then continued. In this way the abundant transformation of ammonium carbonate into urea is readily shown. It has also been found that certain amino acids are converted rather readily in going through the liver. Experiments have shown that in the course of a few hours several grams of leucine, glycocoll or aspartic acid may be transformed into urea under these unfavorable conditions.

The importance of this observation will be recognized. It is well known that the amino acids are among the most important of the disintegration products of the proteins; by hydrolytic and other cleavage reactions these amino complexes result, and we see here the possibility of further destruction with ultimate formation of urea. It is possible that in this reaction carbamates are concerned, as the formation of urea by alternate oxidations and reductions of ammonium carbamate has been shown by Drechsel. These reactions illustrate the relations

$$NH_4O \cdot CO \cdot NH_2 + O = NH_2O \cdot CO \cdot NH_2 + H_2O$$

$$NH_2O \cdot CO \cdot NH_2 + H_2 = NH_2 \cdot CO \cdot NH_2 + H_2O$$

It has been shown that the carbamic acid salt frequently appears in urine, and perhaps normally. This relation is also apparent:

$$\begin{array}{c} NH_4-O \\ | \\ CO \\ | \\ NH_4-O \end{array} \rightarrow \begin{array}{c} NH_4-O \\ | \\ CO + H_2O \\ | \\ NH_2- \end{array} \rightarrow \begin{array}{c} NH_2 \\ | \\ CO + H_2O \\ | \\ NH_2 \end{array}$$

There is one ferment reaction which is known to lead to the formation of urea under definite conditions, and this is the production from the diaminic acid arginine. The liver and other organs contain an enzyme, known as arginase, which has the property of splitting arginine into urea and ornithine, or diamino valeric acid. As arginine is known to be produced normally by the erepsin digestion we have here a source, for a small part at least, of the urea formation.

The Synthesis of Uric Acid. The mode and place of the formation of uric acid in the animal organism have been the subjects of numerous investigations. In birds, serpents and some of the mammals the excretion of nitrogen is largely in the form of uric acid, and experiments have shown that it is, in part at least, of synthetic origin. The excretion of uric acid in birds is increased by doses of ammonium salts; with the livers extirpated there is a decrease in the elimination of uric acid and increase in excretion of ammonium compounds. In a number of such observations the liver has been connected with uric acid formation, and transfusion experiments, in which blood containing ammonium lactate and certain other compounds has been forced through the livers of geese, pointed to the same kind of a synthetic conversion. For the higher animals, however, a different formation has usually been assumed, the oxidation of the purine bodies coming from the breaking down of nucleins being looked upon as the principal formative reaction.

Later, in a chapter on the urine, the relations of the purines to uric acid will be pointed out. It is sufficient to state here that the enzymic production of uric acid from other purines has been clearly shown by recent observers. These enzymes are contained not only in the liver, but in the spleen and elsewhere, and it seems likely that other enzymes, which have been called *nucleases,* must begin the cleavage of the nucleins or parent substances.

Comparatively recent experiments by several authors suggest synthetic reactions as likewise possible. Wiener, for example, mixed chopped beef liver with physiologic salt solution and allowed the mixture to stand at the body temperature an hour. The liquid was then pressed out and the uric acid in it determined after some time in a given volume. To the same volume of liver extract definite weights of urea and various ammonium and sodium salts were added and the mixture allowed to stand as before. In certain cases a very marked increase in the uric acid resulted, pointing to the presence in the liver extract of some agent capable of effecting the combination. The best results were obtained with dialuric acid salts and tartronic acid and urea.

It is fair to state that another interpretation of these results has been given. While admitting the formation of uric acid in this way it is claimed by other physiologists who have repeated the experiments that the purines in the organic mixture are alone converted, the nonnitrogenous bodies used acting merely as accelerators in the reactions. The organs used are all rich in the parent substances of the purines.

The Formation of Ethereal Sulphates. Another reaction of farreaching importance in the body is the production of organic sulphates. The oxidation of the sulphur of proteins leads finally, mainly, to the formation of sulphuric acid which is eliminated in the urine in the form of the ordinary mineral sulphates and the ethereal sulphates. The mineral sulphates are readily formed directly by combinations in the blood, but for the union of the organic groups with sulphuric acid some active agent is required. The addition seems to take place in the liver where it is probable that the oxidation of the sulphur-containing complex, furnished by protein disintegration also occurs. This complex seems to be cystin, $C_6H_{12}O_4N_2S_2$, which undergoes nearly complete oxidation to yield sulphuric acid from the sulphur. A small portion reaches the urine finally in other forms, the so-called "neutral" sulphur.

Several attempts have been made to determine the seat of the reaction by irrigation tests, and comparatively recently it has been shown that blood containing phenol and cystin and led through the liver, freshly dissected, discloses a very considerable oxidation of the sulphur compound with production of aromatic sulphate. It appears that other organs are not much concerned, if at all, in the reaction.

The aromatic radicles which join with sulphuric acid in this way are mainly products of intestinal putrefactive changes, and by absorption finally reach the liver. In addition to sulphuric acid glucoronic acid acts to hold the phenol bodies; it is usually present in traces in normal urine and is often greatly increased pathologically. The glucoronates may be formed in the liver along with the aromatic sulphates. In experiments which have been carried out on the passage of the blood through a liver the conjugate phenol bodies produced have frequently been in excess of the amount called for by the sulphuric acid found; this excess may correspond in the main with the glucoronic acid.

It has been shown recently that the aromatic complex from the intestine and the sulphur body unite in the liver or other organ only when the sulphur group is not yet completely oxidized. In other words, sulphite sulphur and not sulphate sulphur is here concerned. After the union the final oxidation takes place. More will be said about these combinations under the head of urine products.

THE BILE.

The formation of bile is one of the important functions of the liver and the amount secreted in man is several hundred grams daily. Some of the uses of the bile have been referred to in earlier chapters under the head of digestion phenomena. Other functions will be discussed presently.

AMOUNT AND COMPOSITION.

The volume of the bile secreted seems to be subject to variations which are not well understood. Through the aid of a biliary fistula it is possible to collect the total excretion in dogs and other animals which are easily experimented upon and determine the rate of flow and the whole amount. The volumes reported by different observers are not in good agreement. The amounts secreted by different animals in 24 hours for each kilogram of body weight vary between 12 grams for the goose and 137 grams for the rabbit. For man the amounts observed have varied between about 150 and 1,000 grams daily.

The flow of the bile is increased, as far as volume is concerned at any rate, by the administration of certain remedies. These are known as cholagogues and among them calomel, certain resins, rhubarb and oil of turpentine are perhaps best known. That the solids of the secretion are increased is a disputed question. It is proper to state here that many of the older data on this subject were obtained by methods which are open to serious objection.

Composition of Bile. Qualitatively bile is characterized by the presence of certain acids and coloring matters which are not found elsewhere in the body. The acids are *taurocholic* and *glycocholic,* and the coloring matters are *bilirubin* and *biliverdin,* which have been referred to already in their relation to the coloring matter of blood from which they are derived. In addition to these substances several others are present which, while important, are not characteristic. These include cholesterol, fats, soaps, inorganic salts and mucin. The quantitative composition is extremely variable as shown by the analyses below, by Hammarsten, which are frequently quoted. The results are in parts per 1,000:

	1	2	3
Water	974.80	964.74	974.60
Solids	25.20	35.26	25.40
Coloring matters and mucin.	5.29	4.29	5.15
Taurocholates	3.03	2.08	2.18
Glycocholates	6.28	16.16	6.86
Acids in soaps	1.23	1.36	1.01
Cholesterol	0.63	1.60	1.50
Lecithin	0.22	0.57	0.65
Fat	0.22	0.96	0.61
Soluble salts	8.07	6.76	7.25
Insoluble salts	0.25	0.49	0.21

Many of the older analyses quoted were made from bile from the gall bladder. The solids in the bladder bile are always much higher than those given above, because of a concentration which takes place in that receptable. Some results for bladder bile are given below:

	1	2	3	4
Water	860.0	859.2	822.7	898.1
Solids	140.0	140.8	177.3	101.9
Biliary salts	72.2	91.4	107.9	56.5
Mucin and pigments	26.6	29.8	22.1	14.5
Cholesterol	1.6	2.6	47.3	30.9
Lecithin	—	—		
Fat	3.2	9.2		
Soaps	—	—	—	—
Inorganic salts	6.5	7.7	10.8	6.2

Analyses have been made of bile from different animals with the object of connecting composition with the food of the animal or its habits. The results are not very definite. Human bile contains more glycocholic than taurocholic acid, while in carnivorous mammals, birds and fishes taurocholic acid is the more abundant. Hog bile contains largely glycocholic acid, but in ox bile the relation is variable. The amounts of the pigments are small and not accurately known.

Glycocholic Acid. This is a complex substance made up of a combination of glycocoll or glycine with cholalic acid. The constitution of the acid is not known, but the empirical formula $C_{26}H_{43}NO_6$ has been given to it. In the bile it exists in the form of a sodium or potassium salt, which is readily soluble in water or alcohol. The free acid is but slightly soluble; hence the addition of mineral acids to bile produces a precipitate. On boiling a solution of glycocholic acid with weak acids or alkalies a cleavage follows, and glycocoll and the nitrogen-free cholalic acid separate. Water is taken up at the same time. This is a reaction analogous to the separation of glycocoll and benzoic acid from hippuric acid by the same manner of treatment. There appear to be several cholalic acids, but with the common one the reaction would be represented, probably, in this way:

$$C_{26}H_{43}NO_6 + H_2O = C_2H_5O_2NH_2 + C_{24}H_{40}O_5.$$

Taurocholic Acid. To this substance the empirical formula $C_{26}H_{45}NSO_7$ is given. With weak acids it undergoes likewise a hydrolytic cleavage from which taurin and cholalic acid result. Taurin appears to be aminoethylsulphonic acid, $C_2H_4NH_2 \cdot HSO_3$, and the cleavage would be represented in this way:

$$C_{26}H_{45}NSO_7 + H_2O = C_{24}H_{40}O_5 + C_2H_4NH_2HSO_3.$$

The free acid has a bitter-sweet taste; it is much more soluble in water than the glycocholic acid and somewhat soluble in alcohol. The free acid has the property of holding glycocholic acid in aqueous solution, which is shown by the difficulty in precipitating the mixed acids from ox bile. The free acid is but slightly soluble in ether. The alkali salts are soluble in water and alcohol.

Cholalic Acid. Although many investigations have been carried out with this substance its constitution is not clear. The above empirical formula, $C_{24}H_{40}O_5$, is that of a monobasic acid to which Mylius has given this possible structure,

$$C_{20}H_{31}\begin{cases} CHOH \\ CH_2OH \\ CH_2OH \\ COOH \end{cases}$$

The free acid is very slightly soluble in water, but the alkali salts are readily soluble. The free acid is somewhat soluble in ether; hence it is found as a decomposition product of the bile acids in the crude fat extracted from feces. By oxidation cholalic acid yields several new acids which have been much studied with the hope of gaining an insight into the structure of the original acid. Among the various derived acids these may be mentioned: *Choleic acid*, $C_{25}H_{42}O_4$, *dehydrocholeic acid*, $C_{24}H_{34}O_4$, *cholanic acid*, $C_{24}H_{34}O_8$. *Fellic acid*, $C_{23}H_{40}O_4$, and *lithofellic acid*, $C_{20}H_{36}O_4$, are found in some kinds of bile.

Preparation of Acids from Ox Bile. This may be illustrated by the following. Evaporate 200 to 300 cc. of the bile to dryness, or as near to dryness as possible, on the water-bath with the addition of about 60 grams of bone-black. After cooling the mass rub it up thoroughly, transfer to a flask and extract with alcohol by heating over a water-bath. The two bile salts are soluble in the alcohol, while the mucin and inorganic salts present are not. Therefore cool the extracted mixture and filter. The filtrate contains the bile salts along with cholesterol, some fat and traces of other substances. There is also some water present. Evaporate the filtrate to dryness, take up with absolute alcohol and filter again. Taking advantage of the practical insolubility of the bile salts in ether they may be precipitated in this way: Add to the strong alcoholic solution an excess of dry ether, or enough to cloud the mixture, and allow to stand. After some hours or days a crystalline precipitate of the bile salts separates. The crystals may be used for preparation of other substances, or for tests. In the mother liquor cholesterol may be detected by the tests given in an earlier part of this work.

Preparation of Glycocholic Acid. Use the larger part of the above described crystalline precipitate for this purpose. Dissolve in water and add enough dilute sulphuric acid to produce a marked turbidity. Add a little ether, shake the mixture well and allow to stand in a cold place. The glycocholic acid separates in the form of fine silky needles. Press out the mother liquor, redissolve in hot water and allow to crystallize a second time. A nearly pure product may be so obtained.

Preparation of Taurocholic Acid. The separation of this acid from the glycocholic acid is extremely difficult, hence in preparing it, it is best to start with a bile which contains essentially only the one salt. Dog's bile should therefore be employed. Treat it as described for the mixed salts, and decompose finally with dilute sulphuric acid in presence of ether.

Preparation of Cholalic Acid. Dissolve 200 grams of barium hydroxide in 6 liters of water. In this solution saponify 50 gm. of glycocholic acid, by boiling ten to twenty hours, replacing the water lost by evaporation. Filter hot and to the cooled liquid add enough hydrochloric acid to decompose the barium salt. The cholalic acid separates in the form of a granular precipitate. Wash with water and crystallize from hot, strong alcohol.

Optical Properties of These Acids. The three acids and their sodium salts are characterized by rather strong rotating power, which under some circumstances may be used for measurement or identification. The following specific rotations have been found:

	For Aqueous Solution.		For Alcohol Solution.	
	c	$[\alpha]_D$	c	$[\alpha]_D$
Glycocholic acid			9.504	+29.0°
sodium salt	24.928	+20.8°	20.143	+25.7
Taurocholic acid, sodium salt	8.856	+21.5	9.898	+24.5
Cholalic acid, anhydrous			2.942	+47.6
sodium salt	19.049	+26.0	2.230	+31.4

CHEMICAL TEST FOR THE BILE SALTS. The three acids are characterized by giving a certain reaction with furfuraldehyde, or sugar yielding furfuraldehyde, in presence of acid. The test is commonly made by adding to a dilute solution of the salts, say 5 cubic centimeters, a few drops of a dilute cane sugar solution and strong sulphuric acid in volume about half that of the mixture. Let the acid flow down the side of the test-tube so as to form a layer below the lighter liquid. A deep purple band appears at the line of contact. On slowly mixing the liquids in the test-tube the color becomes purple throughout. In this test any excess of sugar must be avoided.

With a trace of pure furfuraldehyde in place of sugar the reaction is sharper, but certain proportions must be observed. A good mixture is 1 cubic centimeter of weak alcoholic solution of the bile acid, 1 drop of 0.1 per cent furfuraldehyde solution and 1 cubic centimeter of strong sulphuric acid. The original test was devised by Pettenkofer; later it was recognized that the reaction belongs to the group of "furfurol" reactions, and the aldehyde was recommended in place of the sugar. The test cannot be used with bile directly because of the presence of other substances, which would give a strong color with the sulphuric acid.

Preparation of Taurin. Use several hundred cubic centimeters of ox bile. Add to it an excess of strong hydrochloric acid, about one-third of the volume of the bile, and boil on the water-bath. A resinous mass separates and when this becomes stringy enough to solidify, when a little is taken up on a rod and allowed to cool, the reaction has gone far enough. Decant from this mass and evaporate the liquid resulting until a crystallization of salt forms. Filter and evaporate to a small volume. If salt separates filter again and pour the liquid finally into a large excess of alcohol. This causes the taurin to separate; wash the crude substance with strong alcohol, and recrystallize from hot water. In a successful separation large plates or prisms of taurin are obtained. The substance may be recognized by several tests. On heating it chars and gives off an odor of sulphurous acid. When fused with sodium carbonate the sulphur is converted into sulphide, from which hydrogen sulphide may be separated and identified by the usual tests.

THE BILE PIGMENTS.

The two substances, biliverdin and bilirubin, are related to hematin from hemoglobin, as pointed out above, and as may be illustrated by these formulas:

Hematin	$C_{32}H_{32}N_4O_4Fe$
Hematoporphyrin	$C_{16}H_{18}N_2O_3$, or $C_{32}H_{36}N_4O_6$
Bilirubin	$C_{16}H_{18}N_2O_3$, or $C_{32}H_{36}N_4O_6$
Biliverdin	$C_{16}H_{18}N_2O_4$, or $C_{32}H_{36}N_4O_8$

The two bile pigments are formed in the liver and normally, apparently, only in the liver, but by what kind of reaction is not clearly known. Hematoporphyrin may be produced from hematin and it is isomeric with bilirubin, though not identical. The relation of bilirubin to blood is perhaps best shown by this observation: in old blood extravasations the blood color appears to be gradually decomposed and in its place the new coloring matter is found, which was called hematoidin by its discoverer. Later studies have apparently shown the identity of this with bilirubin.

Bilirubin is practically insoluble in water, but it seems to act as an acid, the alkali salts of which are soluble. In this form it exists in bile. The solution is reddish yellow and in the air, or by treatment with oxidizing agents, it takes up oxygen and becomes biliverdin, which gives a green solution. The bile always contains the two pigments, from which the greenish yellow color follows. The amount of the two substances in the bile is normally very small, but as the reactions are sharp recognition is easy. The total weight of the two pigments produced in one day is not over 200 milligrams probably; the physiological meaning of the formation is not known. The iron of the original hematin is largely retained by the substance of the liver cells.

Preparation of Bilirubin. The pigment cannot be easily obtained from bile because of the small amount present, but may be obtained from the pathological concretions known as gall-stones, which will be described later. Powder several grams of these stones from cattle very fine and exhaust thoroughly with ether, then repeatedly with boiling water to take out cholesterol and bile acids. In the residue the bilirubin exists as an insoluble calcium compound; this is decomposed by the addition of a little dilute hydrochloric acid, after which what is left is washed thoroughly with hot water, and then with alcohol to leave the pigment in a still better condition for extraction. Finally extract with chloroform in which the substance is relatively soluble. On evaporating the chloroform crude bilirubin is secured, which after washing with alcohol may be recrystallized from hot chloroform or from dimethylaniline, in which it dissolves in the proportion of about 1 to 100 cold or 1 to 30 hot. By several crystallizations it is possible to obtain a product pure enough to employ as a standard for spectroscopic measurements.

By exposing an alkaline solution to the air or by treating with a little acid and sodium peroxide, bilirubin is converted into biliverdin. The latter free substance is not soluble in water, chloroform or ether.

The Bile Pigment Tests. Some of these are extremely delicate and have long been used for the recognition of bile, especially in urine. For the test to be given the bilirubin alkali in very dilute solution may be used, or a diluted bile.

GMELIN'S TEST. In a test-tube take a few cubic centimeters of nitric acid containing some nitrous acid. Over this pour carefully the weak bile solution to be tested. At the junction point colored rings appear which result from the formation of oxidation products of the bilirubin. The colors appear in this order from above down: green, blue, violet, red and yellowish. Of these the green is the most characteristic; the other shades represent more advanced stages in the oxidation. For success in the test the bile solution must not be too strong, and the amount of nitrous acid in the nitric acid must be small.

HAMMARSTEN'S TEST. Use as reagent a mixture of strong nitric acid and strong hydrochloric acid in the proportion of about 1 to 50 by volume. This mixture must stand some time before use, or until it becomes yellow. It keeps a long time. For the practical test mix 1 cubic centimeter of the acid with 4 cubic centimeters of alcohol and add a drop or two of the bilirubin solution to be tested. A permanent green color appears, but if strong oxidation is secured by adding more of the acid mixture the colors change as in the Gmelin test. The reaction can be well applied to urine.

FUNCTIONS AND BEHAVIOR OF BILE.

The bile as a whole has a number of functions to perform in the body, some of which have been referred to in the discussion of digestive processes. It represents also the avenue of escape of a number of by-products formed by the katabolic processes in the liver. Many of these processes are doubtless very complex and in them a variety of secondary or side reactions occur which furnish matters of no further use apparently in the body. These are collected in the gall bladder and finally discharged into the small intestine, where escape from the body with the feces is possible for the constituents having no further value. If the escape of these products from the liver is hindered, some form of icterus results, as the bodies in question must pass more or less directly into the blood.

In part, therefore, the bile must be regarded as an excretion like the urine, but that the parallelism is not complete is shown by the fact that a considerable absorption takes place from the intestine, and products are returned which find further application in the organism. There is evidence to show that this portion returned from the intestine serves as a cholagogue to stimulate new secretion in the liver. It is likely that this free secretion and flow of bile in the liver is necessary for the successful completion of certain metabolic processes going on there, so that it may be regarded not merely as an end but also as a means toward an end.

The one digestive process in which the bile seems to play a practically necessary part is in the splitting and absorption of fats; here its action is partly mechanical as in some way it aids the passage of the finely divided fat through the intestinal walls. The general behavior of bile in this respect may be illustrated by a simple experiment.

Experiment. Moisten two similar filter papers in funnels, one with water and the other with bile. Into each filter pour some fatty oil, such as cotton-seed oil or olive oil. Note that while the oil will not pass through the paper moistened with water a small amount passes slowly through the bile-moistened filter. Similar experiments have been made with animal membranes.

A more important action with fat, however, is shown in the power of bile to form fat emulsions, which depends on the behavior of the bile salts with the fat splitting ferment, as already pointed out, and on the formation of soaps directly. This is now looked upon as the one reaction in the intestine in which the presence of bile is actually practically essential, since the old views of the antiseptic value of the bile in preventing excessive intestinal putrefaction have been shown to be without foundation. In a diet rich in fats the emulsifying behavior of the bile unquestionably comes into play as a leading factor in the final absorption. Experiments have been made on animals in which the flow of the bile could be diverted from the natural outlet into the intestine by means of a fistula. In such cases the digestion of proteins and carbohydrates seemed to suffer no change but the digestion of fats was always imperfect and a large portion ultimately escaped with the feces. Indirectly there may be also a loss in protein if the fat in the food in such cases has a rather low melting point and is abundant. A fatty layer encloses portions of the partly digested proteins and prevents access of the digestive fluids until the lower stretches of the intestine are reached, where bacterial changes soon get the upper hand and rob the protein of any further food value. The action of bile in producing an emulsion with fatty oils may be illustrated by experiment. In an earlier chapter the formation of emulsions by other methods was shown.

Experiment. In a slightly warmed mortar pour about 5 cc. of bile and add to it one cc. of cottonseed oil. Rub the two thoroughly together for several minutes, and then add another small portion of the fatty oil. An emulsion forms slowly, and becomes more persistent as the working with the pestle is prolonged. The amount of oil which can be brought into the form of a stable emulsion with the 5 cc. of bile depends largely on the character of the oil. The presence of a small amount of free fatty acid in the cottonseed oil aids materially in producing the emulsion. The weak alkalinity of the bile is doubtless an important point here, as through the alkali a little soap is formed and this may be the chief factor in producing the emulsion.

In the intestines the stimulating action of the bile salts is probably more important than this last reaction. At the present time these salts are prepared in comparatively pure form as medicinal agents.

Bile contains a large amount of mucin as the analytical table above shows. The stringy character of the secretion is due to this substance which may be recognized by several precipitation tests. The addition of alcohol in excess throws down a flocculent mass which may be separated by the centrifuge. The addition of a little acetic acid produces likewise a precipitate. It is, however, practically impossible to secure pure mucin in this way as other bodies are carried down with the precipitates and their subsequent separation is difficult. The mucin of human bile is said to be nearly pure, while that of other animals is mixed with nucleo albumins.

BILE CONCRETIONS. GALL STONES.

Under conditions not well understood a precipitation of certain constituents of the bile may occur in the gall bladder. These precipitations take the form of solid masses which sometimes grow to considerable size, by gradual surface additions. In every case the deposited material is built up in layers, often well defined, around some body as a nucleus. Three general classes of such calculi are recognized. In man balls of cholesterol, more or less pure, are the most abundant while pigment stones are also frequently found. These pigment stones contain essentially bilirubin in combination with calcium, the alkali earth salts of the pigments being insoluble. The center of the cholesterol stone may be a nucleus of bilirubin calcium. Pigment stones are common in the gall bladders of cattle. Finally we have stones consisting of calcium phosphate or carbonate, which, however, are not usual in man.

The following analyses made of gall-stones of very different appearance illustrate the composition of the cholesterol stones in man:

Water (at 100°)	4.60		4.50	
Cholesterol (and trace of fat)	90.87		90.08	
Bilirubin ($CHCl_3$ extraction)	0.81	} 3.05	0.19	} 1.77
Biliverdin (C_2H_6O extraction)	2.24		1.58	
Mucin and soluble extractives	0.14		1.53	
Total ash	0.88		2.72	
Total P_2O_5	0.20		1.00	

These concretions frequently give rise to serious pathological conditions and they must then be removed by surgical operations. In addition to the above constituents the stones contain small amounts of iron and often traces of copper. But the iron found is far from accounting for the amount which must be separated from the hematin in the formation of bilirubin. In a former chapter the preparation of cholesterol from gall-stones was described, also the general chemical behavior of the substance. The character of a stone is most easily recognized by its behavior toward boiling alcohol, in which cholesterol is rather readily soluble, to crystallize in large thin plates on cooling.

The solutions of cholesterol have a marked action on polarized light, which property may be employed sometimes in the identification and estimation. The specific rotations below have been found.

Ether solution	$c = 2$	$[a]_D^{18} =$	$-31.12°$
Chloroform solution	" 2	"	$-37.02°$
	" 5	"	$-37.81°$
	" 8	"	$-38.63°$

In feces a modified cholesterol is found which has been called *koprosterin* and also *stercorin*. This new substance is a reduction product with the probable formula $C_{n}H_{u}O$ and is dextrorotatory, $[a] = +24°$.

Besides the two principal pigments several derived substances have been obtained from the gall-stones. The following have been described: *bilifuscin, biliprasin, bilihumin, bilicyanin*. These substances exist in small amount and are without practical importance. Their relations to the others are not clearly established.

CHAPTER XVIII.

CHEMISTRY OF THE PANCREAS AND OTHER GLANDS. MUSCLE, BONE, THE HAIR AND OTHER TISSUES.

In this chapter a number of substances will be briefly discussed, the chemical relations of which in some cases are unimportant, or sometimes, when important, not well understood. In regard to the pancreas, it will be recalled that in the discussion of digestive phenomena the behavior of active enzymes in the liquid secreted by the organ was rather fully considered. In the so-called pancreatic juice the three most important enzymes are active in the digestion of carbohydrates, fats and proteins, but in addition to these functions others must be mentioned.

THE PANCREAS.

The organ is relatively poor in solids, containing only about 100 parts per 1,000. The solid substance consists largely of nucleo-proteids with but comparatively small amounts of the other protein bodies. Besides producing the digestive enzymes, or their zymogens, the pancreas cells have an important function to perform in connection with the oxidation of sugar in the body. It has long been known that a kind of diabetes results on the extirpation of the pancreas. Something seems to be produced there which is apparently essential in the oxidation process. Experiments with animals have shown that the oxidation takes place if even a small portion of the organ is left. Of the nature of the active ferment here or of its mode of action practically nothing is known; but it has been pointed out recently by several writers that in this sugar oxidation, taking place in the muscles probably, two things are concerned. The pancreas may furnish one of these and an enzyme formed in the muscle cells is the other. Cell-free extracts from the organs taken separately have been found to be practically inert toward sugar, while in presence of a mixture of the two extracts oxidation follows readily. It has been suggested that one of these organs furnishes an enzyme which is the *catalyzer* for the other, and attempts have been recently made to produce the pancreas enzyme on a large scale for use therapeutically.

Autolysis. The pancreas readily undergoes autolytic digestion under the aseptic treatment or when preserved by toluene. A large number of products may be separated from the altered mass, which in a general way resemble those produced in the liver, as already referred to. Ammonia, leucine, tyrosine, aspartic acid, glutaminic acid and the hexone bases have been recognized; also, the somewhat unusual oxyphenylethylamine, $HO \cdot C_6H_4 \cdot CH_2 \cdot CH_2NH_2$, which may be derived from tyrosine by splitting off of CO_2.

On account of the relatively high content of nucleo-proteids, and the constituent nucleic acids, a marked amount of sugar in the form of pentose is liberated. No other organ subjected to prolonged autolysis seems to yield as much. In certain pathological conditions involving the pancreas, the urine contains a complex which yields a pentose on treatment with acid at the boiling temperature. The pentose is identified through its phenyl hydrazine compounds.

THE SUPRARENAL BODIES.

A soluble substance contained in the capsules, because of its important property of raising the blood pressure, has attracted a great deal of attention in the last ten years. This soluble substance was first recognized as a chromogen which, on account of its oxygen-absorbing power, was assumed to be related to pyrocatechol. An aqueous extract of the capsules becomes dark on exposure to the air and produces a dark green color when treated with ferric chloride. It also reduces Fehling's solution strongly and shows the same behavior toward other metallic salts. The oxygen-absorbing power of the extract had been known about thirty years before the important relation to blood pressure was discovered. It was soon found that the two properties seem to reside in the same constituent of the extract, since the destruction of one is followed by the disappearance of the other. Numerous investigations have been carried out on the isolation of the active principle, especially by Abel, v. Fürth and Takamine, who have given the names *epinephrin, suprarenin* and *adrenalin* respectively to active extracts which they have separated by different processes. Some idea of the nature of the substance may be obtained from considering a method given by Takamine for separating it.

The minced capsules are extracted by weakly acidulated water in an atmosphere of carbon dioxide to prevent oxidation. The temperature of the extraction is at first 50°-60° and finally 90°-95° to coagulate proteins. The extract is concentrated *in vacuo* and precipitated with strong alcohol; the filtrate is concentrated—the alcohol distilled off—*in vacuo*, and to the aqueous residue ammonia is added. This produces a precipitate of the active principle in crude form, which crystallizes in time. The precipitate is redissolved with a little acid in alcohol, and certain impurities are thrown out by addition of ether. The filtrate is concentrated *in vacuo* again and a new precipitation effected by ammonia. By repeating this treatment several times a much purer product is obtained.

A light yellowish crystalline powder is secured, which is somewhat soluble in water. It combines with hydrochloric acid to form the stable salt commonly used in medicine. The empirical formula is $C_9H_{13}NO_3$ and for this several constitutional formulas have been suggested.

The other constituents of the suprarenal capsules have no importance at the present time that can be clearly defined, but as is well known, complete removal of the bodies is usually attended with fatal results, and Addison's disease is associated with certain pathological conditions in the organs. Lecithin bodies and a glucose-furnishing complex are present in small amount, as well as the mass of protein substance which has not yet been fully investigated.

THE THYROID GLAND.

The relation of this gland to certain pathological conditions which sometimes appear in man and which may be induced in animals has been a subject of study for many years. Attempts to isolate the active principle or principles on which the functions of the gland depend have been in a measure successful. In the course of investigations a number of basic bodies have been separated, but these may have no connection with the observed physiological behavior.

From the investigations of Oswald, who has made the fullest contributions to the literature, there are two peculiar protein bodies present, one of which is a globulin and the other a nucleo-proteid. To the first he has given the name *thyreo-globulin;* this exists frequently combined with iodine, and it is the latter complex which is assumed to be theoretically and practically important. It has been called *iodothyreo-globulin* and appears to be found only in those glands which contain colloid, and the amount of iodine present is proportional to the amount of colloid. The normal gland weighs usually 30–45 grams, in which the thyreoglobulin fraction is in the mean about ten per cent. The amount of iodine is usually less than one tenth of one per cent of the whole. In case of enlarged glands—goitre—the whole organ may weigh up to several hundred grams. If the goitre is rich in colloid the iodine appears to be absolutely, but not relatively, increased. In the thyreo-globulin from a normal gland over 0.3 per cent of iodine has been found, while in the preparation from colloid goitres the amount in the mean is 0.06–0.07 per cent.

By treatment with acids the gland, or the thyreo-globulin from it, undergoes a cleavage in which a residue rich in iodine remains. The organic iodine compound so obtained which may be the true active principle is called *iodothyrin* or *thyroiodine*. In earlier experiments Baumann, the discoverer of this compound, found an iodine content of about 9 per cent, but Oswald, starting with pure iodothyreo-globulin which was secured by a salting-out process with ammonium sulphate, obtained finally iodothyrin with over 14 per cent of combined iodine. This iodothyrin is not a protein substance; the analyses of different preparations are not in very good accord, from which it appears that the pure substance has not yet been actually secured. The crude product at present known has been used in medicine and attempts have been made to duplicate or replace it by other iodine compounds.

It is now generally recognized that the physiological activity of the dried thyroid on the market in powdered form is proportional to the iodine content. No exact method of valuation is known.

The smaller glands associated with the thyroid and known as the *parathyroids* are possibly even more important. Both sets of glands have apparently much to do with the general metabolic functions of the body, and the *complete* removal of the parathyroids is usually followed by death. How they act is not clearly known.

THE REPRODUCTIVE GLANDS.

Of the chemical composition of the testicles and their secretion not much can be said. The testicles contain several proteins and extractives, but their investigation has been extremely limited. The most complete examinations of the spermatic fluid are probably those reported by Slowtzoff, from whose work the following figures are taken. The specific gravity of the fluid varies from 1.02 to 1.04; the reaction is alkaline and as measured by the aid of rosolic acid corresponds to 0.15 per cent sodium hydroxide. As a mean of five analyses the following results may be given:

SPERMATIC FLUID.

Specific gravity	1.0299
Water	90.32 per cent.
Dry substance	9.68 "
Salts	0.90 "
Proteins	2.09 "
Ether extract	0.17 "
Water and alcohol extracts	6.11 "

The tables below show the calculations for dry substance and the character of the ash:

For Dry Substance.		Ash.	
Organic	90.81 per cent.	NaCl	29.05 per cent.
Inorganic	9.19 "	KCl	3.12 "
Proteins	24.48 "	SO₃	11.72 "
Ether extract	2.15 "	CaO	22.40 "
Water and alcohol extract	59.36 "	P₂O₅	28.79 "

The ash is peculiar in containing a large amount of sodium chloride and calcium phosphate. The phosphoric acid is present in larger amount than corresponds to the nuclein substances.

The proteins are made up approximately as follows:

Albumins	68.5
Albumose-like bodies	21.6
Nucleins	9.9

A characteristic basic body known as *spermine* is present in small amount. The empirical formula C_2H_5N has been given to it. This substance forms a combination with phosphoric acid which sometimes separates in crystalline form on evaporation of the fluid. The characteristic odor of the discharged secretion is said to be due to partial decomposition of the base.

The *spermatozoa* are relatively stable bodies and resist the action of chemical reagents to a remarkable degree. The heads of spermatozoa consist largely of nuclein compounds while the tails contain other proteins, cholesterol, fat and lecithin. The ash content of the whole is relatively high and is rich in potassium phosphate.

BRAIN AND NERVE SUBSTANCES. CEREBRO-SPINAL LIQUID.

These tissues contain several peculiar compounds of which our knowledge is limited, largely because of the great difficulty in separation. The solid matter of the brain contains globulins, nucleo-proteids, cholesterol, lecithin, fatty bodies and complex compounds not found elsewhere. Various soluble extractives, somewhat similar to those from muscular tissue, are also present.

Protagon. This has been assumed to be an important constituent of the white substance of the brain, which has this elementary composition, according to Gamgee: C 66.4, H 1.07, N 2.4, P 1.07. But as others writers report rather widely different figures it is likely that the pure substance has not yet been isolated. As extracted by means of 85 per cent alcohol at 45° from the minced brain, and purified by crystallization and washing with ether, it is obtained as a white powder practically insoluble in cold ether or alcohol and not properly soluble in water. With much water it finally yields a gelatinous liquid, which suffers decomposition readily.

Notwithstanding the bulky literature which has accumulated in the discussion of this substance, its exact nature is not yet known. All recent investigations seem to show that it is a mixture of a number of bodies. By treatment with certain solvents, or by gentle cleavage, it is possible to separate a group of phosphatides, similar to some of the lecithin bodies, and a group of substances free from phosphorus, but containing nitrogen. Cerebrin and cerebron are names given to two of these products. Of the functions of these little is known.

In the white substance of the spinal marrow the so-called protagon is abundant. In degeneration changes in the tissues of the nervous system it is probably this compound which suffers the greatest alteration, with the production of neurine with marked toxic properties. It is likely that the neurine comes from a lecithin body as one of the groups in the protagon complex, and that these reactions will prove of great importance in pathological study.

It is also known that complex sulphur compounds are present in the brain tissue, but little is known of their reactions.

Cerebro-spinal Liquid. This is a thin, watery liquid of which only a few partial analyses have been recorded. Its general character is shown by these figures recently given by Zdorek:

1,000 parts by weight contain

Dry substance	10.45
Organic	2.09
Inorganic	8.36
Proteins	0.77
Chlorine	4.24
Sodium oxide	4.29

The organic substance includes traces of fats, lecithin, cholesterol and, pathologically, choline or neurine. Common salt, however, is the main solid substance in solution.

MUSCLE AND ITS EXTRACTIVES.

A large part of the solid portion of the body is made up of muscular tissue. A knowledge of the composition of this tissue is of the highest importance, especially since some of the fundamental chemical reactions of the animal organism take place within the cells of the muscles. Fortunately we have fairly satisfactory information on some of the points of interest here, as numerous analyses have been made of the muscles and of the liquid which may be extracted in various ways from them.

The dry part of the muscle is made up largely of proteins of which several are present; in the muscle plasma there are at least five according to Halliburton. In addition to these bodies there are a number of so-called extractives which play an important part.

GENERAL COMPOSITION OF MUSCLE.

The following figures represent approximately the average composition of the fresh muscle dissected free from visible fat.

Water	76 per cent.
Solids	24 "
Proteins (true)	17.6 "
Collagen substance	3.0 "
Fat, interstitial	1.5 "
Flesh bases	0.2 "
N-free extractives	0.4 "
Salts	1.3 "

The Muscle Proteins. It is not possible to give a perfectly clear account of all these bodies at the present time, as the products obtained by different investigators vary with the details of the extraction methods employed. The more important constituents commonly recognized are indicated in the following paragraphs. By washing out the blood from living muscle by physiological salt solution (transfusion), dissecting it, grinding it to a pulp and pressing very strongly a clear yellowish liquid is obtained which is called *muscle plasma*. The ordinary dead muscle treated in the same manner yields a different liquid which may be called *muscle serum*. The plasma has an alkaline reaction and is distinguished by the property of spontaneous coagulation.

The term *myosin* was formerly applied to the solidified or coagulated body as a whole, but experiment shows that two things at least are here present. One of these is called *musculin*, or by some authors, myosin proper, while the other product is known as *myogen*. The musculin, or myosin, coagulates at about 47°, while for myogen the coagulating temperature is about 56°.

The two substances, musculin and myogen, differ also in their precipitation properties. The first is precipitated from solution by adding ammonium sulphate to make up 28 per cent; from the filtrate the myogen may be thrown down by adding the sulphate to saturation, and is found to make up about 80 per cent of the plasma protein.

The serum left after the formation of the plasma coagulum usually contains a little soluble albumin. This may be normal to the muscle substance, or it may be due to the blood not perfectly removed by the preliminary washing. At any rate the plasma consists essentially of the two myosin bodies.

After separation of the plasma what may be called the *stroma* remains. This is mainly albuminous, but its exact nature is not known. The sarcolemma portion of the muscle fiber, which by weight makes up but a small part of the whole, appears to belong to the albumoid group of proteins, resembling elastin. It has been shown in an earlier chapter that from ordinary dead muscle, as represented by lean meat, a considerable amount of "myosin" may be separated by extracting with a weak solution of ammonium chloride. What remains does not agree fully with the stroma left on pressing out the plasma of the fresh muscle, but contains approximately the same substances. By this method of separation the insoluble stroma portion is much larger than the soluble or "myosin" portion. The latter may amount to 7 or 8 per cent of the weight of the muscle in the mean.

Collagen. As given in the above table this refers to the binding substance holding the muscle fibers together and includes the sarcolemma. It is insoluble in cold water, but swells and disintegrates finally in boiling water.

Fat. After removing all visible fat from the dissected muscle, analyses still show a small amount remaining. This must therefore be associated with the minute structure of the fibrils.

Flesh Bases. A number of very remarkable substances are included here. They are sometimes described as the nitrogenous extractives. The most abundant of these bodies is *creatine* or methyl-guanidine acetic acid; some of the purine bases are also present. A brief description of these substances may be given.

CREATINE, $C_4H_9N_3O_2$, may be represented structurally by the formula

$$H-N=C\begin{cases} NH_2 \\ N\begin{cases} CH_3 \\ CH_2 \cdot COOH \end{cases} \end{cases}$$

It is found in all muscles and is probably a product of metabolism, but the method of its formation is not yet known. Being readily soluble in warm water, and in about 75 parts of water at the ordinary temperature its extraction from muscle is easy. When the solution is boiled with dilute hydrochloric acid, through a long period, a molecule of water is split off and the anhydride *creatinine* is left. This is a normal urinary constituent and will be described later. When boiled with alkali solution, especially baryta water, creatinine undergoes a complete cleavage into urea and sarcosine, which relation is an interesting one and has suggested a possible derivation of the urinary urea. Creatinine may be readily crystallized from water solution. It was formerly made for experiment directly from meat. It is best secured from certain crystalline residues occurring as by-products in the manufacture of "beef extract," referred to below.

CARNINE. The amount of this in muscle is very small, but it may be recognized in beef extract. It bears some relation in structure to hypoxanthine, and has been given the formula $C_7H_8N_4O_3$.

Comparatively recently several other crystalline products have been isolated from meat extracts. Among these *carnosine* and *carnitine* are perhaps the most important.

THE XANTHINE BODIES. These constitute a peculiar group of great importance because of their relation to uric acid and other products of metabolism. Traces of several of them have been recognized in the muscular juices; in a later chapter the structure and properties of the substances will be discussed in connection with uric acid. Traces of urea are also found in the muscles.

The Nitrogen-Free Extractives. The muscular juices hold dissolved a number of compounds which contain no nitrogen, some of which are very important. The chief of these are glycogen, inosite, glucose and lactic acid.

GLYCOGEN. The chemical relations of glycogen have been discussed already in earlier chapters. The glycogen as found in the muscles comes from the liver, being transported there by the blood, and in part is probably formed in the muscles by the same kind of an enzymic action which leads to its synthesis in the liver. The liver is capable of storing up a large weight of the reserve substance in a small space. The amount stored in an equal weight of muscle is small, but taking the muscles of the body as a whole the glycogen content is considerable, reaching a hundred grams or more.

It is probably through this glycogen that the muscle is capable of doing its work. Through enzymic hydration the glycogen becomes sugar, possibly maltose and then glucose, and the potential energy of this is liberated by oxidation to water and carbon dioxide ultimately. The oxidation may not be direct; in all probability there are several transformation products before the final stages are reached. But the energy transformation is the same whatever the intermediate steps may be. The importance of the glycogen and related bodies in this direction will be pointed out in a following chapter. It may be recalled that in these oxidation processes, where sugar is concerned, a muscle enzyme and a pancreas enzyme seem to be both necessary.

While glycogen in the muscles must come mostly from sugars, either directly or through the liver, there is also some evidence that it may come in part from other substances, especially from proteins. Animal experiments have shown apparently a storing of glycogen from a protein diet after previous starvation had exhausted the reserve in store. In the breaking down of some proteins it has been shown that certain carbohydrate groups are liberated; it is doubtless these which undergo synthesis to form at least part of the glycogen, and from this standpoint the behavior of protein as a glycogen factor is not so hard to understand.

The glycogen content of the muscles of different animals is variable; in the flesh of the horse it is relatively high, amounting often to over 1 per cent. As the muscle glycogen is not altered rapidly in the dead organ, as is the liver glycogen, the presence of the substance in horse-flesh sausage may be quite readily recognized. Methods have been devised for the identification of horse-flesh, sold for food, based on these facts. Glycogen may be extracted from the muscles by the general method given for the liver in an earlier chapter; the chemical and optical properties may be used for the final identification.

INOSITE. This substance has the empirical formula $C_6H_{12}O_6 + H_2O$ and was long spoken of as *muscle sugar*. It is not a true carbohydrate, however, but an aromatic product $C_6H_6(OH)_6$, that is, hexahydroxybenzene. The amount found in muscle is very small and how it is derived is not known; but it is not peculiar to these tissues, as it occurs in other organs of the body and also in many vegetable substances. It may be extracted from muscles without much trouble and when pure is found to be a white crystalline powder melting at about 220°. It is very soluble in water, to which a sweetish taste is given, and in presence of alkali is not a reducing agent for metallic solutions. Although the usual structural formula does not show an asymmetric carbon atom the substance is optically active and exhibits a strong rotation, both right and left forms being known.

GLUCOSE. From what was said above about the transformation of glycogen it is not surprising that a small amount of sugar should be found in the muscles; both maltose and glucose have been detected.

LACTIC ACID. Several forms of this acid are known, but that occurring in the muscle is the dextrorotatory *paralactic* or *sarcolactic* acid, $C_3H_6O_3$. It is one of the α-hydroxypropionic acids. There has been much speculation as to the source of this acid in the body, but it seems most rational to regard it as derived from the glycogen or sugar by a comparatively simple cleavage. It is also possible that in the katabolic reactions of proteins lactic acid may result from a splitting of the carbohydrate group. The acid is not very readily detected in the living muscle because it is probably oxidized or removed too rapidly by the fluid circulation. In the dead muscle, however, it may accumulate to the extent of half a per cent or more. The living muscle shows a neutral or slightly alkaline reaction, while in the dead muscle the increase of lactic acid changes the reaction.

The lactic acid of the muscle probably results from an enzymic cleavage. In the aseptic autolysis of liver paralactic acid has been recognized among the products, and this fact shows, at least, the possibility of such a formation. The amount of lactic acid formed in the muscle seems to be greatest during working periods, which is true also of the final products of katabolism. The acid may simply represent a stage in the gradual breaking down, whether we consider a carbohydrate or protein as the parent substance. We should expect therefore an increase in the muscle acid if the oxidation processes of the body are hindered or retarded, while at the same time protein or sugar decomposition is increased, or, at any rate, not diminished. In the dead muscle the enzymic formation of lactic acid doubtless continues long after the oxidation reaction ceases, and this is probably the main reason for the ready detection in the muscle after death.

The pure acid occurs as a thickish liquid miscible with water. It forms salts which are mostly readily soluble. The zinc and calcium salts crystallize well and are hence prepared for identification. The pure liquid shows a right hand optical rotation, with $[\alpha]_D =$ about $3°$. The result is not constant because of the difficulty of preparing concentrated solutions free from anhydride or lactide. The rotation of the salts, on the contrary, is to the left.

The Inorganic Salts. Although making up not much over 1 per cent of the weight of the moist muscle, these salts are extremely important. Of dry substance the salts constitute 5 per cent or more. The salts are usually estimated from the ash left in burning the muscle; this gives of course no correct idea of how they are combined in the living muscle, but is the only method available. In the living muscle many of the inorganic elements are doubtless in chemical union with proteins or other organic groups, while in the derived ash we have chlorides, phosphates, sulphates or carbonates. A carbonate is probably formed during the combustion of organic acids and corresponds to no simple preëxisting compound. Phosphorus and sulphur of proteins furnish phosphates and sulphates. The analyses of ash made disclose very different results, but mean values may be given to show the general approximate composition. In the calculation carbonic acid is not considered. The table below is from the König collection.

K_2O	37.04
Na_2O	10.14
CaO	2.42
MgO	3.23
Fe_2O_3	0.44
P_2O_5	41.20
SO_3	0.98
Cl	4.66
SiO_2	0.69

From the table it appears that potassium phosphate is the most abundant substance in the ash. Much of this doubtless preëxists in the muscle juices, while a small portion is of oxidation origin. The small sulphate content is probably due to protein sulphur fully oxidized in the combustion. In the past too little attention has been given to the mineral constituents of the body, it being commonly assumed that they represent "waste" or "ash" only. But the newer applications of chemistry, especially physical chemistry, to physiology have disclosed the fact that the inorganic salts are especially concerned in the proper maintenance of many of the body functions. The balanced osmotic pressure of the body fluids is largely a function of the salt content, and variations here are of great importance. The mineral salts are the carriers of electric charges in the body and as such seem to have important duties to perform.

EXTRACT OF MEAT.

By boiling lean meat with water the soluble constituents are dissolved, producing an extract. When this is concentrated to a paste the article known commercially as "Extract of Meat" results. The article was first made in quantity in South America to utilize the carcasses of cattle slaughtered for the hides, but later the manufacture was introduced elsewhere, and generally to utilize certain waste or by-products in the meat industries. At first the extract was assumed to possess food value in a high degree, but after a time, as the chemistry of the proteins and their derivatives became better understood, this notion was gradually abandoned. Lean meat, muscle, is employed practically in the process; hence little or no fat can be present. At the boiling temperature nearly the whole of the proteins are coagulated and are filtered out. A little gelatin remains, but the food value of this is of minor importance. Unless the boiling is greatly prolonged the extract must therefore contain essentially the meat bases and other extractives referred to above, and the actual nutritive value of these is low, in the case of the bases being nil. On prolonged boiling, however, a small portion of the original protein seems to pass over into the soluble form of albumose, which is therefore found in some extracts. Finally, the phosphates and other inorganic salts, being largely soluble, pass into the extract and constitute a considerable part of the finished pasty product.

In this country "extract" is made by concentrating the broth resulting from the boiling of beef as a step in the canning process. Large quantities of meat being boiled in the same water, it becomes rich in the "extractives" and is finally boiled down to the usual pasty condition. Before the concentration is complete the liquid is filtered and skimmed and therefore leaves a residue free from fat or fiber Roughly speaking the paste extract has about this composition:

```
Water .................................................... 20
Salts ..................................................... 20
Organic substances ....................................... 60
```

Numerous analyses have been made of some of the commercial extracts, but the methods employed have not always been delicate enough to furnish trustworthy information. This is especially true as regards the amounts of so-called peptone and albumose present, for which the definitions have not been fairly uniform until comparatively recently. The recognized relations of these substances are explained in the chapter on protein compounds. Analyses made by the older methods were generally reported as showing more or less "peptone" when, according to the present views, "albumose" is meant. The following figures may be taken as representing approximately the average composition of typical samples of American meat extract:

```
Water .................................................... 20.0
Inorganic salts (ash) .................................... 22.5
Albumose (and gelatin) ................................... 16.5
Flesh bases, etc. ........................................ 26.4
N-free extractives ....................................... 14.6
```

According to these results the food value of the extract would be measured by the nitrogen-free extractives and the albumose and gelatin fractions. In some kinds of extract the flesh bases and related bodies are much higher than here given, with corresponding diminution in the other organic constituents. The real value of these extracts lies mainly in other directions, however. They contain the flavoring and stimulating portions of the meat, and should not be considered so much as foods as additions to foods. Added to vegetables they impart an agreeable taste and doubtless serve a very useful purpose in stimulating appetite for substances not in themselves possessing much flavor. In their action the basic and similar substances in the meat extracts may be perhaps fairly compared with the alkaloids in tea and coffee, which, experience shows, have a real value. Large amounts of the extracts cannot be used, however, as foods, because of the presence of the large percentages of alkali phosphates and other salts. A few simple experiments may be made to show some of the properties of the common commercial extracts.

Experiment. Heat a little of the solid extract on a piece of porcelain until it is reduced to a char. Extract this with dilute nitric acid, filter and divide the filtrate into two portions. In one test for phosphates by the addition of ammonium molybdate and in the other for potassium salts by the flame test. Both reactions should be very distinct.

Experiment. Dissolve 20 grams of extract in water to make about 200 cubic centimeters. A nearly clear solution should be obtained, showing absence of fat or coagulated protein. To a few cubic centimeters add enough weak acetic acid to give a slight reaction, and boil. If a precipitate forms, which is rarely the case, albumin is shown.

With 50 cc. of the liquid make the albumose test. Add to it finely powdered zinc sulphate as long as it dissolves on stirring. On saturating the solution completely a flocculent precipitate gradually settles. This is essentially the "albumose" fraction and may contain a little gelatin. After 24 hours filter, and test the filtrate for peptone by the biuret reaction; this is generally negative.

Use the remainder of the original solution for the recognition of creatine. Add to it carefully a solution of basic acetate of lead as long as a precipitate forms. This will carry down phosphates, sulphates and other compounds forming insoluble combinations with it, but not creatine. A *slight* excess of the lead must be added to insure complete precipitation. This can be determined by allowing the first formed precipitate to settle and adding more reagent as necessary. Finally filter, and remove the excess of lead by passing in hydrogen sulphide. Filter again, and remove as much as possible of the excess of sulphide used, by shaking. Then concentrate the liquid to a small volume by *slow* evaporation on the water-bath and allow it to stand a day or more in a cool place for crystallization of the creatine. Pour off the supernatant liquid and wash the fine crystals obtained with a little strong alcohol in which creatine is but slightly soluble.

Experiment. Dissolve the creatine in a little hydrochloric acid and evaporate the solution slowly to dryness on the water-bath. This action converts creatine into creatinine. Dissolve the residue in a little water and divide the solution into two parts. To one add a solution of zinc chloride, which produces a white crystalline precipitate containing the creatinine-zinc chloride, $(C_4H_7N_3O)_2ZnCl_2$. The character of the crystals can be seen under the microscope. To the other part of the solution add a few drops of a dilute solution of sodium nitroprusside and then, *drop by drop*, dilute solution of sodium hydroxide. This gives a ruby red color which fades to yellow. Add enough acetic acid to change the reaction and warm. The color becomes green and finally blue. This is known as Weyl's reaction. The blue color finally obtained is Prussian blue.

A further very delicate reaction for creatinine is given later, in the chapter on urine analysis.

Experiment. The mother liquor left after crystallizing the creatine contains traces of xanthine bases. Add enough ammonia to give an alkaline reaction and filter. Then add a few drops of ammoniacal solution of silver nitrate which precipitates the several substances in flocculent form.

BONE AND GELATIN.

In the moist bone as it exists in the body the water and solids are, in the mean, in about the proportion of one to two. In very young persons, however, the water is in greater excess, while with age the solids increase. The solid matter consists roughly of 1 part of organic matter to 2 of mineral.

THE ORGANIC MATTER OR OSSEIN.

The crude organic substance in the bone is commonly called ossein; it may be extracted with hot water and forms a gelatinous mass on cooling. But fuller investigations show that this ossein is not a single substance, as several different constituents may be separated by proper solvents. These are, however, closely related substances and for our present purpose they may all be considered as practically identical with the collagen or glue-forming substance of the connective tissues. The conversion of the ossein or collagen into gelatin appears to be a hydration process, as at a higher temperature the reverse operation takes place. The preparation and properties of bone gelatin may be illustrated experimentally:

Experiment. Clean a long, slender bone (best, a rib), and immerse it in dilute hydrochloric acid of about ten per cent strength. Let it remain several days. At the end of this time remove the bone from the acid and observe that it has lost its rigidity and has become very flexible. It may be even possible to tie it in a knot. Wash the elastic mass several times in fresh water to remove all the hydrochloric acid, then with a little dilute sodium carbonate solution followed by more water, and finally boil it with a small amount of pure water. By heating it long enough the ossein becomes converted into gelatin, which solidifies, on cooling, to a jelly.

By boiling the bone ossein under pressure the formation of the gelatin is very much hastened.

The solution as obtained above may be used for tests such as were described in Chapter V, under Gelatin.

THE MINERAL MATTER IN BONES.

We are not able to say exactly how the mineral elements are combined in the moist fresh bone. Our knowledge of these combinations is practically limited to what we can learn by a study of the residue left on burning the bone completely, known as boneash. This is a white powder containing the non-volatile compounds, of which calcium phosphate is the most important. The following table shows the average composition of human boneash:

Calcium phosphate	85.7	per cent.
Magnesium phosphate	1.5	"
Calcium carbonate	11.0	"
Calcium fluoride and chloride	1.0	"
Ferric oxide	0.8	"
	100.0	

The presence of calcium, magnesium and phosphoric acid may be shown in the weak hydrochloric acid extract of the bone described above.

Experiment. To a few cubic centimeters of the filtered solution add some ammonium molybdate solution. In a short time a yellow precipitate appears, indicating presence of a phosphate, as familiar to the student from the reactions of qualitative analysis.

Experiment. To a few cubic centimeters of the solution add solution of sodium acetate until a distinct odor of acetic acid persists. Then add some solution of ammonium oxalate, which produces a white precipitate of calcium oxalate.

Experiment. To another portion of the hydrochloric acid solution add ammonia until a good alkaline reaction is obtained. A white precipitate of calcium and magnesium phosphates settles out. Filter, and to the filtrate add some ammonium oxalate solution. A further precipitate appears. This is calcium oxalate and proves that the original solution contains calcium in excess of that combined as phosphate. The calcium of the carbonate, fluoride and chloride appears here.

Experiment. To detect the small amount of magnesium requires greater care. To another and relatively large portion of the acid solution add enough ammonia to give an alkaline reaction, and then acidify slightly with acetic acid. This dissolves everything except ferric phosphate, which may be filtered off and tested for iron. To the filtrate add enough ammonium oxalate to precipitate all the calcium as oxalate. Separate this after long standing by means of close-grained filter paper. In the clear filtrate the magnesium may be thrown down with the phosphoric acid still present, by the addition of ammonia water in slight excess.

Bone Marrow. The pure marrow consists largely of fat in which olein is abundant; cholesterol is present and some nitrogenous extractive substances, which, however, have not been very thoroughly examined.

CARTILAGE.

Collagen is probably the most abundant substance in the cartilaginous tissue where it exists mixed or combined with several other bodies, of which these have been described: *chondromucoid, chondroitin-sulphuric acid* and an *albuminoid*. The nature of crude collagen has been explained, and in Chapter V the somewhat obscure chemistry of the chondroitin-sulphuric acid has been outlined. Of the nature of the chondromucoid little is known definitely; it has been held by some writers to be merely a combination of part of the collagen with the salts of the complex ethereal sulphuric acid mentioned, while Mörner, who first described it, held it for a distinct body somewhat allied to mucin. His analyses showed C 47.30, H 6.42, N 12.58, S 2.42, O 31.28. The sulphur is probably all in the ethereal combination and on incineration of the cartilage the ash is found to contain a very large amount of alkali sulphate.

Chondromucoid as separated is insoluble in water alone, but with a little alkali forms a thick solution, which is precipitated by acids.

Stronger acids bring about a cleavage with separation of the chondroitin-sulphuric acid. The weak alkali solutions are precipitated by metallic salts, but most of the other protein reactions fail. The ethereal sulphate group seems to prevent the ordinary precipitations.

The albuminoid substance is not well characterized but is insoluble in water, and in weak acids or alkalies.. It undergoes gastric digestion. This protein is said to be found in old cartilage only, and is absent in young cartilage.

KERATIN BODIES.

Compounds of the keratin group occur in hair, the finger nails and horn. They resemble the proteins but contain rather large amounts of sulphur, as shown by these analyses, which are of keratin from several sources:

	Hair.	Nails.	Horn.
C	50.65	51.00	51.03
H	6.36	6.94	6.80
N	17.14	17.51	16.24
O	20.85	21.75	22.51
S	5.00	2.80	3.42

The sulphur in hair is in part loosely combined and may be split off easily by reagents, alkalies for example. The ash of hair is rich in sulphates and contains also silica and other mineral substances. Much of the ash may be removed by washing the hair with weak acids, following treatment with ether and alcohol to remove fatty and other soluble substances. The purified "keratin" thus secured gives results like the above on analysis.

Horn and nails contain along with the insoluble keratin insoluble salts, mainly phosphate of calcium, which stiffen them. From very fine horn shavings these salts may be dissolved out by acids, leaving a soft flexible keratin.

SECTION IV.

THE END PRODUCTS OF METABOLISM. EXCRETIONS. ENERGY BALANCE.

CHAPTER XIX.

THE EXCRETION OF NITROGEN, SULPHUR AND PHOSPHORUS. THE URINE.

Having considered in the foregoing pages the substances used in the nutrition of the body, the agencies of nutrition, and the general character of the products formed, we come now to a short study of the waste products rejected by the body after it has assimilated and used the nutrients furnished to it. The food-stuffs which the animal can utilize are comparatively complex, but consist essentially of the members of the three groups, the fats, carbohydrates and proteins. The theoretically simplest waste or oxidation products of these are nitrogen, carbon dioxide and water, but in the animal organism the breaking down does not go so far. While from fats and carbohydrates essentially only water and carbon dioxide are formed, the protein metabolism is not carried to the elimination of nitrogen, but ends with the formation, largely, of urea, a body in a way related to the theoretical end products, but which would call for three more atoms of oxygen to complete oxidation.

The nitrogen metabolism involves some extremely interesting problems which are still far from complete solution. From the older point of view urea was considered the one normal end point in the chain of katabolic reactions, and the other nitrogenous bodies found in the urine, such as uric acid and creatinine, were looked upon as substances which in some way had accidentally escaped the fate due them. This view is doubtless incorrect, as we have good reason to believe that uric acid is not a step in the ordinary protein metabolism, but is a derivative of certain substances only, which break down to a limited degree. The amount of uric acid which could be formed in this way would not be very large at most. In the metabolism of nitrogen, therefore, a number of normal end products must be considered and these will be discussed in the next few pages.

The question of the fundamental changes in protein before the recognizable end products are reached is one in which there has been a great deal of discussion. In a general way Pflueger assumed that all protein actually katabolized must first be built up into a part of the living tissues, from the absorbed products of protein digestion. The cells of this living tissue must, therefore, undergo constant and far-reaching changes, since the body is able to dispose of some hundreds of grams daily of protein in forced feeding. The somewhat older theory of Voit assumes that the absorbed protein, in the form of complex molecules, from the intestinal tract, is carried along by the blood in dissolved or suspended condition to certain cells or tissues, and is then broken down through the influence of forces residing in, or emanating from, these tissues. This protein is described as circulating protein, and before destruction does not become an integral part of the actual tissues of the body. Both of these theories, following the older views of the conditions under which protein is absorbed after digestion, assume that only the highly complex protein structures are capable of beginning the katabolic change. But in late years the facts brought out by the investigations of Cohnheim, Abderhalden and others, on the fate of protein in the digestive operations, have suggested very different views regarding the general course of this nitrogenous metabolism. It appears probable that the greater part of the protein of the food, broken down, as it largely is, into the component amino acid complexes, and absorbed as such, may not be built up again into structures like the original, but may be at once hydrolyzed and oxidized. A nitrogenous fraction may be separated in the form of ammonia by a hydrolytic cleavage, to be further converted into urea, while the residue, rich in carbon and hydrogen, would suffer ultimate oxidation like a fat or sugar.

This general view, which has found expression notably by Cohnheim and Folin, does not call for the building up of great masses of tissue protein, or even for the circulating protein of Voit. There is reconstruction of protein only insofar as it is needed for the repair of wasted or worn out tissues, and of the extent of this we know but little. It is probable that the protein of the tissues, in its final katabolism may yield some products different from those produced in the hydrolysis of the simply absorbed complexes. A study of the urine gives us some ideas on this subject, which will appear in what follows.

It will be well to begin with the consideration of the urine as a whole, as all these substances are eliminated through that channel.

THE GENERAL COMPOSITION OF URINE.

The work of the kidneys in the discharge of the urine, or more properly the separation of its constituents from the blood, is usually spoken of as one of excretion. But something more than simple elimination of worthless products is here concerned; the work done by these organs is in part secretory, as certain synthetic reactions are beyond question carried out here. Years ago Bunge and Schmiedeberg demonstrated the synthesis of hippuric acid from benzoic acid and glycocoll in the kidney, and since then other changes have been brought to light. Further than this, the peculiar mechanism of the kidney accomplishes another very remarkable thing. The blood circulating through the kidney contains valuable material to be saved as well as worthless substances to be rejected. Toward all these constituents the epithelial cells of the kidney tubules exercise a sort of selective treatment. The proteins, which are colloids, are retained by the blood, but the sugar, which is a crystalloid, and very soluble, is retained also unless its concentration passes a certain limit. The soluble salts are in part passed through the kidneys and in part retained by the blood, with the final result of maintaining a very nearly constant osmotic pressure in that fluid. How this is done we cannot say. It is indeed a problem of physiology and histology rather than of chemistry. We know only this, that the selective absorption and control of the blood concentration are perfectly automatic. When the osmotic pressure of certain constituents is increased beyond a pretty definite limit, the filtering mechanism in the kidney for those constituents becomes active and the excess is allowed to pass. The simple laws of diffusion and osmotic pressure do not help us greatly in explaining the actions of the kidneys where the flow of excreted substances is usually from a level of low concentration to one of higher. Attempts have been made to compare the separating medium between the urine and the blood to a semipermeable membrane, but the comparison is very imperfect unless the degree of impermeability be specially limited for each substance passing from the blood to the urine. The limitation would have to account for a concentration of salt from about 0.6 per cent in the blood to over 1.0 per cent in the urine, while for urea the concentration would change from about 0.05 per cent or lower to over 2.0 per cent, that is, forty fold. Limitations as wide as these render the comparison of little practical service.

Percentage Variations. It is not possible to speak of the mean strength of normal urine since the variations are extremely irregular, depending in health on a great many factors. The volume excreted daily, as stated in the books, is usually given much too high for the conditions obtaining in the United States. In place of the 1,500 cc. as found in most of the foreign works we should take 1,150 to 1,200 cc. as nearer the average excretion for 24 hours. In some hundreds of examinations made by the writer in the last few years on people of both sexes engaged in various occupations the average volume comes within these limits.

A number of complete analyses of urine are found in the literature, but in most of them the uric acid content is placed too low because of the faulty methods of determination formerly employed. In the following table are given some results obtained in the author's laboratory in which the recognized sources of error have been avoided as far as possible. It expresses the mean values obtained in the analysis of the urine of six well nourished men. The daily excretion is taken as 1200 cc., with a specific gravity of 1.023, at 20° referred to water at 4° as 1.000. In grams per 24 hours we have:

Potassium, K	2.82
Sodium, Na	4.87
Calcium, Ca	0.13
Magnesium, Mg	0.15
Ammonium, NH_4	1.13
Chlorine, Cl	8.90
Phosphoric acid, $(PO_4)'''$	2.41
Sulphuric acid, $(SO_4)''$	2.73
Urea, CON_2H_4	33.72
Uric acid, $(C_5H_3N_4O_3)''$	0.88
Creatinine, $C_4H_7N_3O$	1.98
Hippuric acid, $(C_9H_8NO_3)'$	1.00

These figures are merely suggestive, as diet makes, naturally, a great change in the excretion.

Color. In health the straw-yellow color of the urine is characteristic, the depth of shade depending largely on the concentration. With the same solid excretion in 24 hours the color may be light if the volume of water consumed is large, or it may be a deep yellow if the water consumption is deficient. These facts must be kept in mind.

Various darker shades of the urine may be observed after consumption of certain foods or certain chemical substances. Rhubarb, senna, santonin, salicylates and many other aromatic bodies produce highly colored urines. In some cases a marked smoky shade is observed, and this is usually due to the oxidation of more or less complex phenols. With a number of fruits and berries a bright yellowish or yellowish-red color is noticed in the urine.

In diseases the urine may be colored from the presence of substances from the blood, the bile, or from absorbed products of intestinal putrefaction.

Odor. The odor of urine in health is aromatic and absolutely characteristic. On standing it usually changes rapidly from the action of bacteria, and then an ammoniacal odor is ordinarily developed, through the alteration of the urea. Later, other organic matters begin to break down, resulting in the development of putrefactive or other disagreeable odors.

THE URINE.

Certain remedies impart very peculiar odors to the urine, and the same is true of several vegetable foods. The behavior of asparagus and turpentine in this regard is marked.

In disease a great variety of organic substances may be carried into the urine in traces, and the presence of these is often accompanied by some peculiar odor. This may be marked enough to be of importance in diagnosis.

Reaction. The urine for the 24 hours is normally acid to litmus paper. This acidity is due ordinarily, to the presence of acid salts, rather than of free acid; among the acid salts the di-hydrogen sodium phosphate is probably the most important.

Under normal conditions the urine may become temporarily alkaline, usually from the elimination of traces of alkali carbonates due to the combustion of certain organic salts of the diet. This occasional alkalinity must not be confounded with that which is very commonly observed in urine which has been passed some time. In this case the alkaline reaction is due to the presence of ammonium carbonate coming from the bacterial decomposition of the normal urea.

In the practical examination of urine litmus papers are commonly used in preference to other indicators. The measurement of the degree of acidity is uncertain. Occasionally urine shows the so-called amphoteric reaction; that is, it turns blue litmus paper red, and red litmus paper blue. Very sensitive paper is necessary to show this.

The Excretion of Alkali Salts. The alkali salts found in the urine come from the sodium chloride consumed as such in salted food, and in part from potassium salts in the juices of meat and in vegetables. In the analysis of the ash of muscle given some pages back chlorine as well as potassium is shown. Chlorine is found, although usually in small amount, in the ash of all vegetable substances. In the latter, however, especially in the cereals, potassium phosphate is the characteristic constituent of the ash. On a cereal diet we should expect the urine, in consequence, to show a relatively high potash and phosphoric acid content. The ash of potatoes contains in the mean over 60 per cent of potassium oxide while the chlorine is in excess of the sodium. With a mixed diet, therefore, the composition of the alkali salts in the urine must be variable and difficult of explanation. As the alkali compounds are practically all soluble, they are excreted almost solely by the urine and to a small extent only by the feces. The analysis of the urine gives us then, in ordinary cases, a fairly accurate measure of the alkali metals taken in with our food and drink; in normal condition there is no accumulation of alkali salts in the body.

Calcium and Magnesium Compounds. The full significance of these in the urine we can not explain, since without complete analyses of the feces we do not know the relation of the excreted to the ingested alkali-earths. Our natural waters contain usually appreciable amounts of these salts, with those of calcium in excess as a rule. In Lake Michigan water, for example, we have about 125 milligrams per liter of these salts as carbonates, but in our common animal and vegetable foods we consume daily much greater quantities than we could get from water. The ash of wheat contains about 12 per cent of magnesia and 3 per cent of lime, while in the ash of muscle we have over 3 per cent of magnesia and between 2 and 3 per cent of lime. Five hundred grams of lean meat would furnish us then with over 150 milligrams of magnesia and with something less than that amount of lime.

But only fractions of these compounds find their way into the urine. In the original foods they exist, in part at least, in insoluble forms. While some of these substances may be dissolved in the stomach, the conditions are reversed in the intestines, and insoluble phosphates, carbonates and sulphates are lost with the feces. There has been much discussion as to the exact nature of the calcium and magnesium salts excreted. In a measure the discussion is fruitless, as we must certainly admit the free exchange of ions in solution. Under ordinary conditions the acid ions of the urine appear to be in slight excess of the metals, which prevents precipitation of insoluble phosphates, for example. Temperature plays a very important part in the problem of the stability of the calcium and magnesium compounds in the urine, and the problem is further complicated by the presence of uric acid, the peculiar behavior of which will be touched upon below.

THE NITROGEN EXCRETION.

For many reasons this excretion is the most important which we have to consider in connection with the urine, as it gives us an insight into some of the fundamental problems in metabolism. The largest part of it leaves the body as urea, but the proportion excreted in other compounds cannot be neglected. We have pretty accurate methods for the estimation of urea, ammonia, uric acid, creatinine and purine nitrogen as they are found in the urine. Hippuric acid, which is found in urine, is not as readily measured, and for several other compounds which contain nitrogen our methods are far from exact. The following table shows the distribution of the nitrogen in the urine of six men on whose complete excretion daily tests were made in the author's laboratory through a period of four months. The figures are the mean values for the whole period, and are in percents of the total nitrogen excretion, as measured by the Kjeldahl process. The general mean represents 720 determinations for each constituent.

Under the head of undetermined nitrogen, shown in the table, there is included the nitrogen of hippuric acid, oxyproteic acid, alloxypro-

No.	Urea Nitrogen.	Ammonia Nitrogen.	Purine Nitrogen.	Uric Acid Nitrogen.	Creatinine Nitrogen.	Undetermined Nitrogen.
1	83.26	4.39	0.67	1.70	5.38	4.60
2	84.50	3.56	0.61	1.69	5.52	4.12
3	82.43	5.55	0.36	1.63	5.64	4.39
4	85.05	4.56	0.41	1.23	5.50	3.25
5	81.46	4.71	0.61	1.94	6.29	4.99
6	84.17	4.26	0.51	1.69	4.94	4.43
Mean.	83.48	4.50	0.53	1.65	5.54	4.30

teic acid and traces of other bodies of obscure composition. A brief discussion of each one of the important constituents will follow.

UREA.

The relation of this substance to ammonium carbonate has been referred to many times, but especially in discussing the enzymic processes of the liver. The nutrient proteins contain many amino groups which seem to be split off in the general combustion or hydrolytic processes going on in the body; also a great excess of groups which oxidize more completely and yield carbon dioxide. The large part of this escapes by way of the lungs, while another part is evidently taken care of in the liver through combination with the amino groups to form urea. It is also true that normally some of this amino nitrogen fails to take this simple course, because of the presence of strong acid radicles, which have great tenacity in their combining reactions. The ammonium salts so formed are stable and cannot be worked over into urea.

It appears, also, that the nitrogen of some other groups in addition fails to reach the urea stage. Creatinine and uric acid nitrogen are not included here, as these substances seem to have an independent origin which will be discussed below. But there are obscure compounds in the urine in small amount of which we know but little, and some of these contain nitrogen. The *oxyproteic* acid referred to above is an illustration. What the relation of this is to urea we cannot say, but an idea of this kind suggests itself: the original protein complex may contain certain groups which do not fall an easy prey to the work of the oxidation enzymes in the body; they do not break down to amino compounds and carbon dioxide, but remain intact as very resistant residues, and hence when the liver is reached they are not in condition to pass into the urea stage. In the katabolic changes of protein it is possible that a number of such resistant groups may be produced, and it is likely that the amount of nitrogen or other element which so escapes the normal end reaction depends largely on the strength of the enzymic functions. These must vary in different individuals, and hence sometimes more and sometimes less of these resistant, or left over, residues will find their way into the urine.

From this point of view urea represents that part of the original body nitrogen, aside from the creatine and nuclein derivatives, which takes the normal course. It represents no store of practically realizable energy, while with some of the other bodies which escape in the urine this is not the case; under more favorable conditions they might be expected to suffer further oxidation with liberation of more heat. Such ideal conditions are realized in some individuals more than in others.

Urea may be built up outside of the body by many synthetic processes, but is most easily prepared by the conversion of ammonium cyanate, NH_4OCN, into the isomer. On evaporation of a solution of this salt the transformation into urea is complete. Urea is very soluble in water, from which it may be obtained easily in crystalline form. Its solutions are easily decomposed by many oxidizing agents with formation of water, carbon dioxide and free nitrogen, on which behavior several of the processes for determining it are based. This change is brought about by hypochlorites, for example, in this manner:

$$CON_2H_4 + 3NaOCl = 3NaCl + 2H_2O + CO_2 + N_2.$$

The amino groups in urea may be completely converted into ammonia in many ways, and this reaction, also, is applied in estimating urea, as will be shown in the next chapter.

On the other hand, urea may take part in synthetic reactions and may be combined to form complex substances, in certain cases, as will be shown below.

AMMONIA.

This represents a portion of the protein disintegration which for a number of reasons has not been converted into urea. The ammonia passing into the urine takes that course ordinarily through combination with mineral or other acids, which are not destroyed, or may not be destroyed, by oxidation. In any pathological increase of such acids, if there is not enough fixed alkali in the blood to combine with them, ammonia is split off from protein derivatives in quantity sufficient to complete the neutralization. This may be shown also by the injection of free mineral acids either directly or with the food; an increased elimination of ammonia results. It should be expected, therefore, that the proportion of ammonia in the urine would be subject to marked fluctuations, which is indeed the case. Taken with other determinations the estimation of ammonia may possess considerable diagnostic value, as it measures to some extent the excessive acid excretion. In advanced stages of diabetes, with marked elimination of acid, the ammonia content of the urine may increase to several grams daily. The normal amount is usually a gram or less.

Ammonia must be determined in fresh urine only, since in old urine fermentation changes soon produce large quantities of the substance from the breaking down of urea.

URIC ACID AND THE PURINE BODIES.

Few topics in physiological chemistry have attracted more attention than the relations of uric acid to other nitrogenous products excreted in the urine, and its behavior in relation to disease. The importance of the substance in this point of view has undoubtedly been very frequently over-estimated and even at the present time clinicians are much divided as to the part it plays in certain diseases. This much may be said with truth, however, that many of the fine-spun theories which have been advanced by medical men on the uric acid question, and which have held our attention for a longer or shorter period, have been founded on very weak *chemical* evidence, and this, it should be mentioned, is the real factor in the case.

Under the older view, as explained already, uric acid was supposed to be but a step in the formation of urea, the normal end product in protein metabolism, and numerous disorders were attributed to the accumulation of uric acid in the blood through some failure in the final oxidation processes. But it appears now from the evidence available that uric acid is *not* a natural step in the oxidation of the simple proteins; it does result, however, from the breaking down of the complex nucleo-proteids which are represented to a limited extent only in the body, as compared with the muscle proteins, for example. The glandular organs rich in cells furnish the chief amount of the nuclein complexes. In the katabolism of these, true proteins and the residues rich in phosphorus known as nucleic acids result; the proteins undergo the usual further oxidation probably, while the nucleic acids break down into a variety of products of which the purine bases, the pyrimidine bases, phosphoric acid and carbohydrate groups are the most important. The purine bodies in turn doubtless give rise to uric acid. As pointed out in Chapter V several nucleic acids exist; their structural formulas are not known, but empirically these formulas have been given to acids from different sources.

$C_{40}H_{32}N_{14}O_{22}P_4$	Salmon milt
$C_{40}H_{54}N_{14}O_{24}P_4$	Salmon milt
$C_{20}H_{43}N_{14}O_{26}P_4$	Yeast cells
$C_{41}H_{61}N_{16}O_{21}P_4$	Wheat embryo

The cleavage products of these acids are not constant, since from different acids different purine bases have been made. Those found in the animal body are the following: *xanthine, hypoxanthine, guanine, adenine, heteroxanthine, paraxanthine* and *epiguanine*. In order to show the relations of these compounds to uric acid, E. Fischer proposed to consider them all as derivatives of a nucleus group which he called *purine*.

As the chemistry of these bodies is complex it may be well to illustrate their relations by the structural formulas worked out or confirmed by Fischer. Starting with the assumed purine nucleus we have these formulas, with the nucleus atoms numbered, as suggested by Fischer:

```
 1 N—C 6              N=CH                HN—CO
   |   |   7           |  | H             |   |
 2 C 5 C—N             HC C—N             OC C—NH
   |   |   \C 8        ||  ||  \CH        |   ||  \CH
 3 N—C—N  /            N—C—N /            HN—C—N /
       4 9
Purine nucleus, C₅N₄   Purine, C₅H₄N₄     Xanthine, C₅H₄N₄O₂


 HN—CO                N=C—NH₂             HN—CO
 |   |                |   |               |   |
 OC C—NH              HC C—N              HC C—NH
 |   ||  \CO          ||  ||  \CH         ||  ||  \CH
 HN—C—NH              N—C—N—H             N—C—N /
Uric acid, C₅H₄N₄O₃   Adenine, C₅H₅N₅     Hypoxanthine, C₅H₄N₄O
```

Employing the Fischer nomenclature these bodies have the following names:

Adenine	6-aminopurine
Hypoxanthine	6-oxypurine
Xanthine	2, 6-dioxypurine
Uric acid	2, 6, 8-trioxypurine
Guanine	2-amino-6-oxypurine

As their relations have been shown by various syntheses and other transformations, and as further, the xanthine and hypoxanthine, adenine and guanine have been directly derived from the nucleic acids, the relation of uric acid to the latter bodies is not far to seek.

Not all of the nucleic acid destroyed can be assumed to come from body cell structures; many of our foods contain nucleins and these must give rise to the same derivatives on oxidation without passing through, becoming part of, the cells of the glandular organs of the body. Accordingly we distinguish between *endogenous* and *exogenous* purines and uric acid. With the food nucleins eliminated as far as possible, it has been found that the uric acid excreted becomes nearly constant and bears a more uniform relation to the urea. This indicates that the destruction of cell substance in the body leads as regularly to uric acid as does that of muscle proteins to urea. The use of rich protein foods does not necessarily occasion greater elimination of uric acid. It is only when they contain appreciable amounts of the nucleins that this is the case. In addition to these facts it has been found experimentally that the oxidation of nucleins outside the body leads to the production of uric acid in small amount.

Uric acid may be obtained synthetically by combining urea with glycocoll, and at a high temperature it may be decomposed with production of urea, ammonia, prussic acid and other bodies, under different conditions. But little importance is attached to these facts at the present time, but formerly they were supposed to support the view that uric acid is a stage in the urea formation through which all the katabolic nitrogen should pass. Of greater interest is this fact that when uric acid is introduced into the circulation of certain animals some of it appears to be destroyed, and with the production of a little urea. Such observations suggest that possibly a small part of our urea *may* come from uric acid, but they have no bearing on the proposition that the acid in turn has its origin in the nucleins and not in the common proteins.

According to the structural formula above given uric acid appears to have four hydrogen atoms of equal value in the formation of salts. But apparently only two classes of salts may be formed: neutral salts, in which two hydrogens are replaced, and acid salts, in which but one hydrogen is replaced. We have therefore salts of the types $MC_5H_3N_4O_3$ and $M_2C_5H_2N_4O_3$. In addition to these, so-called *quadriurates* are known as urine sediments. These salts are of the type $MC_5H_3N_4O_3 \cdot C_5H_4N_4O_3$. The pure acid requires nearly 40,000 parts of water for solution; the neutral salts of the alkali metals are much more soluble; while the acid salts are but slightly soluble. The data given by different observers are very contradictory. The salts of barium, strontium and magnesium are nearly insoluble in water.

Solutions of urates in presence of alkali exhibit a reducing action toward copper, silver and certain other salts, which fact possesses an importance which will be explained later.

XANTHINE AND THE OTHER PURINES.

The amount of these purines not changed to uric acid is small, the sum of all of the nitrogen so held not being over one-third of the uric acid nitrogen, according to most observers. They are but slightly soluble in pure water, but dissolve readily with alkali hydroxides to form compounds like the urates. With the heavy metals the combinations are mostly insoluble.

PYRIMIDINE DERIVATIVES.

Among the decomposition products of nucleic acids the so-called pyrimidines should be referred to, as they must bear some relation to the urinary nitrogen. Pyrimidine itself represents a nucleus like purine from which various derivatives may be formed. The relations as outlined by Kossel are these:

```
 1 N=C—H 6         NH—C=O           NH—CO
   |   |            |   |            |   |
 2 CH  C—H 5       C=O  C—H         C=O  C—CH₃
   ||  ||           |   ||           |   |
 3 N—C—H 4         NH—C—H           NH—CH
 Pyrimidine, C₄H₄N₂   Uracil, C₄H₄N₂O₂    Thymine, C₅H₆N₂O₂
                    2, 6-dioxy pyrimidine   5-methyl uracil
```

We have no definite knowledge of the occurrence of these bodies in urine, but they are among the compounds which should be expected to result from the nuclein metabolism and possibly form a part of the nitrogen residue not fully accounted for. Thymine has been obtained as a well-crystallized compound, soluble readily in warm water. It is identical with the body formerly described as nucleosin.

CREATININE.

Creatinine, $C_4H_7N_3O$, is the anhydride of the creatine described under the head of the muscle extractives, and is always present in normal urine. The amount in which it occurs is indicated by the analyses given above. Much has been written on the question of the relation of creatine to urea, but the evidence for the view held by some writers that the latter is derived from the former is not very convincing. In a laboratory way by boiling creatine with baryta water, urea, sarkosine and several other things result. But no corresponding reaction appears to take place in transfusion experiments with creatine, and it has been held until recently that creatine introduced with the food appears in the urine, not as urea, but as creatinine.

Of this transformation there is now considerable doubt, and we have at present no very satisfactory theory concerning the origin of the urinary creatinine. Much of our information on this subject has come from the investigations of Folin, who pointed out a few years ago that the elimination of creatinine is fairly constant for any given individual from day to day, but varies in different individuals. The total elimination seems to be independent of muscular exertion, and varies with the body weight, amounting to about 16 to 20 milligrams per kilo daily. The excretion is not essentially changed by a high protein diet, and is not lowered by fasting. It may be, then, the measure of some peculiar phase of nitrogenous metabolism characteristic for each individual.

The conversion of creatine into creatinine by boiling with weak acids follows quantitatively according to the following reaction, but the change is not rapid, as formerly assumed.

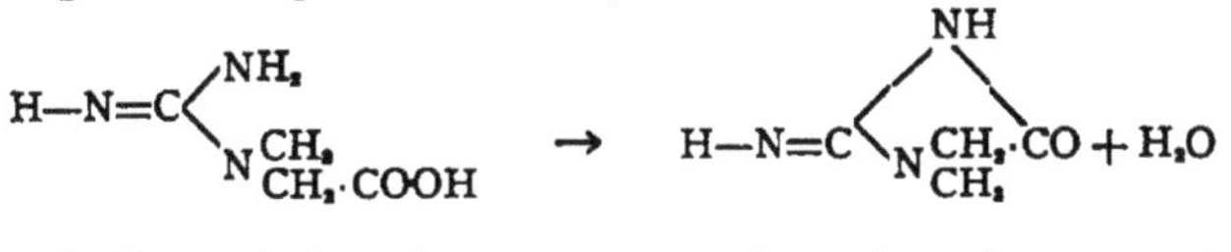

$$H-N=C\diagdown_{N}^{NH_2} {}_{CH_3}^{CH_2 \cdot COOH} \rightarrow H-N=C\diagdown_{N}^{NH} {}_{CH_3}^{CH_2 \cdot CO} + H_2O$$

But in alkaline solution the reverse reaction takes place, creatine being slowly formed. Creatinine may be separated from the urine most easily with zinc chloride as a crystalline double salt which is yellow because of the coprecipitation of pigments; when purified the crystals are colorless. Creatinine is found in normal amount in the urine of vegetarians. Toward copper and some other metallic solutions it behaves as a reducing agent when alkali is present.

OTHER NITROGENOUS BODIES.

It has been intimated that other compounds of nitrogen, besides those just mentioned are excreted with the urine. Of the nature and amount of these much remains to be shown by experiment. It was pointed out that 4 to 5 per cent of the total urine nitrogen in the table given above was "undetermined." It did not belong to the urea, uric acid or other purines, ammonia or creatinine. Among the bodies which contain some of this nitrogen the following may be mentioned as the most important.

HIPPURIC ACID. This is benzoyl glycine, $C_6H_5CO \cdot NHCH_2COOH$, and is always present in some amount in normal urine. It is formed by synthesis from the glycine group of metabolism, and the benzoic acid ingested with many aromatic food substances, certain acid fruits and spices, for example. Some aromatic products of protein metabolism may also give rise to it. Moderate doses of benzoates, or benzoic acid, appear in the urine wholly in the form of hippuric acid, as the glycine groups seem to be always present in sufficient amount to complete the synthesis. These groups, otherwise, go to form urea, apparently. The hippuric acid content of the urine is far from constant, as it depends so largely on the diet, but a gram or more daily is often present. The nitrogen in the acid amounts to 7.8 per cent.

OXYPROTEIC ACID. This product was first described by Bondzynski and Gottlieb, and has been studied by others. The exact formula is not known, but the percentage composition is about C, 39.62; H, 5.64; N, 18.08; S, 1.12; O, 35.54. Under the names of *antoxyproteic acid* and *alloxyproteic acid* Bondzynski and other colleagues have described somewhat related substances. As these, and still other bodies which have been found in urine contain both sulphur and nitrogen, it is evident that they represent a peculiar small fraction of the original protein which has failed to be hydrolyzed and oxidized in the general metabolism. It has been estimated that 2 or 3 per cent of the total urine nitrogen is found in these compounds, and the sulphur in them would be included in the so-called "neutral" sulphur, to be referred to later.

PATHOLOGICALLY the relations in the nitrogen excretion may be very much changed, and protein, as such, may occur in the urine in considerable amount. The detection of this protein will be discussed in the following chapter, along with the methods for the measurement of the normal constituents.

THE SULPHUR EXCRETION.

The sulphur in the urine is found in several compounds and comes from different sources. A small part has its origin in traces of sulphates taken directly in food and natural waters; some comes from the sulphur existing in peculiar combinations in certain vegetables, while the largest part has its origin in the sulphur of proteins, which undergoes more or less complete oxidation before elimination through the kidneys. Most of this sulphuric acid of oxidation combines with alkalies for elimination; if fixed alkalies are deficient ammonium sulphate is formed and this ammonia therefore escapes the natural oxidation to urea.

It has been explained in an earlier chapter that in the putrefactive changes taking place in the intestines certain phenol bodies are split off from proteins there remaining, or perhaps in most cases from the unabsorbed protein derivatives. A very considerable part of these phenol bodies escapes with the feces, but another portion, often much increased in disease, is absorbed by the blood vessels from the lower intestine and carried to the liver where combination with sulphuric acid is effected, probably through some kind of enzymic action. From recent observations it appears likely that the combination is effected before the sulphur is completely oxidized, that is, while it is in the sulphite condition. The final oxidation then follows. The ethereal sulphate so formed is discharged finally with the urine. The fraction of the sulphuric acid so voided is extremely variable, reaching sometimes 20 per cent of the whole. The most abundant of these combinations are salts of phenyl-, cresyl-, indoxyl-, and skatoxyl-sulphuric acid, the structural relations of which are shown by the following formulas:

$\begin{matrix} C_6H_5O \\ HO \end{matrix} \} SO_2$. $\begin{matrix} CH_3C_6H_4O \\ HO \end{matrix} \} SO_2$.

Phenyl-sulphuric acid Cresyl-sulphuric acid Indoxyl-sulphuric acid

Skatoxyl-sulphuric acid is the methyl derivative of indoxyl-sulphuric acid. Phenyl- and cresyl-sulphuric acids are frequently called phenol- and cresol-sulphuric acids, but the former names are preferable. The alkali salts of indoxyl-sulphuric acid are known as indican, the appearance of which in the urine in quantities above minute traces is an indication of the existence of excessive putrefactive reactions in the intestines. The observation is therefore of value in diagnosis. Accurate methods are known for the recognition of the substance in the urine.

In addition to the oxidized or sulphate sulphur in the urine there are always traces of so-called "neutral" sulphur compounds present. According to various authorities the sulphur in these compounds, which are all complex organic bodies, may make up 12 to 25 per cent of the total sulphur. The sulphur bodies here in question are not easily recognized in most cases and the existence of several of them described in the journals is yet to be demonstrated. The fact of the excess of sulphur over that contained in the mineral and ethereal sulphates may be easily shown by determination of the total sulphates directly, and then after complete oxidation of the urine by sodium peroxide, or other agent. This unoxidized or neutral sulphur is found in such bodies as the oxyproteic acid mentioned above, and occasionally in cystin or taurin present.

The following table gives a good idea of the sulphur excretion

No.	Inorganic Sulphur.	Ethereal Sulphur.	Neutral Sulphur.
1	72.30	9.71	17.98
2	74.96	9.42	15.62
3	75.93	8.48	15.59
4	78.13	6.55	15.32
5	71.84	9.13	19.03
6	76.80	6.39	16.81
Mean.	74.99	8.28	16.73

through a long period, in the urine of six men in normal health. The values represent fractions of the total sulphur, and in each case are the means of daily observations on the whole urine, through four months. The observations were made in the author's laboratory.

THE PHOSPHORUS EXCRETION.

As mentioned in an earlier chapter, some of our foods are very rich in phosphates, or phosphate-furnishing material. This is especially true of the cereals, of a few animal proteins and of the lecithins. The phosphates formed by oxidation of these substances are mainly eliminated with the urine, and in traces with the feces. A very considerable portion of the urinary phosphoric acid comes from this source, which may be described as the exogenous phosphoric acid. Another portion comes from the breaking down of the nucleic acids of the cell substances and from the so-called phospho-globulins or nucleo-albumins also found in the body. These are the *endogenous* sources. How the phosphorus is held in these last-named bodies is not known, but in the nucleic acids it appears to be in oxidized form, according to some authorities as metaphosphoric acid. The excreted product is orthophosphoric acid, combined to form salts of the type MH_2PO_4 or M_2HPO_4. The alkali salts are soluble readily, while those of calcium and magnesium are only in part soluble in water. The conditions of solubility in urine are complicated by the presence of other salts.

The amount of phosphoric acid, as P_2O_5, excreted daily varies within wide limits; according to the above analysis the mean may be about 2.5 gm. If the urine becomes alkaline through the fermentation of urea a very considerable part of the phosphate may be precipitated in the form of calcium and magnesium salts. One of the commonest of these is the so-called *triple phosphate*, $NH_4MgPO_4 \cdot 6H_2O$. The determination of the amount of phosphoric acid in the urine will be discussed in the next chapter.

CARBOHYDRATES.

Normally no large amount of any carbohydrate passes from the blood into the urine, but with increased sugar concentration in the blood from any cause, the urine may contain sugar in more than traces. When sugars are consumed in more than the "assimilation limit" a part of the excess may be found in the urine. This is true of glucose as well as of cane sugar and maltose.

But under normal conditions there is evidence that traces of several carbohydrates and of bodies related to them pass also into the urine. It is well known that the α-naphthol reaction is given with ordinary normal urines, and this points to some carbohydrate derivative which can yield furfuraldehyde. The other delicate sugar tests give the same indication. Some of the carbohydrate doubtless appears as such, but a portion of it probably exists in the form of glucoproteids or other complex groups. Part of the carbohydrate can be readily fermented, while another portion resists the action of yeast. Such observations may be interpreted as suggesting the presence of other carbohydrates than the common glucose. More will be said about this below.

Pathologically, the passage of even large quantities of sugar into the urine is a common phenomenon. This is most frequently observed in *diabetes mellitus,* a disease in which the power of oxidizing sugar in the muscles seems to be wholly or partly lost. It has been already mentioned that this oxidation is possibly an enzymic operation in which certain muscle and pancreas enzymes are at the same time active. In this form of diabetes several hundred grams daily of sugar may be excreted.

In another, "artificial," form of diabetes sugar may be caused to appear in considerable quantity temporarily in the urine. The administration of small doses of phloridzin, a glucoside, is followed by the appearance of sugar in the urine, but this condition is now believed to depend on a disturbance of the proper functions of the kidneys rather than on any alteration of the sugar-oxidizing power. The power of retaining the sugar in the blood from which it should be taken up and oxidized by the muscular tissues depends on the maintenance of the integrity of the membranous structures of the kidneys. If these suffer strain, which seems to be the case after administration of phloridzin, a little sugar more than the normal may be able to diffuse or pass through in some manner. It is possible that many other substances may have a somewhat similar action, but in much smaller degree, and so occasion sugar excretions still within what may be called "normal" limits.

PROTEINS.

As in the case of the sugars, so with the proteins; there is good evidence that traces of these nitrogen compounds are normally present in the urine. The amount which may be present is, however, very small, in the mean not over 40 to 50 milligrams per liter. With these traces it is not possible to say how many different kinds of proteins may occur, but the serum albumin is doubtless the most abundant. With greatly increased amounts of albumin in the diet the excreted albumin is also increased, with no visible impairment of the kidney mechanism.

It is also true that foreign albumins injected into the blood circulation, white of egg, for example, are speedily eliminated by the kidneys.

Pathologically, however, the serum albumin and serum globulin of the blood may pass through the kidneys in considerable quantity. This is occasionally due to disturbances in the circulation which may bring about an increase of blood pressure, but ordinarily is due to structural changes in the kidneys themselves, through which the power of perfectly retaining albumins of the blood is lost. A discussion of the nature of these changes is not within the scope of this book, and for an explanation of the tests which are employed in recognizing the proteins of the urine the reader is referred to the next chapter. Many of these tests are but modifications of the delicate protein tests described in one of the earlier chapters.

URINARY SEDIMENTS.

The urine when passed is usually clear, but frequently it soon becomes cloudy and deposits a precipitate. This precipitate may consist of a variety of constituents and may be due to several causes. Sooner or later all urines, unless specially protected by preservatives, undergo the so-called ammoniacal fermentation in which urea is converted into ammonium carbonate by certain bacteria. When this alkaline condition is reached the condition of equilibrium in which the various salts exist together is destroyed and insoluble products are commonly formed which appear as precipitates. The alkali-earth phosphates are the bodies usually thrown out in this way.

But disturbances in the equilibrium may result from changes of temperature also, and precipitation occur because of the simple cooling of the warm voided urine. Many of the peculiar urate sediments are formed in this way. Sometimes the separation of sediment begins in the bladder or ureters and this sediment may take the form of hard concretions or calculi. In years past numerous attempts have been made to explain the formation of these deposits, especially as they occur within the body. The theoretical explanations given have been in general far from satisfactory, and the most recent studies have only gone to show the complexity of the problem. It is now coming to be recognized that we have here one of the most difficult problems of physical chemistry, which like other questions of chemical equilibrium in solution may be approached only through elaborate studies. The beginning of such studies may be seen in some of the valuable papers which have been published in the last few years on the solubility of uric acid and its salts, and the degrees of dissociation which obtain in the various solutions. As yet these matters are scarcely in condition for elementary presentation.

The recognition of the general character of the sediments from urine is most readily effected by the aid of the microscope, which is explained in the next chapter.

THE REDUCING POWER OF NORMAL URINE.

On account of the presence of some of the bodies described in the last few pages, normal urine exhibits a certain reducing action toward metallic solutions. At one time this was ascribed to traces of carbohydrates present, but later doubt was thrown on this conclusion and the presence of even traces of sugar-like bodies was denied by most writers concerned with the question. With the development of greater accuracy in the methods of detecting sugars, especially through the aid of the phenylhydrazine combination and the formation of benzoic esters, the question of the passage of traces of carbohydrates into the urine seems to be finally settled in the affirmative, but the amount of sugar which may be so identified is too small to account for the total reduction easily measured by copper solutions. This normal reduction cannot be quantitatively followed with the ordinary Fehling solution, but if a dilute Pavy solution, as described in Chapter III, is used, a very satisfactory determination may be made. This modified Pavy solution is given such a strength that 1 cc. oxidizes 1 mg. of glucose in very dilute solution, approximating 0.2 per cent strength or less.

In a large number of tests made in the author's laboratory a few years ago it was found that to reduce 50 cc. of such a solution urine volumes ranging between 14.9 cc. and 58 cc. were required. The mean of all the determinations on these normal urines was 23 cc. In hundreds of normal urines examined since similar results have been obtained. In all these urines the creatinine and uric acid present were accurately determined and an attempt was made to connect these bodies with the reduction.

The Reducing Power of Creatinine. The reducing power of creatinine may be easily found by the method referred to, using the weak ammoniacal copper solution. A liter of this solution contains 8.166 gm. of $CuSO_4 \cdot 5H_2O$, corresponding to 2.6042 gm. of CuO. A measured volume of this, usually 25 or 50 cc., is reduced by a creatinine solution of definite strength and the volume required noted. The details of some experiments are given in the following table. The copper solution was diluted to 100 cc. before the titration:

Creatinine in 100 cc.	Copper Solution Taken	CuO Equivalent.	Creatinine Solution Used.	Creatinine to 130.2 mg. CuO.	Mols. CuO to 1 Mol. $C_4H_7N_3O$.
50 mg.	25 cc.	65.1 mg.	92.5 cc.	92.5 mg.	1.998
50	25	65.1	94.0	94.0	1.967
120	50	130.2	76.0	91.2	2.026
120	50	130.2	77.0	92.4	2.000
				Mean,	1.998

It appears, therefore, that in this kind of solution 2 molecules of copper oxide are required to oxidize 1 molecule of creatinine. The 2 molecules of copper oxide yield 1 atom of oxygen. It will be recalled that 5 molecules of copper oxide are used up in oxidizing 1 molecule of dextrose. For equal weights the reducing power of dextrose is about twice as great as is that of creatinine.

Reducing Power of Uric Acid. With the same reagent the reducing power of uric acid may be measured, the uric acid being dissolved with a little alkali to form a soluble urate. The table below illustrates the relation of the reducing and oxidizing solutions:

Uric Acid in 100 cc.	Copper Solution Taken.	CuO Equivalent.	Uric Acid Solution Used.	Uric Acid to 130.2 mg. CuO.	Mols. CuO to 1 Mol. $C_5H_4N_4O_3$.
80 mg.	15.3 cc.	39.8 mg.	36 cc.	94.2 mg.	2.92
80	25	65.1	58	92.8	2.96
120	50	130.2	76.8	91.8	2.99
120	25	65.1	37.8	90.8	3.03
					Mean, 2.98

From this it appears that 1 molecule of uric acid requires 3 molecules of copper oxide for oxidation under the conditions of the test, or 1.5 atoms of oxygen. This is a larger amount of oxygen than would be required for the oxidation of uric acid to urea and alloxan. This requires 1 atom:

$$C_5H_4N_4O_3 + O + H_2O = CON_2H_4 + C_4H_2N_2O_4.$$

But secondary reactions also take place, and a partial oxidation to parabanic acid may be represented in this way:

$$2C_5H_4N_4O_3 + 3O + 2H_2O = 2CON_2H_4 + C_4H_2N_2O_4 + C_3H_2N_2O_3 + CO_2.$$

This possibly represents the course of the reaction with such a solution.

With these reducing values established it is possible to estimate the fraction of the total urine reduction which is not due to these two most important substances besides sugar. To do this it is necessary to find the amount of uric acid and creatinine in a given volume of urine and determine its reducing value in terms of CuO with the same ammoniacal solution. If the reducing power of the creatinine and uric acid be subtracted from the total, the reduction due to glucose and other bodies is arrived at. To illustrate, in the examination of a large number of urines these values were found:

> Total reducing power of 1 cc. of urine in mg. of CuO = 6.204
> Reducing power of the creatinine present in same terms = 1.961
> Reducing power of the uric acid present in same terms = 0.935
> Reducing power of the sum of uric acid and creatinine = 2.896

Nearly one-half of the total reduction then corresponds to the action of these two nitrogen bodies, while the reduction of the other substances is equivalent to 3.308 mg. of CuO per cc. If this is calculated as glucose it represents 1.27 mg. per cc. As several other bodies contribute to this reducing action the value of any saccharine substance present must be still smaller. It has been found by earlier investigations that on concentration urine loses a part of its reducing power. This suggests that some volatile substances may be responsible for part of it. It should be mentioned in addition, that the relative extent of the reducing action varies with the reagent employed. Knapp's mercury solution has been used for the purpose but it is not as convenient as the ammoniacal copper solutions.

A knowledge of this normal reducing action is not without value, since very frequently the mistake has been made of assuming the presence of sugar in suspicious quantities in the urine merely from a reduction test. In some of the extreme cases quoted in one of the statements above, the normal reduction was equivalent to over 0.3 per cent of glucose, when in reality it was largely due to uric acid and creatinine. This point has importance in clinical observations. In another direction also the question is important, as the reducing power of the urine is a measure of the extent of certain kinds of excretion. Creatinine and uric acid are probably perfectly normal end products of metabolism, and when large in amount the reduction is high. It will be recalled further that, as mentioned in a former chapter, strongly reducing substances are produced in the autolytic changes taking place in the liver and pancreas. In cases then, this " normal " reduction may become excessive.

THE ELECTRICAL CONDUCTIVITY OF URINE.

In one of the chapters on the blood the nature of electrical conductivity in the fluids of the body was explained and the methods of determination outlined. As this conductivity depends mainly on the sum of the inorganic constituents present, and as sodium chloride is the most abundant of these, the determination in itself has but a limited importance. In some cases the value of the conductivity would be merely a rough measure of the salt consumed with the food, and the salt consumption is extremely variable.

Nearly all the other substances found in the urine have a significance very different from that of the salt. The latter is consumed and excreted as such, while the other important urinary constituents are products of metabolism, that is, of the breaking down of the digested and absorbed food materials. The organic products of metabolism are practically non-electrolytes or bodies with a very low conducting power; indeed the conductivity of a weak salt solution is materially lowered by the addition of urea and the effect of the purine bodies is practically in the same direction. Aside from the chlorides, the inorganic salts of the urine are mainly phosphates and sulphates of the alkali or alkali-earth metals, and these are made up largely from the oxidation of sulphur and phosphorus of protein foods. We consume a certain amount of phosphoric and sulphuric acids in complex organic combination, in the lecithins and chondroitins, for example, and small amounts of mineral sulphates and phosphates are also found in some of our foods, but these amounts are not large enough to vitiate the truth of the general proposition that the sulphuric and phosphoric acids as detected in the urine are results of certain kinds of metabolism. Now, the conductivity measures the combined effect of these products of oxidation and, if it could be determined apart from the effect of the chlorides, a factor of considerable practical importance would be secured.

Approximately, the conductivity due to metabolic products may be found by subtracting from the observed conductivity that due to sodium chloride in the same solution. The chlorine may be accurately determined and calculated as chloride; then from tables the conductivity for a solution of this concentration may be found and used as a correction to be taken from the total observed conductivity, leaving

A	B	Period.	A			B		
			Volume Passed, cc.	Specific Gravity $\frac{20°}{4°}$	Conductivity, κ.	Volume Passed, cc.	Specific Gravity $\frac{20°}{4°}$	Conductivity, κ.
Observations of first day; Excretion = 1,162 cc. Liquid consumption = 1,750 cc.	Excretion = 861 cc. Liquid consumption = 1,815 cc.	6– 9 A. M.	145	1.024	0.02589	112	1.022	0.02264
		9–12	178	1.023	0.03021	220	1.019	0.02519
		12– 3	135	1.024	0.02834	77	1.026	0.02226
		3– 6	118	1.027	0.02790	70	1.027	0.02614
		6– 9 P. M.	128	1.026	0.02694	198	1.021	0.02178
		9– 6	458	1.023	0.02421	184	1.026	0.02008
		Means, 3-hr. period		1.024	0.02650		1.024	0.02228
Observations of second day. Excretion = 1,265 cc. Liquid consumption = 2,100 cc.	Excretion = 1,131 cc. Liquid consumption = 2,565.	6– 9 A. M.	192	1.020	0.02645	410	1.007	0.01039
		9–12	245	1.020	0.02926	122	1.017	0.02408
		12– 3	155	1.026	0.02803	96	1.023	0.02422
		3– 6	113	1.026	0.02702	138	1.024	0.02355
		6– 9 P. M.	155	1.025	0.02702	180	1.021	0.02144
		9– 6	405	1.025	0.02122	285	1.024	0.02369
		Means, 3-hr. period		1.024	0.02518		1.020	0.02184
Observations of third day. Excretion = 1,324 cc. Liquid consumption = 1,670 cc.	Excretion = 1,099 cc. Liquid consumption = 1,330 cc.	6– 9 A. M.	230	1.022	0.02683	278	1.015	0.02629
		9–12	260	1.021	0.02939	158	1.022	0.03037
		12– 3	160	1.026	0.02792	92	1.029	0.02720
		3– 6	134	1.026	0.02755	71	1.030	0.02566
		6– 9 P. M.	146	1.025	0.02580	112	1.027	0.02577
		9– 6	394	1.023	0.01997	388	1.014	0.01492
		Means, 3-hr. period		1.024	0.02468		1.021	0.02251

the desired residual or metabolic conductivity. In this plan, however, an error is involved, because the conductivity of the chloride taken from the tables directly is that found in pure aqueous solution in absence of other salts, and is larger than the true conductivity of the chloride as it exists in the urine. It is necessary, therefore, to use as a correction the value of the salt conductivity, not in aqueous solution, but in a solution of a concentration corresponding to that of urine.

Several attempts have been made to measure the conductivity of mixtures of electrolytes and formulate the results. Most of the experiments have been made with dilute solutions and can not be used well in cases like the present one. The author has investigated the question with special reference to mixtures like the urine and has considered the effect of urea as a non-electrolyte also. The conductivity varies from individual to individual, and from hour to hour according to the kind and amount of food metabolized, but is seldom above $\kappa = 0.03$. The table (page 310) illustrates the variations in the urine of two men, both well nourished on mixed diet, with the water consumption intentionally high.

By making complete urine analyses and duplicating the results by mixing the inorganic and organic ions and the non-electrolytes in proper proportion, it is possible to reach almost exactly the corresponding urine conductivity as shown, for example, with six urines, the complete analyses of which were employed in reaching the mean values given at the beginning of the chapter.

No. of Urine.	1	2	3	4	5	6
κ as observed.	0.02372	0.02402	0.02793	0.01984	0.02898	0.02251
κ from mixtures.	0.02332	0.02400	0.02768	0.01944	0.02886	0.02158

But if an attempt were made to calculate the urine conductivity by adding together the individual conductivities of the various salts for the concentrations as found by analysis the result would be too high, as the conductivity of each substance is lowered somewhat by the presence of the others. It is possible, however, by experimenting with known artificial mixtures, to determine the extent of this modification and so be able to introduce a correction. For the present purpose the disturbing action of the other urinary constituents on sodium chloride is all that is called for. This has been done, but the details of the investigation can not be given here.

In a long series of experiments it was found that a mean correction for the effect of sodium chloride may be made in this way. The chlorine is accurately determined in the urine and calculated as sodium chloride. From conductivity tables the value of κ for the calculated concentrations is found and this value is diminished by 3 per cent which is the average correction due to the presence of other substances, organic and inorganic, in the urine. This corrected salt conductivity in turn must be subtracted from the observed total conductivity to find the fraction due to metabolic products. This correction gives a result which is near the truth in ordinary normal urines and which may be taken as furnishing a peculiar kind of measure of the total metabolism, that will be found to have a value in certain calculations.

THE FREEZING POINT OF URINE. CRYOSCOPY.

In the thirteenth chapter the application of cryoscopic methods to blood examinations was discussed, and the apparatus used described. In urine investigations also freezing point determinations have become important and a very considerable literature has accumulated. The Beckmann apparatus may be employed, as with the blood, but the Zikel modification has been found to give good results and is somewhat simpler in manipulation.

Conductivity and cryoscopic methods do not yield exactly parallel results; the conductivity power of the urine depends essentially on the number of inorganic molecules or ions present, while the freezing point depression depends on the sum of all the dissolved substances. Urea is therefore important in the one instance, but not in the other, as it is practically a non-conductor. This being the case it is evident that valuable information may be obtained by a combination of the two methods, as it is possible to determine the fraction of the osmotic pressure of the urine due to electrolytes and non-electrolytes. Such applications are frequently made.

While the osmotic pressure of the blood is nearly constant, that of the urine is extremely variable. Ordinarily the limits are between $\Delta = -1.3°$ and $-2.0°$, but after great water consumption on the one hand, or consumption of much nitrogenous food, or salt, without sufficient liquid, on the other, the freezing point of the urine may vary from $\Delta = -0.1°$ to $-3.0°$. That is, the urine concentration may range from one-fifth that of the blood to over five times the blood concentration, expressed in active molecules. It will be remembered that a freezing point depression of $1°$ C. corresponds to an osmotic pressure of 12.1 atmospheres.

The applications of this cryoscopic method to urine are mainly in the direction of diagnosis. Since it is possible to collect the urine from each kidney separately, by ureter catheterization or equivalent means, a test of the two portions will disclose any difference in the performance of the two organs. Normally, the secretions from the two kidneys, for a given time, should be the same. A freezing point determination is easily made and will show if one kidney is doing more work than the other. By including a conductivity test it may be found that the difficulty in excretion is more pronounced for one class of substances than for another.

CHAPTER XX.

SOME PRACTICAL URINE TESTS.

In this chapter brief directions will be given for the routine examination of urine, such as is called for in clinical observations. In such work the color and reaction, already referred to, are always noted, and the specific gravity.

SPECIFIC GRAVITY.

For very accurate work the specific gravity of the urine is determined by means of the pycnometer or weighing bottle, but in the ordinary clinical practice it is customary to employ the urinometer, which may furnish very accurate results if the instrument is properly made and graduated. Formerly the reading was based on a temperature of about $15.5°$ in the urine, referred to water of the same temperature as unity. A much more rational plan is to assume a standard temperature of $25°$ C. in the urine and refer the reading to water at $4°$ C. as the unit.

From a test of the specific gravity it is possible to make an approximate estimate of the total solids present. From a large number of observations made in the author's laboratory it has been found that by multiplying the last two figures of the specific gravity at $25°$ C., by 2.6 a product is secured which represents the weight of urinary solids in 1000 cc. This 2.6 is **Long's coefficient**, and takes the place of the **Haeser coefficient** based on tests at a lower temperature. Thus, if the specific gravity at $25°$ C. is 1.023 (Sp. gr. $\frac{25}{4} = 1.023$) we have $23 \times 2.6 = 59.8$, or the number of grams of solids per liter. The direct determination of total solids by evaporation and weighing is not perfectly accurate, because of losses through partial decomposition of some of the constituents.

The specific gravity of the urine varies, naturally, with the food consumption and volume of liquid consumed. It may, normally, be as low as 1.010 and as high as 1.030 or perhaps higher. A diet rich in meats gives rise to a urine with high specific gravity because of the resultant urea excretion; much salt in the food has the same effect. On the other hand, a diet rich in fats and carbohydrates, with low proteins, gives rise to a urine with low specific gravity, because in such cases little is formed to be excreted by the kidneys. With ordinary mixed diet a specific gravity of 1.020 to 1.025 is generally observed.

In diabetes mellitus the specific gravity of the urine may be very high, from the excretion of sugar, as will be pointed out below. But it must be kept in mind that a specific gravity high enough to **suggest** sugar may be reached through high salt and protein and low water consumption.

The clinician is not so much interested in the normal constituents of the urine as he is in the possible presence of bodies pointing to a pathological condition in his patient. A marked variation in some of the normal constituents of the urine may also have an important bearing on the diagnosis, and he must know how to make these tests. Following the usual routine, we shall take up albumin and sugar first.

THE TESTS FOR ALBUMINS.

Albumins are present in normal urine in traces only, and such traces are not detected by the usual reagents described below. Albumin in the urine in any appreciable amount is indicative of a pathological condition, and may be due to a number of causes. A great number of tests have been suggested for the recognition of this pathological albumin, and these all depend, practically, on the fact that the soluble, invisible protein may be easily coagulated in the urine and so rendered visible. The ordinary serum albumin will be considered first.

Heat Test. At a temperature near 70° C. the serum albumin becomes coagulated and opaque. Heat a few cc. of the urine in a test-tube nearly to the boiling point. If the liquid becomes cloudy albumin is suggested. But the urine usually contains enough so-called earthy phosphates to become cloudy on heating, and this cloud resembles the albumin coagulation very closely. To avoid a mistake here the urine must be *very weakly* acidified with dilute acetic acid before warming. This prevents the precipitation of phosphates but does not interfere with the albumin coagulation. Avoid using much acid.

Nitric Acid Test. In contact with strong nitric acid the albumin of the urine is immediately coagulated. To make the test pour two or three cc. of strong nitric acid into a narrow test-tube, warm slightly and by means of a dropping-tube introduce over the acid a layer of the clear urine (filtered previously, if not clear). In presence of albumin a cloudiness, or even a heavy precipitate, appears at the junction of the two liquids. This is an extremely delicate test and is very commonly employed clinically. The nitric acid is warmed to prevent the possible precipitation of urea nitrate, which occasionally happens with cold, very concentrated urine. A simple change of color at the junction layer is not due to albumin. The latter yields an actual coagulation.

Picric Acid Test. Many reagents besides nitric acid have the power of forming precipitates with albumin. A saturated aqueous solution of picric acid is very useful for this purpose, and when added to albuminous urine produces a yellowish white cloud or precipitate. It is best to acidify the urine slightly with acetic acid first. If this causes a precipitate, of mucin possibly, filter it off and then add the picric acid, drop by drop. The test is extremely delicate, but is not better than the preceding one.

Double Iodide Test. This is known also as Tanret's Test and the Mercuro-Potassium Iodide Test. The reagent used is made by dissolving 33.1 gm. of potassium iodide in distilled water and adding to it gradually 13.5 gm. of mercuric chloride. The mixture is stirred until all is dissolved, and diluted to make 800 cc. To this 100 cc. of pure, strong acetic acid is added. If a slight precipitate remains decant carefully and dilute to one liter with distilled water. The solution contains approximately 4 KI to $HgCl_2$.

To make the test filter the urine, if not clear, and add enough acetic acid to give a sharp reaction. If a precipitate or turbidity now appears, possibly due to mucin bodies, filter again and to 10 cc. of the clear filtrate add a few drops of the reagent. In presence of proteins a white cloud is formed, or even a precipitate if much of the albumin substance is in the solution.

The test is extremely delicate, but it must be remembered that the same reagent is employed in testing for alkaloids. If the urine happens to contain traces of those bodies, from previous medication, a reaction will certainly follow. But such precipitates may be distinguished by their solubility in alcohol. A precipitate may be formed also in presence of excess of urates. Such a precipitate is dissipated by heating; it may be avoided in the first place by proper dilution of the urine.

The Amount of Albumin. It is not alone sufficient that we are able to detect the presence of albumin in urine; we often need to know its amount to determine the practical value of a line of treatment pursued from day to day. To be of the greatest possible service, a method must be so easy of execution that approximately correct results may be obtained by it by the use of simple apparatus and in a short time. Several methods are known by which the amount of albumin in urine can be found. One of these, and the best, may be called the gravimetric method, as by it the albumin is precipitated, collected, and weighed. In another, the albumin is precipitated and its volume measured, while in a third process the amount of albumin is estimated from the degree of turbidity caused by its precipitation in the urine.

The gravimetric method consists essentially in coagulating the albumin, collecting it on some form of filter, and, after proper washing, weighing the precipitate. This method may be made to give very good results but is too difficult and tedious for the clinical laboratory.

Volume Methods. One of the simplest of these is the one proposed by Esbach. In this a special tube is used, called the **Esbach albuminometer**, and a special solution or reagent made by dissolving 10 grams of pure picric acid and 20 grams of pure citric acid in a liter of distilled water. The solution must be filtered if it is not perfectly clear, and is the same as the one used for the qualitative test. The principle involved in the employment of the method is this: The precipitate of albumin and picric acid settles in coherent manner and in a compact volume proportional to its weight, provided certain definite amounts of the reagent and urine are taken. The albuminometer, or measuring tube used, resembles a test-tube of heavy glass about six inches long and is graduated, empirically, to show how much urine and reagent to take and the amount of albumin obtained in grams per liter, or tenths of 1 per cent.

The test is carried out in this manner:

Urine is poured in, to the mark, and then the reagent, described above, to its proper level. The tube is closed with the thumb and tipped backward and forward eight or ten times until the liquids are thoroughly mixed. It is then closed with a rubber stopper and allowed to stand in a perpendicular position twenty-four hours. This will give the precipitate time to settle thoroughly after which the amount can be read off on the scale. The results are accurate enough for clinical purposes and by practice can be made to agree moderately well with those found by the gravimetric method. But the tube must not be violently shaken, and, as the test is an empirical one, it should not be allowed to stand much longer than one day before reading the result.

In any case in applying this test the urine should not be highly concentrated. The best results are obtained with urine of low specific gravity, and with the albumin not over 0.3 per cent. When the test shows an amount in excess of this, the urine should be diluted and a new trial made.

A yellowish-red precipitate which sometimes separates on long standing must not confuse the analyst. It consists of uric acid.

The Tests for Globulins. These protein bodies resemble the albumins in many respects, and like them occur only pathologically in the urine. In all the preceding tests, globulins react as do the albumins. Among the distinctive tests one only need be mentioned here.

Dilution Test. Globulin is insoluble in water, but soluble in dilute salt solutions; —hence its solubility in urine. If the latter is diluted until the specific gravity is 1.002 or 1.003 the globulin may separate out. At any rate the addition of a few drops of dilute acetic acid will produce the desired result. A current of carbon dioxide passed into the diluted liquid for several hours accomplishes the same end.

The test may be modified in this manner. Filter the urine if it is not perfectly clear, and then pour it, drop by drop, into a tall, narrow beaker of distilled water. If globulin is present it is thrown out as a white cloud which shows as the drops pass down through and mix with the lighter, clear water. The globulin may afterward be confirmed by adding a small amount of salt solution which will cause the precipitate to disappear.

PEPTONES AND PROTEOSES.

True peptones are very rarely found in urine. The bodies described as peptones are probably highly converted proteoses, or albumoses of the deuteroalbumose type.

The recognition of albumose is not a matter of difficulty, as it is distinguished from the other protein compounds sometimes found in the urine by several well-marked characteristics. It is not coagulated by heat or by the addition of acetic or warm nitric acid, and is very soluble in hot water. It is much less soluble in cold water, but the presence of small amounts of salts seems to increase its solubility here in marked degree.

In presence of albumin or globulin it can be found by the following process unless it is in very small amount. The urine is saturated with pure sodium chloride, which will precipitate albumose if present, and then enough acetic acid is added to give a very strong acid reaction. The mixture is boiled and filtered hot. This treatment throws out both albumin and globulin, and redissolves a precipitate of albumose which may have formed. The latter would therefore be found in the clear filtrate and sometimes in amount sufficient to precipitate as this cools. The filtrate should therefore be allowed to remain at rest until quite cool. If much albumose is present it will appear as a white cloud. Sometimes, however, it will be necessary to concentrate the filtrate before looking for this reaction, and this is done by evaporating slowly on the water-bath to half the volume. On now cooling, salt will quickly settle out while the albumose precipitates later in flocculent form. After the salt has separated the liquid may be tested by the biuret reaction, described in earlier chapters.

Another test is this: Separate the albumin and globulin by boiling with a small amount of acetic acid without the salt. Filter while warm and concentrate the filtrate to a volume of one-third. Allow to cool thoroughly and add a large excess of saturated solution of ammonium sulphate. This gives a white flocculent precipitate of albumose, if present. The precipitate can be collected on a filter and washed with the saturated ammonium sulphate solution and then dissolved in a little distilled water, poured on the filter. This filtrate gives tests with picric acid, potassium ferrocyanide and acetic acid, and other albumin reagents.

If the original urine shows no reactions for albumin or globulin the albumose tests can be applied directly after concentration. The method by precipitation by means of picric acid gives good results. The biuret test is also delicate if the urine is clear and of light color.

THE TESTS FOR SO-CALLED MUCIN.

Much confusion exists at the present time as to the exact nature of the protein bodies in the urine which have long been known as mucins. The true mucins belong to the group of gluco-proteids and are seldom found in the urine, but certain other protein bodies are practically always present normally, and with them several acid substances which, under the proper conditions, are able to precipitate these normal traces of protein. Among these acids chondroitin-sulphuric acid is the most important, and in acetic acid solution it combines with and precipitates the normal protein. Sometimes the trace of protein normally present is so small that no precipitate forms on slightly acidifying. In such cases the addition of a little albumin solution to the acidified urine usually produces a precipitate, as the albumin-precipitating agent is ordinarily present in sufficient amount.

The kind of albumin which may be present and give this reaction is not clearly defined. In some cases it appears to be a nucleo-albumin, and occasionally a gluco-proteid, which yields a reducing substance on boiling with acids. In working with a large volume of urine it is sometimes possible to separate from the so-called mucin precipitate a reducing substance suggesting the gluco-proteid; in other cases this precipitate may be free from sulphate or reducing substance, but contain a phosphate-yielding complex which can point to a nucleo-albumin or nucleo-proteid. These reactions are too complex for ordinary clinical demonstration.

Acetic Acid Test. The addition of an excess of acetic acid, that is enough to give a strong acid reaction and make up about 0.2 per cent. of the whole volume, produces a flocculent precipitate in presence of the "mucin" substance. Heat is not applied in this test, and its delicacy may be increased by dilution of the urine.

Citric Acid Test. In this test a layer of the urine is poured over a concentrated solution of citric acid. The "mucin" appears as a cloud at the junction of the two liquids.

In pouring urine over nitric acid a mucin cloud or band sometimes makes its appearance about a centimeter **above** the junction point. In this test albumin shows as a cloud at the junction point. In these acid tests the precipitation of uric acid is prevented by the dilution with two or three volumes of water.

THE TESTS FOR SUGARS.

Normal urine contains a trace of carbohydrate, probably glucose, which may be recognized by very delicate tests. Pathologically glucose, and sometimes other sugars, occur in quantity and may be identified by a great number of tests. We have to distinguish then, between what is known as a *physiological glycosuria*, and a pathological condition most commonly characteristic of *diabetes mellitus*. In the normal urine the trace of sugar may not be over one-twentieth of one per cent., while in advanced diabetes it may amount to five per cent., or even more, with a volume of five or six liters daily.

The most convenient of our sugar tests are based on the fact that the usually-occurring glucose is an aldehyde body and responds to certain characteristic tests. Fructose, sometimes present, is a ketone sugar, and responds to the same tests in general. Among these we have, first, the so-called reduction tests, the best of which are as follows:

Moore's Test. This depends on the reaction between reducing sugars and strong alkali solutions. When a solution of sugar or diabetic urine is mixed, without heating, with a solution of sodium or potassium hydroxide, no change is at first apparent unless the amount of sugar present is large or the alkali very strong. But on application of heat, even with weak sugar solutions, a yellow color soon appears which grows darker, becoming yellowish brown, brown, and finally almost black, while an odor of caramel is quite apparent. The strong alkali-sugar solution absorbs atmospheric oxygen, giving rise to a number of products among which lactic acid, formic acid, pyrocatechol and others have been recognized. The brown color is due to other unknown decomposition products.

This is a good reaction for all but traces of sugar, as the intense dark brown color and strong odor are not given by other substances liable to be present in urine.

But traces of sugar cannot be recognized by this test with certainty, as the color of normal urine even is darkened to some extent by the action of alkalies.

Urine containing much mucin becomes perceptibly darker when heated with sodium, potassium, or calcium hydroxide solutions.

The Trommer Test. This is one of the oldest and best known of the tests for the recognition of sugar in urine, and is a typical reduction test. It is performed by adding to the urine an equal volume of 10 per cent. solution of sodium or potassium hydroxide and then a *very few drops* (three or four to begin with) of dilute solution of copper sulphate as described in Chapter III.

The test must be used with certain precautions. If albumin is present it must be coagulated and filtered out. The amount of copper sulphate used must be small, because if only a trace of sugar is present and much copper is used the latter will give a blue precipitate which can not redissolve, and which turns black on boiling, thus obscuring a sugar reaction which may be given at the same time. In making this test if a black precipitate is thus found it must be repeated, using less of the copper.

The active body in producing the reaction is copper hydroxide, but this must be in solution to act as a good oxidizing agent with sugar; and the test, therefore, becomes uncertain or unsatisfactory if so much copper is added that the hydroxide formed cannot be dissolved by the sugar which may be present. In doubtful cases it becomes necessary to make several trials before the right proportion between urine, alkali, and copper solution is found. In the solution, on completion of the reaction, several oxidation products of sugar are found, among which are formic acid, oxalic acid, tartronic acid, etc. But the complete reaction is obscure. In order to avoid the indicated uncertainty of the Trommer test when used for small amounts of sugar the next one was proposed.

The Fehling Solution Test. Fehling suggested the use of a solution containing along with the copper sulphate and alkali a tartrate to dissolve the copper hydroxide formed by the first two. Many substances besides sugars have the power of dissolving copper hydroxide with a deep blue color. Among these may be mentioned tartaric acid and the tartrates, glycerol, mannitol and others of less value.

A solution prepared by mixing certain quantities of alkali, copper sulphate, and either one of these bodies, with water in definite proportions, remains perfectly clear when boiled. But if a trace of glucose (or several other sugars) is present the usual yellow precipitate forms.

In performing the test with the Fehling solution described in Chapter III three or four cubic centimeters of the solution are poured into a test-tube, diluted with an equal volume of water, and boiled. The solution must remain clear. Then the urine is poured in, a few drops at a time, and the mixture is boiled. If sugar is present in the urine in an amount above one-fifth of one per cent, it should show at once with this treatment. If but a trace of sugar is present, more urine must be added and the boiling repeated.

When normal urine is heated with Fehling solution a greenish flocculent precipitate usually makes its appearance. This has no significance, as it is due to the phosphates normally present, which come down when the reaction is made alkaline. Many urines produce a clear dark green solution when heated with the Fehling solution. This is a partial reduction reaction and like the other has no special importance, as urines free from sugar give it. At other times urines free from sugar yield an almost colorless mixture when boiled with the Fehling solution. These peculiar reduction effects are due to the presence of uric acid, creatinine, pyrocatechol and several other substances and are generally characterized by discharge of the deep blue color of the solution without precipitation of the copper suboxide. Certain substances taken as remedies give rise to products in the urine which exert a similar action. Occasionally, however, the amount of uric acid is so large that the reduction is accompanied by actual precipitation of the copper as red oxide. This fact is of interest as it makes the test, at times, somewhat uncertain, but it is a very simple matter to determine whether or not a great excess of uric acid is present, as will be pointed out later. The liability to error in the Trommer test from these causes is less than in the Fehling test, but notwithstanding this the latter must still be regarded as the better test practically, because of its great convenience and the sharpness of the reaction with even traces of sugar. The ingredients of the Fehling test are best kept in separate bottles closed with rubber stoppers, as the mixed solution does not keep well, as ordinarily prepared. It is customary to keep the copper sulphate in one bottle and the alkali and tartrate in another.

Several other copper test solutions are in use; the Loewe solution contains glycerol in place of sodium-potassium tartrate of the Fehling solution, while mannitol is employed for the same purpose in the Schmiedeberg solution. The reduction is the same as in the Fehling test.

The Bismuth Test. Böttger found (1856) that in the presence of alkali, bismuth subnitrate is reduced to the metallic condition by the action of glucose in hot solution. As a urine test he recommended to make it strongly alkaline with sodium carbonate, and then add a very small amount, what can be held on the point of a penknife, of the pure bismuth subnitrate. On boiling the mixture the insoluble bismuth compound, which settles to the bottom, turns dark if sugar is present.

The test is at present carried out by adding to the urine in a test-tube an equal volume of 10 per cent. solution of sodium or potassium hydroxide, and then the subnitrate. Boiling gives the reaction as before. In absence of sugar (or albumin) the bismuth compound remains white.

In performing this test only a very small amount of the subnitrate should be taken. This is absolutely necessary in the detection of traces of sugar. In this case the reduction is but slight, and not much black powder of bismuth or its oxide can be formed. If a great excess of the white subnitrate is taken it may be sufficient to completely obscure the reduction product. It is frequently well to use not more than four or five milligrams of the subnitrate.

The black precipitate formed was at one time supposed to be finely divided metallic bismuth. Later investigations seem to show that it consists essentially of lower oxides of bismuth. This test has certain advantages over the copper tests. It is easily made, and with materials everywhere obtainable in condition of sufficient purity. Furthermore, the reaction is not given with uric acid, which it will be remembered may act on the Fehling solution if excessive.

Albumin, however, interferes with the test, as it gives, also, a black precipitate when boiled with alkali and the bismuth subnitrate. In this case the albumin gives up sulphur and forms bismuth sulphide; if it is present in a urine it should be coagulated and filtered out before trying the test.

Fallacies in the Reduction Tests. It has been shown above that several bodies normally found in urine are able to reduce the alkaline copper solutions. Some of these interfere with the bismuth reactions also, but not to the same degree; but attention must be called to another source of error which is very important. In warm weather it is often desirable to add something to urine to prevent its rapid decomposition, and several substances have been suggested for the purpose. The best known are chloral, chloroform, salicylic acid, phenol, and formaldehyde. Unfortunately all of these except phenol have a rather marked action on the copper solutions. As a preservative phenol is objectionable from other standpoints. Urine intended for sugar tests should be tested as soon as possible after collection, and no foreign substances should be added as a preservative. Neglect of this very simple and obvious precaution has caused many serious blunders, especially in the examination of urine from applicants for life insurance.

The Phenylhydrazine Test.—In this test a reaction discovered some years ago has been applied by v. Jaksch to the examination of urine. Add to about 10 cc. of urine 0.2 gram of phenylhydrazine hydrochloride and a slightly greater amount of sodium acetate. Warm the mixture gently, and if solution does not take place add half the volume of water and heat half an hour on the water-bath. Then cool the test-tube by placing it in cold water and allow it to stand. If sugar is present a yellow precipitate settles out, which consists of minute needles generally arranged in rosettes, visible under the microscope. Albumin does not obscure this test, but if much is present it is best to coagulate it as well as possible by heating, and then filter. The yellow precipitate is phenyl-glucosozone. The reaction is not much used in urine analysis, but is of great value in the identification of sugar under other conditions.

The Fermentation Test.—When yeast is added to urine containing sugar and the mixture left in a moderately warm place the usual fermentation soon begins which is shown by two principal changes. Carbon dioxide is given off, which may be collected and identified, and the mixture becomes lighter in specific gravity. When only traces of sugar are present the test by collection and identification of the carbon dioxide frequently fails because of the solubility of the gas in the liquid.

The variation in the specific gravity is an indication of greater value, as it can be readily observed with proper appliances. The test has practical value, however, only as a confirmation of some other one. If by the copper solutions, for instance, a strong indication is obtained which it is suspected may be due to an excess of uric acid, the reaction by fermentation may be resorted to because only sugar will respond to it.

The Amount of Sugar.—It is not always sufficient to be able to detect the presence of sugar in urine. A knowledge of the amount is frequently of the greatest importance. A number of methods have been proposed by which a quantitative determination can be made, some of them crude and of little practical value, while others give, when properly carried out, results which are accurate. The methods in general may be divided into four groups, depending on the

(1) Reduction of solutions of heavy metals, and measurement of the amount of reduction.
(2) Change of color produced in organic solutions, by action of sugar, the depth of final color being proportional to the amount of sugar.
(3) Results of fermentation, with measurement of change in specific gravity of the urine, or measurement of evolved carbon dioxide.
(4) Observation of rotary polarization of light.

The reduction methods, which alone need be considered here, are illustrated in the use of the Fehling solution as a qualitative test and in the bismuth tests. The general principles involved in making a quantitative determination of sugar by aid of the Fehling solution are the same as those involved in making other volumetric analyses with standard solutions, and are fully explained in Chapter III. A measured volume of the properly prepared Fehling solution is poured into a flask and brought to the boiling point. Then from a burette the urine is run in slowly, a few cubic centimeters at a time, until the deep blue of the copper solution is just discharged, leaving a pale yellow. At this stage the copper is all reduced to the condition of insoluble red oxide, Cu_2O, and the volume of urine added from the burette contains the amount of sugar measured by the oxidizing power of the Fehling solution taken. The details of the titration are as follows: Use the standard quantitative Fehling solution as described, and dilute an accurately measured volume with exactly 4 volumes of water. That is, 50 cc. should be diluted to 250 cc. or 25 cc. to 125. One cubic centimeter of this diluted liquid will oxidize almost exactly one milligram of glucose, as was shown, provided the sugar is in approximately 1 per cent. solution. For all practical purposes of urine analysis the oxidizing power may be considered the same in a solution of one-half per cent strength, and only very slightly increased in still weaker solutions. Therefore, before beginning the test dilute the urine, if necessary, accurately with four or nine volumes of water. This can be done by making 50 cc. up to 250 or to 500 cc. and mixing well by shaking.

With the diluted urine so prepared proceed as follows: Measure out 50 cc. of the dilute Fehling solution, pour it in a flask and heat to boiling on gauze. Fill a burette with the diluted urine and when the solution in the flask is actively boiling run in about 3 cc. Boil two minutes, remove the lamp, and wait half a minute to observe the color. If blue is still visible, heat to boiling again and run in 3 cc. more. After boiling two minutes as before, wait a short time and observe the color near the surface of the liquid in the flask. If still blue repeat these operations until on waiting it is found that the blue has given place to a yellow. The urine should be so dilute that at least 10 cc. must be run in to reduce all the copper hydroxide.

When the volume required is found to within 2 or 3 cc. a second experiment must be made, the urine being added very gradually now, without interrupting the boiling longer than necessary, until the first of the limits between which the correct result must lie, as shown by the former test, is reached. From this point the addition of the urine is continued, with frequent pauses for observation of color, until the reduction is complete. The volume of urine used contains 50 milligrams of sugar.

If the preliminary experiment shows that the urine is strong in sugar and that the reduction is easy, that is, that the cuprous oxide separates and settles readily, the second test may advantageously be made with 50 cc. of a stronger Fehling solution. With many strong diabetic urines it is possible to use the undiluted copper solution with the oxidizing power of 4.75 milligrams of sugar to each cubic centimeter. The difficulties in this test have been very much overestimated; with a little practice any one can make a good sugar determination in urine. The important point is to find by a few simple preliminary tests the best conditions of dilution of Fehling solution and urine to give a precipitate which settles readily. With this information, and it can be acquired in a few minutes, the actual quantitative experiment can be easily made.

As an illustration of the calculations involved let it be assumed that 50 cc. of the dilute Fehling solution is reduced by 11 cc. of urine. Each cubic centimeter of the urine must therefore contain 4.54 milligrams of sugar. If the urine was undiluted this corresponds to 4.54 grams to the liter. If it had been diluted with 9 volumes of water the result must be multiplied by 10, giving as the original strength 45.4 grams per liter. If the specific gravity of the urine were found to be 1.032, the percentage strength would be

$$\frac{4.54}{1.032} = 4.39.$$

Method by Use of Ammoniacal Copper Solution. This is described fully in Chapter III. The solution can be used with dilute urines only, as it has about one-fifth of the oxidizing value of the standard Fehling solution. For urine work one cc. is considered the equivalent of one milligram of glucose.

The great advantage in the use of this solution rests in the fact that the end point is easily recognized by the disappearance of color of the standard solution when the copper is all reduced. Cuprous oxide dissolves in ammonia without color while the slightest trace of cupric oxide leaves a marked blue in the liquid. To use this solution with advantage the student should learn to judge, from the results of a qualitative test, about how far the urine should be diluted. A very strong diabetic urine can be accurately titrated only after marked dilution.

Other Sugars in Urine. Besides the glucose other sugars are occasionally found in the urine. Fructose and lactose have been frequently described, and also a pentose. The identification of these sugars calls for certain precautions which can not be detailed here. The significance of the pentose has been the subject of much discussion. Fructose is found, as a rule, only when glucose is present.

In addition to the sugars a few other bodies are occasionally found in urine which exhibit some of the reduction and similar reactions. One of the most characteristic of these is *glucoronic acid*, $C_6H_{10}O_7$, which in structure bears some relation to glucose. The acid is a strong reducing agent, reacts with phenylhydrazine and possesses a marked dextro-rotating power. When distilled with hydrochloric acid it yields furfurol, recognized by its reaction with aniline. In urine the glucoronic acid usually occurs in ester-like combinations with phenols and other bodies. These exhibit a levo-rotation, which changes to the other direction on boiling the urine with a little acid. This optical property is of importance in the recognition of the acid.

ACETONE, ACETOACETIC ACID AND OXYBUTYRIC ACID.

The first of these is frequently found in urine in small amount. Indeed, it may be true, as has been asserted, that it is normally always present in traces. This physiological acetonuria has no clinical significance. Under some circumstances, however, it may be found in larger quantity, sometimes in amount sufficient to be detected by the odor alone, which fact first called attention to it. At one time it was supposed to be related to the sugar found in urine, but it is now established that it more generally accompanies albumin and is frequently observed in many febrile conditions.

Acetone in urine is believed to be a decomposition product of albumins, or of bodies which may in turn be looked upon as resulting from protein disintegration, the fatty acids, for example. It has been shown that in health, even, it can be much increased by a diet rich in nitrogenous materials.

But, occurring as it does in fevers and in advanced stages of diabetes mellitus, a certain interest attaches to its detection, and numerous methods have been proposed by which it may be identified in small amount. Those which depend on its direct recognition in the urine are mostly uncertain. It is always safer to distil the liquid and apply the test to a portion of the distillate. Half a liter, or more, of the urine is poured in a retort attached to a Liebig's condenser, and, after addition of a little phosphoric acid, is subjected to distillation. 100 cc. of distillate will be enough. A portion of this can be taken for each test as follows:

Legal's Test.—Add to 25 cc. of the liquid a small amount of a fresh solution of sodium nitroprusside, and a few drops of a strong potassium hydroxide solution. If a ruby-red color appears which slowly gives place to yellow, and if the addition of acetic acid changes this to purple or violet-red, the presence of acetone is indicated.

Creatinine gives a ruby-red color as does acetone when the nitroprusside reaction is directly applied to urine, but after adding acetic acid a green or blue color results.

Lieben's Test.—This depends on the production of iodoform, and is carried out in this manner. To about 5 cc. of the distillate add a few drops of a solution of iodine in potassium iodide (the "compound solution of iodine," Lugol's solution), and then enough sodium hydroxide to make distinctly alkaline. Warm gently. If acetone is present a yellowish white precipitate soon appears, which, on standing, becomes crystalline and more deeply colored. The test is said to be sharper and more characteristic if ammonia is used instead of the fixed alkali. The liquid is first made strongly alkaline with ammonia, and then the iodine solution is added until the brownish precipitate formed at first dissolves very slowly. In a short time the yellowish iodoform precipitate makes its appearance. A rough quantitative measure of the amount of acetone present is given by noting the smallest volume of the distillate with which a distinctive iodoform reaction can be seen. It is said that 0.0001 milligram in 1 cc. can be detected; 0.5 milligram in 10 cc. can be recognized by the nitroprusside reaction.

Acetoacetic Acid. This acid is frequently found with acetone in the urine of fevers and diabetes, and also in many other pathological conditions, but its relation to these disorders is not clearly understood. While acetone may be normally present, the acetoacetic acid is probably always pathological.

As it is extremely unstable tests for it must be made in the fresh urine. As it yields acetone by decomposition, tests for this substance must be made first. If these are negative there is no need in going further, but if they are positive the following direct test may be made for the acid.

Ferric Chloride Test. Add this reagent, a few drops at a time, and look for the formation of a reddish color. Ordinarily there can be nothing in the urine to give a similar reaction, and the color test is a strong indication of the presence of acetoacetic acid. But several coal tar products given as remedies furnish residues in the urine which also give a red or purple color with ferric chloride when added. To detect the acetoacetic acid with certainty under these conditions it is necessary to proceed with greater care. To this end add to the urine, which should be fresh, a few drops of ferric chloride or enough to precipitate the phosphates present. Filter and add a little more of the chloride. A red color indicates the acid. Divide the liquid into two portions; boil one and allow the other to stand a day or more. In the boiled portion the color due to acetoacetic acid should disappear within a few minutes, while in the other it should remain about twenty-four hours.

Acidulate another portion of the urine with dilute sulphuric acid and extract it with ether, which takes up acetoacetic acid. Remove the ethereal layer and shake it with a very dilute aqueous solution of ferric chloride. The red color in the new aqueous layer should appear as before and disappear on boiling, which behavior distinguishes the acid from other substances likely to be present.

β-Oxybutyric Acid. The detection of this acid in the urine by chemical methods is by no means simple, but if present in amounts not too minute it may be found by the aid of the polariscope, inasmuch as its solutions possess a strong negative rotation. In dilute solution the specific rotation is approximately $[\alpha]_D = -23.4°$. As this acid is found associated with sugar in diabetes it is necessary to destroy the sugar by fermentation before making the test. Amounts as high as 200 grams in the day's urine have been reported, but usually, where present at all, the amount is far below this, 15 to 20 grams being nearer the average.

If the urine does not give a test for acetoacetic acid it is useless to look for the β-oxybutyric acid. To test for the latter evaporate about 50 cc. of the urine, freed from sugar if necessary, to a syrup and distil the residue, mixed with an equal volume of strong sulphuric acid, from a small flask. Collect the distillate without a cooler in a test-tube. If the oxybutyric acid was present in the urine the distillate will contain α-crotonic acid, which on strong cooling yields a crystal mass melting at about 72° C. To get characteristic crystals it may be necessary to extract the distillate with ether and allow this to evaporate.

NORMAL COLORING-MATTERS.

Although many investigations have been carried out on the subject of the normal urinary pigments we are yet unable to give a very definite account concerning them. This is partly due to the fact that the coloring substances exist in the urine in minute traces only, which makes their separation and recognition exceedingly difficult, and partly to another fact that some of them are easily altered or destroyed by the action of the reagents employed in their investigation. By proceeding according to different methods, physiologists have obtained very different results indicating the existence of several colors, or at any rate modifications of colors. The difficulty of detecting the normal colors in urine is sometimes increased by the presence of traces of accidental coloring-matters having their origin in peculiar or unusual articles of food consumed. Some of these will be referred to below.

It seems to be settled, however, that in health not merely one but several coloring-bodies must be present. It has not yet been found possible to separate these in the free state.

Uroerythrin is the name given by Thudichum and others to a common reddish coloring-matter which often precipitates with urates and other substances. It is colored green by solution of potassium hydroxide, but the color is not restored by addition of acid.

Urochrome. This seems to be the most important and characteristic coloring matter of normal urine. It is responsible for the ordinary yellow color of the secretion, and its decomposition products appear on treatment of the urine with strong acids.

Urobilin is a coloring matter found in much smaller amount than the last one. When separated it is a reddish brown amorphous substance. In the urine the following test is sufficient for identification. Precipitate 200 cc. of urine with basic lead acetate, collect the precipitate on a filter, wash it with water and dry it, and then wash with alcohol. Finally, digest with alcohol containing a little sulphuric acid, and filter. The filtrate is usually fluorescent. Make it strongly alkaline with ammonia, and add solution of zinc chloride. This will give the fluorescence referred to above if but little is added, while if an excess of the zinc chloride is added, a reddish precipitate falls.

Urophain. This is the name given by Heller to a substance identical with, or similar to, urobilin. Heller gives this test: Take a few cubic centimeters of strong sulphuric acid in a conical glass and pour on it, drop by drop, about twice as much urine. As the two mix, a deep garnet-red is produced.

This reaction is not, however, characteristic, as several other matters may give it.

Urohematin is the name given by Harley to a coloring-matter similar to the above. He applies this test: Dilute or concentrate the urine so that it is equivalent to 1,800 cc. for the twenty-four hours. Take a few cubic centimeters in a test-tube or wine-glass, and add one-fourth of its volume of strong nitric acid. No change of color can be observed if the urohematin is present in normal amount, but if in excess various shades from pink to red may appear.

Indican. This is a normal constituent of urine in small amount, but in many intestinal diseases may be greatly increased. Its detection in the urine is then a matter of considerable clinical importance. It is formed from the oxidation of indol which is a common product of bacterial origin in the intestines. The oxidation product, indoxyl, combines with sulphuric acid and the potassium salt of this is excreted as indican. The formula of the indican is $KSO_4C_8H_6N$. Through further oxidation this yields indigotin, a blue coloring-matter which is the substance finally identified in the test. For further relations consult earlier chapters.

Indican is found in normal urines in very small amount only. It may, under favorable circumstances, be detected as here given: Take about 4 cc. of pure hydrochloric acid in a test-tube and add about half as much urine, shaking well. A blue or violet color shows indican. This test depends on the conversion of the indoxyl compound into indigo, but the oxidizing action of the acid is not always strong enough to bring about the change in presence of other organic bodies in the urine.

A more generally applicable method is this: To 10 cc. of urine and the same volume of strong pure hydrochloric acid, add 2 or 3 cc. of chloroform. Then add, drop by drop, solution of sodium hypochlorite, shaking after each addition. The hypochlorite acts as an oxidizing agent, liberating the coloring-matter, which is then taken up by the chloroform. The oxidation must not be carried too far; that is, too much hypochlorite must not be added, as it would then destroy the color as fast as formed. In fact, small traces of the product sought might be completely overlooked in the process, as the hypochlorite is a very active oxidizer, the effect going far beyond the production of indigo. It has therefore been proposed to use nitric acid as the oxidizing agent. The urine is boiled with an equal volume of hydrochloric acid and then a few drops of nitric acid are added. The mixture is cooled and to it a little chloroform is added and well shaken. In presence of indigo the blue color appears. Bromine water in small amount may be employed in the same manner.

More recently hydrochloric acid containing a little ferric chloride has come into use as an oxidizing agent. This is less destructive than the nitric acid or hypochlorite. The reagent may be made by dissolving 3 or 4 grams of the ferric chloride in a liter of strong hydrochloric acid. It is known as the Obermeyer reagent.

As a comparative quantitative test the following may be employed: To a hundredth part of the whole day's excretion add an equal volume of the above reagent and 5 cc. of chloroform. Shake a minute or more, and allow to settle. Compare the color with that of a strong Fehling solution, taken as 100, as a standard.

ABNORMAL COLORS.

We have here a number of pathological substances which are best recognized through certain color reactions. These substances may not have much significance in themselves but they are often important in pointing to certain conditions, indirectly.

The Bile Pigments. Several coloring-matters originating in the liver may find their way into the urine. Their recognition is frequently important, and several reactions are available for this.

Biliary urine has generally a characteristic greenish yellow color sometimes tinged with brown. The froth from such urine is readily recognized by its yellow color, which is often a sufficient test in itself. Among the chemical tests the following are the best known.

Gmelin's Test. This is easily performed and depends on the oxidation of bilirubin, the pigment commonly present in fresh jaundice urine, by nitrous acid. Pour in a test-tube about 5 cc. of the urine under examination and by means of a pipette introduce below it an equal volume of strong nitric acid mixed with nitrous. This should be carefully done so as to avoid mixing the liquids much. At the junction of the two liquids, if bile is present, several colored rings appear of which the green due to biliverdin is most characteristic. The bands or rings appear above the acid in this order, yellowish red, red, violet, blue, and green. The last is essential. It must be remembered that nitric acid gives the other colors at times with urine free from bile, but green is characteristic of the latter.

Fleischl modified this test by mixing the urine with a strong solution of sodium nitrate and then adding strong sulphuric acid carefully. This settles below the urine and decomposes the nitrate at the point of contact, liberating the necessary nitric and nitrous acids for the oxidation as before. This method is a very good one.

Trousseau's Test. Add to some urine in a test-tube a few drops of tincture of iodine, allowing this to float over the urine. If the bile pigments are present a greenish color appears at the junction of the two liquids and may remain some hours. An excess of iodine must not be used.

The Diazo Reaction. Ehrlich and others have called attention to the behavior of many urines with solution of diazobenzene sulphonic acid, which often has importance in diagnosis. Normal urine treated with a weak solution of this reagent shows no marked change, but in several pathological conditions after adding the acid and saturating with ammonia a deep carmine or scarlet-red color appears, followed by greenish or violet precipitation. After the addition of ammonia the foam should show a distinct pink color.

This reaction depends on the combination of the sulphonic acid of diazobenzene with some aromatic compound found in the urine in pathological condition. At one time it was supposed to have special significance in the diagnosis of typhoid fever, but it now appears that in many diseases of the intestinal tract the urine receives traces of complex aromatic products of bacterial origin which respond to the test. It has, therefore, general rather than special significance.

As the diazobenzene sulphonic acid is not very stable, it is not convenient to use, and a reagent is made which, in its application, is its chemical equivalent. The reagent is prepared by dissolving 1 gram of sulphanilic acid in 200 cc. of water with the addition of 10 cc. of pure hydrochloric acid. Another solution is made by dissolving 1 gram of sodium nitrite in 200 cc. of water. To make the test take 50 cc. of the first solution, add 5 cc. of the nitrite solution and then 50 cc. of urine. Ammonia is then added in sufficient quantity to impart a strong alkaline reaction after thoroughly shaking. A scarlet-red color is the result if the urine in question contains the abnormal products referred to.

It must be remembered that normal urines usually give *some* color. It is only the strong reactions which have significance.

Ehrlich has introduced another test which discloses the presence of products of intestinal origin. This is a 2 per cent solution of dimethyl amino benzaldehyde in 15 per cent hydrochloric acid. A few drops of this reagent added to 5 cc. of urine strikes a red color, weak in normal urine, but deep in certain pathological urines. The exact significance of the reaction is not known.

BLOOD COLORING MATTERS.

These may appear in the urine from several sources. We may have color due to the presence of corpuscles themselves and we may have the color due to dissolved hemoglobin. It is important to distinguish between these conditions. The presence of the corpuscles, which are best recognized by the microscope, points to a lesion of the kidney or some part of the urinary tract, and to a fresh lesion if the corpuscles are sharp in outline. The term **hematuria** is applied to the condition when corpuscles themselves are present, while **hemoglobinuria** is the condition in which few or no corpuscles are found, but the free pigment is present. Hemoglobinuria is the result of the disintegration of corpuscles within the tissues or blood vessels, permitting the oxyhemoglobin or some derivative to pass through the kidney with the urine. With much of the coloring matter present, the urine may have a very dark color. Sometimes the hemoglobin suffers further decomposition, giving rise to hematin and methemoglobin.

The following are the best chemical tests for the recognition of these bodies:

Heller's Test. Treat the urine with solution of sodium or potassium hydroxide, and heat to boiling. This produces a precipitate of the earthy phosphates which in subsiding carry down coloring-matters. If a precipitate does not separate readily it may be hastened by adding two or three drops of magnesia mixture. Hemoglobin, when present, is decomposed by this treatment with separation of hematin, which in turn settles down with the phosphates, imparting a red color to the precipitate.

Struve's Test. Make the urine slightly alkaline with sodium hydroxide solution, and then add enough solution of tannic acid in acetic acid to change the reaction. If hemoglobin is present a dark brown precipitate of hematin tannate settles out. The test is a good one, and easily performed, but is not sufficiently delicate for the detection of small traces of hemoglobin directly. By collecting the precipitate on a filter and washing it, it may be used for two confirmatory tests. One of these depends on the formation of hemin crystals and is made in this manner: Place a small portion of the precipitate on a microscopic glass slide and add a minute crystal of sodium chloride. Then add a large drop of glacial acetic acid and cover with a cover glass. Warm very gently over a small flame. When the acid, salt, and precipitate have become thoroughly mixed allow the slide to cool. Small rhombic crystals of hemin should now appear, which are best seen under a microscope.

The washed precipitate may also be ashed and used for an iron test. The ash should be dissolved in a little pure hydrochloric acid in a porcelain dish and tested by the addition of potassium ferrocyanide and ferricyanide to give the well-known reaction. This test presupposes purity and freedom from traces of iron in the reagents used.

Almen's Guaiacum Test. In a test-tube mix equal volumes of fresh tincture of guaiacum and ozonized turpentine—2 or 3 cc. of each will suffice. The mixture, if made of proper materials, must not show a green or blue color after thoroughly shaking. Now add a few cubic centimeters of the urine to be tested, a drop at a time, and agitate after each addition. If hemoglobin is present it causes the oxidizing material of the ozonized turpentine (probably hydrogen peroxide) to act on the precipitated guaiacum resin, imparting to it first a greenish, and finally a blue color. Old and alkaline urine must be made faintly acid before performing the test. Pus in the urine gives a somewhat similar reaction, and a few other bodies, very seldom present, interfere. The test is very delicate, and if it gives a negative result it is safe to conclude that blood is absent.

Vegetable and Other Colors. It has long been known that many peculiar coloring-matters enter the urine from substances taken as remedies, or with the food. Santonin from wormseed, chrysophanic acid from some kinds of rhubarb, the coloring substances from blueberries, carrots and other fruits and vegetables, are all occasionally met with in the urine, where they may lead to confusion. In most cases the addition of acid produces a yellowish tinge, while alkalies give rise to a red color.

Many aromatic compounds given as remedies pass into the urine in small amount, where they may often be recognized by the peculiar color reactions they yield with certain reagents. Salicylic acid is an illustration, and it may be recognized by the blue color it strikes on addition of ferric chloride.

TOTAL NITROGEN.

In all complete examinations of the urine a determination of the total nitrogen present is necessary. This may be easily made by means of the simple Kjeldahl method, as follows:

Measure accurately 5 cc. of urine into a large Kjeldahl digesting flask and add 25 cc. of pure sulphuric acid and 10 grams of pure potassium sulphate. Heat over a free flame about an hour, or until the mixture has become perfectly colorless. Complete hydrolysis is effected and the nitrogen is all left as ammonium sulphate. Cool the flask, dilute largely with distilled water, neutralize with an excess of pure alkali solution and distill off the ammonia.

This ammonia is caught in 25 cc. of N/4 sulphuric acid, which is titrated afterwards with corresponding alkali, using alizarin red or methyl orange as indicator. The results are accurate.

The total nitrogen determination should always be made as a check on the sum of the nitrogen factors found in the following paragraphs. It will be remembered that the urea nitrogen makes up a large fraction of the total nitrogen.

UREA.

The physiological importance of urea was discussed in the last chapter. As it makes up a large fraction of the nitrogen excretion, which in turn varies with the protein of the diet, the total amount excreted may vary between a few grams and fifty grams daily. By bacterial agency it is rather quickly converted into ammonium carbonate in the voided urine. Quantitative tests should be made, therefore, on fresh urine only.

Recognition of Urea. Because of its extreme solubility urea cannot be easily obtained by evaporation of urine. It may be shown, however, by a simple experiment that by concentrating the urine slowly to a small volume—to one-third or one-fourth—cooling and adding strong nitric acid, a crystalline precipitate of plates of urea nitrate separates, which is characteristic. From this precipitate pure urea can be obtained.

Clinically, this test has no importance, as we are concerned only with a measurement of the amount of urea. This determination can be made in several ways, but in actual practice we employ three essentially different methods. The first depends on the fact that solutions of urea precipitate solutions of certain metals in a definite manner, from which a volumetric process has been derived. The second depends on the fact that solutions of certain oxidizing agents decompose solutions of urea with the liberation of its nitrogen (and carbon dioxide) in gaseous form. From the known relations between weight and volume of the gas, and weight of nitrogen and weight of urea, the absolute amount of the latter may be calculated. The third method depends on the fact that when the urea of urine is heated in solution under certain conditions it is converted quantitatively into ammonium salts from which the ammonia may be distilled and measured.

In the practice of the first method a standard solution of mercuric nitrate is employed, as this precipitates urea quantitatively, but as the reaction is complicated by the presence of other bodies it is seldom used practically at this time. This is the old Liebig method. Ordinarily for clinical purposes we measure urea by the second process; that is from the nitrogen liberated in some oxidation reaction, as will now be described.

Method by Liberation of Nitrogen. A solution of urea is decomposed by a solution of a hypochlorite or hypobromite as illustrated by this equation.

$$CON_2H_4 + 3NaOCl = CO_2 + N_2 + 2H_2O + 3NaCl.$$

That is, nitrogen and carbon dioxide gases are given off. If the reaction is allowed to take place in an alkaline medium the carbon dioxide will be held and the nitrogen alone given off. The volume liberated is a measure of the weight of urea decomposed. From the above equation it is seen that 28 parts by weight of nitrogen correspond to 60 of urea, from which it follows that 1 cc. of pure nitrogen gas, measured at a temperature of 0° C. and under the normal pressure of 760 mm. corresponds to 0.00269 gram of urea. One gram of urea furnishes 371.4 cc. of nitrogen gas.

In employing these principles in practice it is simply necessary to bring together a measured volume of the urine or urea solution and the hypochlorite or hypobromite reagent under such conditions that all of the nitrogen liberated may be collected and accurately measured.

As a reagent, a solution of sodium hypobromite is very commonly employed. As it does not keep well it must be made fresh for use, which is inconvenient

SOME PRACTICAL URINE TESTS.   329

unless many tests have to be made at one time. The reagent may be prepared in this manner:

Dissolve 100 grams of good sodium hydroxide in 250 cc. of water. When cold add 25 cc. of bromine by means of a funnel tube carried to the center of the solution. The bromine must be poured into the funnel a little at a time, with constant agitation, the containing vessel being kept cold. There is an excess of alkali here to hold the carbon dioxide liberated.

A very convenient form of apparatus used in this test is shown in the annexed cut.

The tall jar is filled with water which must stand until it has the air temperature. A 50 cc. burette is inverted in the jar, the delivery end being connected with a bottle holding about 150 cc., by means of a piece of firm rubber tubing. The rubber tube is slipped over a short glass tube passing through the hole in a rubber stopper which must close the bottle accurately. In the bottle is a short stout test-tube, or vial, holding about 10 cc. and which contains the urine to be tested. Into the bottle itself is poured the reagent as above described, which must not reach to the top of the test-tube. On mixing the liquids the urea decomposes, liberating the gas as explained, which passes through the rubber tube and displaces water in the burette so that the volume can be readily determined.

The test is made practically in this manner: Pour about 20 cc. of the strong hypobromite or 40 cc. of the weaker hypochlorite reagent into the bottle. With a pipette measure some exact volume of urine, usually 5 cc., into the small test-tube, and by means of small iron forceps place the latter carefully in the bottle containing the reagent. Insert the stopper which connects the bottle with the burette standing in the jar of water. Allow the apparatus to stand about ten minutes to get the proper air temperature, and then adjust the levels of the liquid in the burette and jar. Note the reading in the burette, and the temperature. Incline the bottle to mix the urine and the reagent, and shake to complete the reaction. The liberated nitrogen, or its equivalent volume of air, passes over into the burette and depresses the water. When there is no change in the gas volume allow the whole apparatus to stand until the contents of the bottle and burette have cooled down to the air temperature again. Then lift the burette, with the clamp as before, to restore the levels and read the gas volume. From this subtract the volume at the first reading. The difference is the volume of nitrogen gas liberated in the reaction, at the observed temperature and atmospheric pressure. If, as sometimes happens, more gas is liberated than the burette will hold, repeat the experiment, using urine diluted with an equal volume of water.

FIG. 30.

Reduce the gas volume to standard conditions by the usual formulas, and for each cubic centimeter calculate 2.7 milligrams of urea as present in the volume taken.

Exact investigations have shown that the whole of the nitrogen is not liberated in the reaction, as at first assumed, but falls short between 7 and 8 per cent. It appears that under some conditions a small part, possibly 3 or 4 per cent., of the nitrogen of the urea is oxidized to nitric acid in the reaction and escapes measurement. Another small portion is left in the ammoniacal condition. Attempts have been made to prevent this abnormal oxidation by adding to the urine a reducing agent to destroy nitric acid, as fast as formed. Dextrose has been used for the purpose, also cane-sugar, and apparently with success. But a great excess of sugar must be added. The results are therefore only approximately correct.

Doremus has introduced a small apparatus for the quick and approximate estimation of urea in which 1 cc. of urine is used. This apparatus is often employed clinically. The same reagent is used.

In the more exact methods of estimation ammonia is formed by hydrolysis reactions, usually aided by hydrochloric acid. Folin has simplified these methods greatly.

The Folin Method. This conversion is accomplished by boiling the urine in a flask with magnesium chloride and hydrochloric acid, the boiling point of the mixture being high enough to effect the hydration. The determination is carried out practically as follows: Measure accurately 5 cc. of urine into a 200 cc. Erlenmeyer flask, add 5 cc. of strong hydrochloric acid and 20 gm. of crystallized magnesium chloride. Add also a small piece of paraffin, about half a gram, and a few drops of alizarin red as indicator. Attach a small reflux condenser to the flask, or the special condensing bulb tube, which Folin recommends, and heat to boiling. Continue the ebullition until the drops returning from the condenser produce a sharp sound when they strike the liquid, now concentrated, remaining in the flask. After this the boiling may be continued more gently through 45 minutes or an hour. But the reaction in the flask must remain strongly acid, as shown by the indicator. The condensing bulb permits the return of evaporated acid, as necessary. Then the mixture is diluted with water, washed into a liter flask with water enough to give a volume of 700 cc., made alkaline with 10 cc. of 20 per cent. sodium hydroxide and distilled. This carries over the ammonia which is best caught in N/10 sulphuric acid, of which about 50 cc. should be taken. At the end of the distillation, which is carried on nearly to dryness, the excess of acid is titrated back, and the difference corresponds to the ammonia. For 17 milligrams of ammonia calculate 30 milligrams of urea.

As urine contains a little ammonia, this must be deducted in the final calculation. It should be determined by a special process. The distillation of the urea ammonia requires about an hour and in the final titration alizarin red is used as indicator. Uric acid and creatinine are not appreciably attacked in this hydrolysis. With practice the method gives very good results. As much of the magnesium chloride on the market contains traces of ammonia this must be separately determined and subtracted from the amount found.

Benedict and Gephart have recently recommended a method in which the hydrolysis of the urea is effected by heating the urine with dilute hydrochloric acid in an autoclave. 5 cc. of urine is mixed with 5 cc. of 8 to 10 per cent acid in a test-tube, which is heated an hour and a half to about 150° in the autoclave. Many tubes may be heated at one time. The results are a trifle higher than in the Folin method, as other constituents appear to be slightly hydrolyzed. The contents of the test-tubes are diluted to about 400 cc., made alkaline, and distilled as in the Folin method.

AMMONIA IN URINE.

The normal amount excreted varies usually between 500 milligrams and one gram daily, and makes up about 5 per cent of the total nitrogenous excretion as shown in the last chapter where, also, the conditions of formation are explained.

Determination of Ammonia. As ammonia is always present in urine in some amount, qualitative tests have little value, and we proceed immediately to quantitative methods. As the ammonia present is in combination as a salt it must be liberated by the action of a strong alkali, but in the choice of one for this purpose we are limited by the fact that the hydroxides of sodium and potassium have a decomposing action on urea and other nitrogenous bodies in urine and cannot, therefore, be used. Milk of lime is free from this objection and may be employed. The experiment is so arranged that the liberated ammonia may be absorbed by a measured volume of standard sulphuric acid, the amount of this neutralized being the measure of the ammonia absorbed:

$$H_2SO_4 + 2NH_3 = (NH_4)_2SO_4.$$

98 parts by weight of the acid correspond to 34 parts of ammonia, NH_3. The old Schloesing method is based on these principles. But it is by no means as accurate or convenient as the later processes in which the ammonia is drawn out from a measured volume of urine by an air current or through the aid of a vacuum pump, absorbed in standard acid, and so titrated. One of the best of these absorption processes is the following.

The Folin Ammonia Determination. In this method 25 cc. of urine is measured into a tall narrow cylinder with about a gram of dry sodium carbonate. Enough light petroleum is poured into the cylinder to form a layer about 5 mm. deep. This is to diminish the frothing in the following operation. The tall cylinder, furnished with a doubly perforated stopper, is connected with an absorption apparatus to wash the air current to be drawn through, and free it from ammonia, and is followed by another absorption apparatus containing a measured volume of N/10 sulphuric acid. The system is connected with a large Chapman pump, capable of drawing several hundred liters of air through in an hour. With a sufficiently rapid air current bubbling through the urine, the ammonia may be completely drawn out in an hour and a half or two hours, and absorbed in the standard acid, of which 25 cc. should be taken and diluted to 200 cc. with water. To aid in the absorption of the ammonia carried along by the air current Folin suggests a special bulb tube through which the air must pass into the standard acid. It is also well to put a Reitmair or Hopkins bulb over the tall cylinder holding the urine, to catch any liquid which may be carried up by the froth produced by the rapid air current. The time required in the determination depends on the air current available, and this should be found by a few preliminary experiments. The results are extremely exact.

With a good air current it is possible to work six or eight of these combinations in series, and complete the tests in the same time. One protecting washing flask at the end of the series is sufficient for the whole, and the tube from this leads to the bottom of the liquid in the first urine cylinder. The cylinder may be 30 to 36 cm. high and 4 or 5 cm. wide, with advantage.

URIC ACID.

The relations of uric acid to the whole nitrogen excretion were explained in the last chapter. The normal variations are between about 0.2 gm. and 1.2 gm. daily.

In the recognition of the acid the following points may be noted: When present in large amount it frequently precipitates from the urine in free form, or as an acid urate; in both cases the precipitate is yellow. When the amount present is small it may be found by acidifying with hydrochloric acid and then allowing the urine to stand some hours in a cool place; uric acid crystals separate. In mixed sediments it may be recognized by this test:

Murexid Test. Throw the sediment on a filter and wash once with water. Place the residue in a porcelain dish, add a drop of strong nitric acid, and evaporate to dryness on the water-bath. A yellow or brown mass is obtained, and this touched with a drop of ammonia water turns purple.

Unless the uric acid or urate is present in the sediment in fine granular form its recognition by the microscope is very simple. Illustrations of the forms of uric acid and certain urates are given in the paragraphs on the sediments.

THE AMOUNT OF URIC ACID.

For determination of the amount of the acid in the urine we have the choice of several methods, not one of which is very convenient or of the greatest accuracy. The first of these depends on the fact referred to above, that hydrochloric acid liberates uric acid from its combination, precipitating it in crystalline form.

Precipitation Test. Measure out 200 cc. of urine and add to it 20 cc. of strong hydrochloric acid. Mix thoroughly and set aside in a cool place for about forty-eight hours. At the end of this time collect the reddish-yellow deposit on a weighed filter, wash it with a little cold water, dry, and weigh. Not over 30 or 40 cc. of water should be used in the washing. The precipitated uric acid is not pure, holding coloring and other substances which increase its weight. On the other hand, it is soluble to some extent in cold acidulated water so that not the whole of it is obtained on the filter and a correction must be made. It is usually recommended to add to the weight obtained 4.8 mg. for each 100 cc. of filtrate and washings.

If the urine under examination contains albumin, the latter must be coagulated by heating with a drop or two of acetic acid and filtered out, before the test is made. If the urine is very cold to begin with and has a sediment of urates, the latter must be brought into solution by warming before beginning the test. To prevent precipitation of phosphates during the warming a few drops of hydrochloric acid may be added. This method gives only approximate results.

Ammonium Sulphate Precipitation. It has been found that the addition of certain ammonium salts to urine produces a precipitate of ammonium urate which is practically complete. Fokker and Hopkins recommended the chloride. Folin suggested the sulphate, which possesses some advantages, as follows:

A special precipitating mixture is employed which in 1000 cc. contains 500 gm. of ammonium sulphate, 5 gm. of uranium acetate, 6 cc. of absolute acetic acid and water to make 1 liter. Measure out 300 cc. of the urine and add 75 cc. of the above reagent. Mix well and allow to stand five minutes. A precipitate containing phosphate and some protein substance, which would interfere with subsequent work if left, separates. Filter off two portions of 125 cc. each, equivalent to 100 cc. of the original urine, and to each add 5 cc. of strong ammonia. Allow the mixtures to stand 24 hours, in which time a complete precipitation of ammonium urate takes place. Collect the precipitates on a small filter and wash practically free from chlorine by aid of 10 per cent. ammonium sulphate solution. Next dissolve the precipitates in 100 cc. of hot distilled water, add 15 cc. of pure strong sulphuric acid, and while still hot titrate the mixture with N/20 potassium permanganate solution, each cubic centimeter of which oxidizes 0.00375 gm. of uric acid. A reduction of the reagent, with loss of color, follows. When the uric acid is fully oxidized a further addition of permanganate leaves a pink tinge in the liquid. The addition of the standard reagent from the burette should cease as soon as a pink tinge is reached, which is permanent two seconds after good shaking. By waiting a longer interval the color fades and more solution must be added from the burette. If the reaction is stopped with the first decided tinge obtained, as explained, for each cubic centimeter of permanganate used from the burette, 3.75 milligrams of uric acid may be calculated as present. In illustration, suppose we start with 300 cc. of urine and precipitate, wash and dissolve as described. If now we run 12.5 cc.

of the twentieth normal permanganate solution into the hot uric acid solution to obtain the pink color, the amount of this acid present is $12.5 \times 3.75 = 46.87$ milligrams in the 100 cc. As ammonium urate is slightly soluble in the mother liquor, a small correction must be added. This amounts to 3 milligrams for each 100 cubic centimeters of the urine carried through to the titration. Urine contains traces of other bodies which are precipitated with the uric acid, but in amount so small that their effect may be practically neglected. The duplicates should agree closely.

THE PURINE BODIES.

These compounds were discussed in the last chapter. Leaving uric acid out of consideration the amount present in urine is not large, but because of their relations to certain diets a determination is often a matter of importance. It is not practically possible to make an accurate separation of the individual purines, and this is, besides, not necessary. Several methods have been suggested for the group precipitation, and of these the following gives good results and is generally employed.

This method depends on the precipitation of the purine bodies as cuprous salts, and may be carried out in this way: measure out 200 cc. of urine into a large casserole, slightly acidify with acetic acid, add 10 grams of sodium acetate and boil about a minute. Add 50 cc. of saturated sodium bisulphite solution, 40 cc. of 10 per cent copper sulphate solution, boil again and filter. Wash the precipitate with hot water on the filter. Return the precipitate to the same casserole by aid of a stream of hot water, add 25 cc. of strong solution of sodium sulphide, and then enough acetic acid to give a distinct acid reaction. Boil five or six minutes to expel the liberated hydrogen sulphide, filter and wash the precipitated copper sulphide with plenty of hot water. Save the filtrate and washings in the same casserole in which the process was begun. This solution contains an approximately pure mixture of the purines, free from other nitrogenous bodies of the original urine. By repeating the precipitation the traces of foreign substances may be removed. Therefore add to the liquid in the casserole copper sulphate and the bisulphite solutions as before, boil several minutes and filter. Wash the copper salts thoroughly with hot water, and finally work down into the bottom of the filter, which should not be large. Allow the precipitate to drain thoroughly, and throw it, with the filter paper, into a Kjeldahl flask for determination of nitrogen by the process already given. It may be well to add 35 to 40 cc. of strong sulphuric acid, 15 gm. of potassium sulphate, about half a gram of copper sulphate and some small flakes of feather tin to aid in the oxidation, which under these conditions follows easily. A clear solution is obtained in less than one hour. This is neutralized with pure alkali and distilled as before into 50 cc. of N/4 sulphuric acid.

In the process as outlined the uric acid is precipitated with the other purines, and the nitrogen determined is the nitrogen of all the purines, including the uric acid. Therefore, the uric acid nitrogen, as determined above, must be subtracted from this result to find the true purine nitrogen. It is absolutely essential to employ a fresh acid sulphite solution in the original precipitation process in order to secure a proper reduction and precipitation of the cuprous salts. The acid sulphite solution may be made by leading a good current of sulphurous oxide (from copper and sulphuric acid) into sodium hydroxide solution until saturation is reached.

CREATININE.

This product, having the formula $C_4H_7N_3O$, occurs normally in urine and is excreted to the amount of 1.5 to 2 grams daily. It is, therefore, more abundant than uric acid, as already pointed out. As it is readily soluble in water and acids it escapes detection, except when looked for by special reagents. In weak solutions it is precipitated by phosphotungstic acid, phosphomolybdic acid, and especially by solutions of several heavy metallic salts. The precipitate given with a neutral solution of zinc chloride is the most characteristic. It gives certain color reactions also. The following test may be applied to urine. If acetone is present it must be expelled by heat. To about 25 cc. of this urine add half a cubic centimeter of a dilute solution of sodium nitroprusside made alkaline with caustic soda. With this the urine gives a ruby-red color, fading to yellow. Then add acetic acid in slight excess and warm. A green color soon appears, deepening finally to blue.

With picric acid creatinine gives a very characteristic reaction, suggested by Jaffé. Add to the urine an aqueous solution of picric acid and a few cc. of dilute sodium hydroxide solution. A deep red color is formed almost immediately, which persists a long time. The reaction is characteristic, as other bodies in the urine do not give it. Folin has made the reaction the basis of a quantitative colorimetric test as follows:

Determination of Creatinine. This is accomplished by aid of a colorimeter in which the color of the above Jaffé reaction is compared with the color from a known creatinine solution, or from a standard dichromate solution. Folin suggests the Duboscq colorimeter for the purpose, but other forms may be used, and employs a half-normal potassium dichromate solution, with 24.54 grams to the liter as the color standard.

Measure 10 cc. of urine into a 500 cc. flask, add 15 cc. of saturated picric acid solution and 5 cc. of 10 per cent. sodium hydroxide solution. Allow the mixture to stand 5 or 6 minutes and dilute to the mark with distilled water. Meanwhile pour dichromate solution into one of the cylinders of the colorimeter and adjust to a depth of exactly 8 millimeters. Pour some of the urine solution into the other cylinder and adjust the depth until the shades are the same, leaving the dichromate side at 8 millimeters. An exact reading of the depth of the urine-picric acid layer is noted, and a calculation of the strength is made from this basis: Folin found that 10 mg. of pure creatinine in 500 cc., under the same conditions, gave such a color that a layer of 8.1 mm. was equivalent to the 8 mm. of the dichromate. The concentration of the solution is inversely proportional to the depth of layer required to match the constant standard. Suppose this is a millimeters. Then $10 \times (8.1/a) = x$. If, for example, we read off a depth of 7.5 mm., $x = 10.8$ mg. of creatinine in the dilute solution. If the calculation shows over 15 mg. or below 5 mg. it is necessary to make a new dilution with a smaller or larger volume of urine, to get the most accurate results, as the comparison is based on a mean content of 10 mg. of creatinine to 500 cc. after the dilution. The colorimetric reading should always be made without delay.

Creatine may be determined in the same manner after conversion into creatinine by prolonged boiling with dilute hydrochloric acid, and subsequent neutralization.

HIPPURIC ACID.

The amount of this acid in the urine is not large, ordinarily, but may be increased by the consumption of certain fruits and vegetables, especially by those containing benzoic acid. The methods of determination proposed are not very exact. The following is the best one.

Measure out 200 to 300 cc. of urine, make it alkaline with sodium carbonate and filter. Evaporate the filtrate nearly to dryness and extract it four or five times by shaking with alcohol. Unite the alcoholic extracts and distill off all the alcohol. The aqueous residue is acidified with hydrochloric acid and extracted by shaking with five or six portions of pure acetic ether. The hippuric acid is dissolved in this way, and the united extracts are partially purified by washing with a little water. The acetic ether is evaporated to dryness at a moderate temperature. A residue of hippuric acid with traces of other substances is left. Most of these impurities may be removed by washing this residue with light petroleum ether, in which fats, benzoic acid and certain other things are soluble.

The hippuric acid left after this treatment may be still further purified by dissolving it in a little *hot* water, heating with well-burned animal charcoal, filtering hot and evaporating the colorless filtrate slowly to dryness. A crystalline residue should now be obtained. When properly carried out 90 per cent of the hippuric acid in the urine may be removed by this general method.

THE PHOSPHATES IN URINE.

Phosphoric acid occurs normally in the urine combined with alkali and alkali-earth metals, of which combinations the alkali phosphates are soluble in water, while the earthy phosphates are insoluble. In the urine, however, they are held in solution through several agencies. The larger part of the earthy phosphates appear to be held here normally in the acid condition; that is, as compounds of the formulas $CaH_4(PO_4)_2$ and $MgH_4(PO_4)_2$. The salts of the type $CaHPO_4$ are present, also, in small amount. As long as the urine maintains its acid reaction these bodies may be expected to remain in solution, but if it becomes alkaline by fermentation, or by the addition of the hydroxides or carbonates of ammonium, sodium, or potassium, the acid phosphates are converted into insoluble, neutral phosphates and precipitated. Most urines contain along with the acid phosphates traces of neutral phosphates which precipitate on boiling. It has been suggested that these phosphates are held by traces of ammonium compounds or by carbonic acid, both of which are driven off by heat, allowing the phosphates to precipitate. It is well known, however, that some urines can be boiled without showing any sign of precipitation. In such cases it is probable that the neutral phosphates are not present.

Part of the phosphoric acid of the urine comes directly from the phosphates of the food and another portion results from the oxidation of the phosphorus-holding tissues. In health, the rate of such oxidation is practically constant, or nearly so, but in disease it may be greatly increased or diminished. Variations in the amount of excreted phosphates may therefore become of considerable clinical importance.

Various statements are found in the books regarding the mean excretion of the alkali and earthy phosphates. Different observers have reported between 2 and 5 grams of phosphoric anhydride (P_2O_5), while 3 grams may be taken, perhaps as the mean.

The recognition of the phosphates is an extremely easy matter. The presence of earthy phosphates may be shown by adding to the urine enough ammonia water to give a faint alkaline reaction and then warming. A flocculent precipitate, resembling albumin, appears and is usually white, or nearly so. But sometimes coloring-matters come down with it in amount sufficient to give it a brownish or reddish shade. It will be recalled that the color of this precipitate was referred to under the head of blood tests.

The alkali phosphates can be detected in the filtrate after separation of the earthy phosphates. To this end, add to the clear alkaline liquid a little more ammonia and some clear magnesia mixture. A fine crystalline precipitate of ammonium-magnesium phosphate separates and settles rapidly. This is very characteristic. The qualitative tests for phosphates have, however, little value in examination of the urine. We are chiefly concerned with the amount, the measurement of which will now be described.

Determination of Phosphates. It is customary to measure the total phosphoric acid, not the alkali or earthy phosphates, separately. We have at our disposal several methods, gravimetric and volumetric, of which the latter are accurate and most convenient. A volumetric process will be described which serves for the measurement of the phosphoric acid as a whole, and which can be used for the separate measurement of the earthy and alkali phosphates by dealing with the precipitate and filtrate described in the qualitative test above. This method depends on the fact that solutions of uranium nitrate or acetate precipitate phosphates in greenish-yellow colored, flocculent form, and that in a solution holding in suspension a precipitate of uranium phosphate any excess of soluble uranium compound may be recognized by the reddish brown precipitate which it gives with a solution of potassium ferrocyanide. The latter substance serves, therefore, as an indicator. If to a phosphate solution in a beaker a dilute uranium solution be added precipitation continues until the whole of the phosphates have gone into combination with the uranium. If, during the precipitation, drops of liquid from the beaker are brought in contact with drops of fresh ferrocyanide solution on a glass plate, no reddish brown precipitate of uranium ferrocyanide appears until the last trace of uranium phosphate has been formed. The production of uranium ferrocyanide is the indication, therefore, of the finished precipitation of the phosphate.

The reaction between uranium and phosphates is shown by the equation:

$$UO_2(NO_3)_2 + KH_2PO_4 = UO_2HPO_4 + KNO_3 + HNO_3$$

From this it follows that 238.5 parts of uranium are required for 71 parts of P_2O_5. In order to have the reaction take place as above it is necessary to neutralize the nitric acid as fast as formed, or dispose of it in some other manner. The best plan is to add to the solution some sodium acetate and acetic acid. The latter brings the phosphates into the form of acid salts while the acetate decomposes with formation of sodium nitrate and free acetic acid, which does not interfere with the reaction.

We need the following reagents:

(a) **Standard Uranium Solution.** This is made by dissolving 36 gm. of the pure crystallized nitrate, $UO_2(NO_3)_2 \cdot 6H_2O$, in water to make one liter. The solution is standardized as below:

(b) **Standard Phosphate Solution.** This is made by dissolving 10.087 gm. of pure crystals of sodium phosphate, $HNa_2PO_4 \cdot 12H_2O$, to make one liter. 50 cc. of the solution contains 100 mg. of P_2O_5.

(c) **Sodium Acetate Solution.** Dissolve 100 grams in 800 cc. of distilled water, add 100 cc. of 30 per cent. acetic acid and then water enough to make one liter.

(d) **Fresh Ferrocyanide Solution.** Dissolve 10 grams of pure potassium ferrocyanide in 100 cc. of distilled water. The solution should be kept in the dark.

The actual value of the uranium solution is determined by the following experiment. Measure out 50 cc. of the phosphate solution, (b), add 5 cc. of the acetate solution, (c), and heat in a beaker in a water-bath to near the boiling temperature. Place several drops of the ferrocyanide solution on a white plate. Fill a burette with the uranium solution and when the solution in the beaker has reached the proper temperature run into it from the burette 18 cc. of the uranium standard. Warm again, and by means of a glass rod bring a drop of the liquid in the beaker in contact with one of the ferrocyanide drops on the plate. If the uranium solution has been properly made no red color should yet appear. Now run in a fifth of 1 cc. more from the burette, warm and test again, and repeat these operations until the first faint reddish shade begins to show on bringing the two drops in contact. With this test as a preliminary one make a second, adding at first one-fifth of a cubic centimeter less than the final result of the preliminary, and finish as before. Something less than 20 cc. should be heeded to complete the reaction. Supposing 19.8 cc. are required for the purpose, the whole solution should be diluted in the proportion,

$$19.8 : 20 :: a : x$$

in which a represents the volume on hand. Each cubic centimeter precipitates exactly 5 mg. of P_2O_5. The liter contains 35.38 gm. of the true uranium nitrate, which, if the salt were absolutely pure, could be weighed out directly.

The test of the urine is made exactly as above. Measure out 50 cc., add 5 cc. of the acetate mixture and finish as before. The 50 cc. of urine, in the mean contains about as much phosphoric acid as was present in the same volume of standard phosphate solution. The titration must be made hot, because the reaction is much quicker and sharper in hot solution than in cold. Make always two tests; the first is an approximation, while the second gives a much closer result.

A separate test of the earthy phosphates may be made by adding to 200 cc. of urine enough ammonia to give an alkaline reaction. The urine then must stand until the precipitated phosphates settle out. The precipitate is collected on a small filter, washed with water containing a very little ammonia, and then allowed to drain. It is next dissolved in a small amount of acetic acid, the solution diluted to 50 cc., mixed with 5 cc. of the sodium acetate solution and titrated as before. The reaction here is not quite as accurate as with the alkali phosphate, but the results are satisfactory for the purpose. The difference between the total phosphates and the earthy phosphates, expressed in terms of P_2O_5, is the amount combined as alkali phosphates.

It is also possible to determine the amount of phosphoric acid combined as monohydrogen salt and that combined as dihydrogen salt, but the determination has at the present time little clinical value.

Instead of finding the end point in the precipitation with uranium solution by means of drops of ferrocyanide as explained, the following process may be followed. Add to the urine the sodium acetate as before and then three or four drops of tincture of cochineal. Heat to boiling and add the uranium solution to the hot liquid. Just as soon as the phosphate is combined and a trace of uranium left in excess it produces a green color or precipitate with the cochineal, which thus serves as an indicator to show the end of the reaction. If the urine is quite warm the color is sharp.

THE CHLORIDES IN URINE.

Practically all the chlorine consumed with the food, mainly in common salt, is eliminated in the urine. The excretion, therefore, varies within wide limits in different individuals, but in the mean amounts to 10 or 15 gm. daily of the salt. A large increase in excreted chlorine points merely to increased consumption, but a marked decrease may point to one of several pathological conditions. Quantitative tests have sometimes considerable value, and are easily made through the reaction between silver nitrate and a chloride, by a volumetric process.

Determination of Chlorine. The reaction between nitrate and chloride is expressed as follows:

$$NaCl + AgNO_3 = AgCl + NaNO_3$$

from which it appears that 5.85 mg. of sodium chloride require for precipitation 16.997 mg. of silver nitrate. A standard silver solution may be made then of this strength, by dissolving 16.997 gm. of the pure fused nitrate to make 1 liter with distilled water.

In one method of determination 10 cc. of urine is evaporated in a platinum or porcelain dish to dryness, after mixing with 2 gm. of potassium nitrate, and 1 gm. of sodium carbonate. The dry residue is carefully fused, and the resulting mass dissolved in water. After exactly neutralizing with nitric acid, and adding a little potassium chromate as indicator the chlorine is titrated in the usual manner. But the process is not generally followed, being replaced by the next one.

Volhard's Method. We have here a method by which the chlorine in urine can be quickly and accurately determined without fusion. The principle involved in the process is this. If to a chloride solution a definite volume of standard silver solution be added, and this in excess of that necessary to precipitate the chloride, the amount of this excess can be found by another reaction, subtracted and leave as the difference the volume actually needed for the chloride. The reaction for the excess depends on these facts. A thiocyanate solution gives with silver nitrate solution a white precipitate of silver thiocyanate, AgSCN. It also gives with a ferric solution a deep red color due to the formation of soluble ferric thiocyanate, $FeS_2(CN)_3$. If the silver and ferric solutions are mixed and the thiocyanate added the second reaction does not begin until the first is completed; that is, the silver must be first thrown down as white thiocyanate before a permanent red shade of ferric thiocyanate appears. The presence of silver chloride interferes but slightly with these reactions. Therefore, if we have a thiocyanate solution of definite strength we can use it with the ferric indicator to measure the *excess* of silver used after precipitating the chlorine solution.

The reaction between silver nitrate and a thiocyanate is expressed by the following equation:

$$AgNO_3 + NH_4SCN = AgSCN + NH_4NO_3.$$

For 16.99 milligrams of the silver nitrate we use 7.6 milligrams of the thiocyanate. In this method the standard solutions required are

(a) **Standard Silver Nitrate Solution,** $N/10$.—Made as before with 16.997 grams of the fused salt to the liter.

(b) **Standard Thiocyanate Solution.** Weigh out about 7.7 gm. of ammonium thiocyanate and dissolve to make 1 liter. Adjust the exact strength as below:

(c) **Ferric Indicator.** Use for this a strong solution of ferric alum, free from chlorine.

To find the exact strength of the thiocyanate solution proceed as follows: Measure into a flask or beaker 25 cc. of the $N/10$ silver nitrate, and add about 3 cc. of the strong ferric alum solution. Add also enough strong, pure nitric acid to make a perfectly clear mixture, for which not over 2 or 3 cc. should be needed. From a burette run in the thiocyanate solution a little at a time, shaking after each addition. A red color appears temporarily, but vanishes on shaking. After a time this color disappears very slowly, which shows that the end point is near. The burette solution is therefore added more carefully, best by drops, until at last a single drop is sufficient to give a permanent reddish tinge. Something less than 25 cc. should be used for this. Repeat the test and if the same result is found dilute the thiocyanate solution so as to make 25 cc. of the volume used in the titration. For instance, if 24.2 cc. were required 900 cc. of the solution may be diluted in this proportion:

$$24.2 : 25 :: 900 : x \qquad \therefore x = 929.8.$$

We have now a standard thiocyanate solution corresponding exactly to the silver solution. To test it further and illustrate its use with chlorides measure out 25 cc. of an N/10 sodium chloride solution, very accurately prepared, and add to it, from a burette, exactly 30 cc. of the silver nitrate solution, then the ferric indicator and the nitric acid as given above. Shake the mixture and filter it through a small filter into a clean flask or beaker. Wash out the vessel in which the precipitate was made with about 20 cc. of pure water, pouring the washings through the filter. Then wash the filter with about 20 cc. more of water, allowing the washings to mix with the first filtrate. This mixed filtrate contains all the silver used in excess of the chloride. Now bring it under the thiocyanate burette and add this solution until a reddish tinge becomes permanent. Exactly 5 cc. should be necessary for this.

The chlorides of the urine may be treated in about the same manner. To a measured volume of the urine, usually 10 cc., an excess of silver nitrate solution is added; 25 cc. with most urines is enough, and then the indicator and acid.

But as the coloring matters in urine interfere somewhat with the sharpness of the titration it is best to destroy them by partial oxidation with sodium peroxide or potassium permanganate. The latter is preferable. To 10 cc. of urine add 3 cc. of pure, strong nitric acid, then the ferric alum. Add also 3 or 4 drops of a saturated, chlorine-free, solution of potassium permanganate. The red color which forms at first, soon disappears. Then add 25 cc. of the N/10 silver nitrate, stir up, filter and titrate the excess of silver with thiocyanate, as above. If the first drops run in from the thiocyanate burette produce a red color it is evidence that there is *no silver in excess*, and that the urine was very strong in chloride. In this case start a new test with urine diluted one-half.

To illustrate the calculation, if we use 10 cc. of urine, 25 cc. of silver nitrate and finally 3.4 cc. of thiocyanate, $25-3.4 = 21.6$ cc., the amount of silver nitrate actually needed for the chloride. Then, $21.6 \times 3.55 = 76.68$ mg., the amount of chlorine in the urine. This is equivalent to 126.36 mg. of sodium chloride, or 12.636 gm. per liter.

THE TOTAL SULPHUR AND SULPHATES IN URINE.

It was shown in the last chapter that sulphur appears in the urine in the ordinary mineral sulphates, in the organic or ethereal sulphates and in the so-called neutral or unoxidized form. The total sulphur is determined by oxidizing everything to the condition of mineral sulphate, and precipitating as barium sulphate. As oxidizing agents which may be used with urine the following have been tried: fuming nitric acid, sodium peroxide, potassium or other nitrate and chlorates. The method given below, which was worked out by Benedict in the author's laboratory yields good results.

Total Sulphur. An oxidizing reagent is made by dissolving 200 grams of *pure* copper nitrate and 50 grams of potassium chlorate in water to make 1 liter. To 10 cc. of urine in a small porcelain dish add 5 cc. of the above reagent and evaporate to dryness over a low flame, then increase the heat gradually and bring up to a high temperature with a good Bunsen flame. Continue the heat five to ten minutes after the mass has fused and solidified. Finally allow the dish to cool, add 10 cc. of 10 per cent hydrochloric acid, and warm until solution takes place. Filter into a small Erlenmeyer flask, wash the filter thoroughly and make the filtrate up to 150 cc. Add now, slowly, 10 cc. of 5 per cent barium chloride solution, best drop by drop, stir gently and allow to stand an hour. At the end of this time collect the precipitate on a weighed Gooch crucible in the usual manner. The increase in weight gives the barium sulphate from the total sulphur.

Total Sulphates. The method recommended by Folin gives uniform results. It is as follows: Measure into an Erlenmeyer flask of 250 cc. capacity, 25 cc. of urine and 20 cc. of 8 per cent hydrochloric acid. Boil gently half an hour, with a watch glass or small beaker over the neck of the flask. At the end of this time cool and dilute with cold water to 150 cc. Add 10 cc. of 5 per cent barium chloride solution, without agitating more than is necessary to mix the liquids. If the barium chloride is added **very slowly** from a dropping tube no further agitation is necessary. At the end of about an hour filter through a weighed Gooch crucible and wash with 250 cc. of cold water. Dry and weigh as usual.

In this method of determination the long boiling of the urine with the hydrochloric acid effects the splitting of the ethereal sulphates, so that in the following barium chloride precipitation they come down with the inorganic sulphates. The latter may be found separately, as recommended by Folin.

Inorganic Sulphates. Dilute 25 cc. of urine with 10 cc. of 8 per cent hydrochloric acid and water to make 150 cc. in an Erlenmeyer flask of 250 cc. capacity. Add **slowly**, from a dropping pipette, 10 cc. of 5 per cent barium chloride solution, without shaking. Allow the mixture to stand an hour and then filter through a weighed Gooch crucible. Wash the precipitate with 250 cc. of cold water, dry and ignite in such a manner that the barium sulphate is protected from reducing gases.

By the conditions of the precipitation the ethereal sulphates are not decomposed, as the liquid was not heated during the operation. From the weight of the barium sulphate the sulphur should be calculated in each case for a given volume. The difference between the two results is the sulphur in the form of ethereal sulphates.

If the difference is taken between the sulphur of the "total sulphur" test and the "total sulphate" test the result is the "neutral" or unoxidized sulphur.

THE SEDIMENT FROM URINE.

Urine is frequently cloudy when passed and on standing deposits a sediment of the substances imparting the cloudiness. Other urines which may appear perfectly clear at first also throw down deposits after a time. This is always the case with urine allowed to stand long enough to undergo alkaline fermentation, when a precipitate of phosphates forms. The deposit is frequently caused by a change of temperature. Warm voided urine holding an excess of urates may be perfectly clear, but becomes cloudy as its temperature goes down with the formation of a light reddish sediment. This is a perfectly normal action, and indeed most sediments may be considered in the same light. Urine containing a deposit is not necessarily pathological.

There are conditions, however, in which the sediment is an indication of abnormality, and its examination becomes important clinically. Certain sediments are pathological because of their origin, others because of their amount. For instance, blood and pus corpuscles, casts of the uriniferous tubules of the kidney and a few other forms are not found normally in urine, and their presence is of importance, whether observed in large or small quantity. Sediments containing phosphates, uric acid and urates, calcium oxalate and other salts, are common enough and usually attract no attention, but if the amount of these deposits is very large there may be attached to them clinical significance and they deserve study.

In the examination of a sediment it is necessary to allow the urine to stand long enough to deposit the important forms it may contain, which may require twenty-four hours or more. For the deposition of a sediment the urine should be left in a place with an even temperature, preferably not above 15° C. A low temperature favors the precipitation of urates, while decomposition may begin if the temperature be allowed to go up. Some of the light organic forms have a specific gravity so little above that of the urine that they may remain a long time in suspension. It is important, therefore, to allow plenty of time for these to settle. If the weather is warm and there is no good means at hand for keeping the temperature of the urine down until the examination can be made, or if for any reason this must be delayed for some days, it is well to add some preservative to the urine; i. e., something to prevent fermentation. Many substances have been suggested for this purpose, some of which are very objectionable inasmuch as they form precipitates which often obscure what is sought for. Chloroform is the simplest and at the same time one of the best substances which can be added.

To 100 cc. of the urine to be set aside for tests add three or four drops of chloroform and dissolve by shaking. It is not well to add more than this, as there is danger of leaving minute droplets undissolved, and these are confusing in the subsequent examination. The chloroform may be applied in the form of aqueous solution. Add about 10 grams of chloroform to a liter of distilled water and shake thoroughly; about three-fourths will dissolve at the ordinary temperature; 25 cc. of this saturated solution may be added to 100 cc. of the urine to be examined, which is then allowed to stand as before.

Recently, formaldehyde has come into use as a urine preservative and is applied as is the chloroform. It must be remembered that both of these substances are reducing agents, and therefore should not be used with urine to be tested for sugar.

All of these methods of preservation are unnecessary if a centrifuge is at hand, with which a deposit from about 10 cc. of urine may be secured in a few minutes. This plan is always preferable, and is now generally followed.

After the deposit has settled pour off the supernatant liquid very carefully and by means of a small pipette with a coarse opening transfer one or two drops to a perfectly clean glass slide. Clean a cover glass with great care and by means of small brass forceps lower it on the drop of liquid in such a manner as to exclude air bubbles. This can be done by lowering it inclined to the slide, not parallel with it, so as to touch the liquid on one side first. In settling down, the cover now pushes the air in front of it and gives a field generally free from bubbles. The slide is then examined under a microscope with a magnifying power of 250 to 300 diameters. Either natural or artificial light may be used, but it must not be very bright. A very common mistake in the examination of urinary sediments by the microscope is to employ so high a degree of illumination that the lighter and nearly transparent bodies are completely overlooked.

Sediments from urine are commonly classed as organized and unorganized, these divisions being then subdivided according to various plans. The important forms under each division are shown in the following schemes:

Organized Sediments.	**Unorganized Sediments.**
Blood corpuscles.	Uric acid.
Mucus and pus corpuscles.	Various urates.
Epithelium from various locations.	Leucine and tyrosine.
Mucin bands, or threads.	Cystin.
Casts of the uriniferous tubules.	Cholesterol.
Spermatozoa.	Fat globules.
Fungi.	Hippuric acid.
Certain other parasites.	Calcium carbonate.
	Calcium phosphate.
	Calcium oxalate.
	Magnesium phosphates.

In addition to these there are often found in the urine certain bodies whose presence must be called accidental; for instance, hairs, fibers of cotton, silk or wool, starch granules, bits of wood, mineral dust, etc. Some of these will be referred to later.

ORGANIZED SEDIMENTS.

Blood Corpuscles. Urine containing blood presents a characteristic appearance easily recognized, unless it be present in very small quantity. If the reaction of the urine is acid the color is generally dark; but if alkaline the shade is inclined to reddish. Blood corpuscles enter the urine from several different sources and their presence is usually a pathological indication, but not always, as they may come, for instance, from menstruation. The kidneys, or their pelves, the ureters, the bladder, the urethra, the vagina, or the uterus may be the seat of the lesion from which the blood starts, and its appearance sometimes gives a clue to its origin.

Fresh blood corpuscles are clear in outline and show distinctly their biconcavity. But corpuscles which have been long in contact with the urine become much swollen, less distinct in outline, often biconvex, or nearly spherical even, and lighter in color. As long as the reaction of the urine is acid the corpuscles

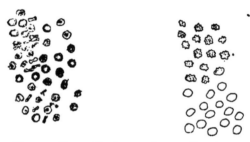

FIG. 31. Human blood corpuscles, 400 diameters.

remain comparatively fresh in appearance, but with the beginning of the alkaline reaction disintegration and loss of color soon set in.

The microscopic recognition of blood in urine is easy enough if it is not too old. The fresh, red corpuscles of human blood have a mean diameter of about 0.0077 mm., but when swollen by absorption of water they are somewhat larger. When seen on edge they appear as shown at the left in the figure above. If presenting the flat side to the eye, they appear as disks whose centers grow alternately light and dark by changing the focus of the instrument. In old urine, especially with alkaline reaction, they appear as granulated spheres, shown at the left of the figure. In all cases the color is more or less yellowish. It is generally assumed that the paler washed-out corpuscles come from lesions higher up, from the pelvis, or kidney even, while the brighter fresh blood suggests a lesion nearer the point of discharge; that is, from the bladder or urethra. This is pretty certain to be the case if the blood is discharged but little mixed with the urine and settles rapidly as a distinct mass.

The *crenated* appearance of the corpuscles, shown at the right and above in the figure, is due to contact with a denser medium, to partial drying and sometimes to certain pathological conditions, while the swollen appearance, from the absorption of water, is illustrated below.

Mucus and Pus Corpuscles. These are white corpuscles somewhat larger than the red blood corpuscles and spherical in outline. The term *leucocyte* is frequently applied to these as well as to the so-called white corpuscles of blood. Their size varies greatly but the average diameter may be given as 0.009 mm. All these corpuscles present, when fresh, a slightly granular appearance and occasionally show one or more nuclei. The addition of a little acetic acid to the sediment brings the nucleus out distinctly so that it may be seen under the microscope as a characteristic appearance.

Mucus corpuscles in small number are normally present in urine, but pus corpuscles enter the urine as a constituent of pus itself which is an albuminous product discharged from suppurating surfaces and not normal. It has been pointed out that the reactions of mucus and albumin are distinct, but urine containing pus always affords reactions for albumin. Pus in urine tends to form a sediment at the bottom of the containing vessel and may be recognized by the following method:

Donne's Test. Pour the urine from the sediment and add to the latter an equal volume of thick potassium hydroxide solution, or a small piece of the solid potassa. Stir with a glass rod. The strong alkali converts the pus into a thick viscid mass closely resembling white of egg. Sometimes this is so thick that the test-tube containing it can be inverted without spilling it. In alkaline urine this glairy mass is sometimes spontaneously formed.

The appearance of the mucus or pus corpuscles in urine depends largely on the concentration of the latter. In urine of low specific gravity the corpuscles absorb water and swell to larger size than normal, while in a highly concentrated urine they may give out water and become reduced in size and shrunken in appearance.

To recognize them under the microscope transfer a few drops of the sediment to a slide, and cover as usual. If the nuclei are not distinct place a drop of diluted acetic acid on the slide at the edge of the cover glass. Part of the acid will flow under the cover and mix with the urine. As it does this the clearing up of the corpuscles, with appearance of the nuclei, can be very easily followed. Urine containing much pus is white and milky. The same appearance is often noticed with an excess of earthy phosphates, but the latter clear up with acids while the pus does not.

Epithelium Cells. Epithelium cells from different sources may appear normally in the urine, and the light cloud which separates from normal urine on standing consists chiefly of these cells. When present in small amount this epithelium has usually no clinical importance, as it easily finds its way into the urine from the bladder, vagina, or urethra. An abundance of cells from these organs would, however, be considered pathological, pointing to a catarrhal condition.

Unfortunately, it is not possible in all cases to determine the source of the cells, as found in urine, partly because cells from different localities have frequently the same general appearance, and partly because, owing to immersion in the urine, they become greatly changed from what they are in the tissue as shown by the microscopic study of sections. It is customary to make three rough divisions of the cells as found in the urine:

1. Spherical cells. 2. Columnar or conical cells. 3. Flat or scaly cells.

The spherical cells are probably normally much flattened, but by absorption of water they become swollen and globular. These cells may be derived from several sources, as from the uriniferous tubules or from the deeper layers of the lining membrane of the pelvis of the kidney, or the bladder, or the male urethra.

These cells have a well-defined nucleus resembling that of a pus cell. But they are much larger, and besides show the nucleus without addition of acid. In nephritis, or other structural diseases of the kidney, these round cells are found along with albumin, and their recognition is then a matter of importance as indicating a breaking down of the tubular walls. Sometimes these cells form a variety of tube cast, to be described later. But it must be remembered that we cannot distinguish with certainty between the cells from the tubules and those from the other localities mentioned.

Conical cells come generally from the pelvis of the kidney, from the ureters and urethra. Some of these cells are furnished with one or two processes, and are broad in the middle and taper toward each end, while the others are broad at the base and taper to a point.

The large flat cells come from the vagina or bladder, and it is generally impossible to distinguish between them. Sometimes they are very nearly circular, sometimes irregularly polygonal in outline. Sometimes the vaginal epithelium is found in layers of scales, which appear thicker and tougher than the cells from the bladder, which occur singly.

What was said about the decomposition of blood or pus cells in urine obtains also for the various epithelium cells. In acid urine they may maintain their distinct outlines many days, but in alkaline secretion they soon undergo disintegra-

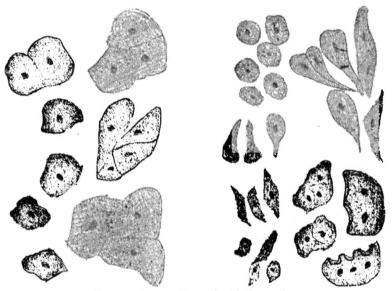

FIG. 32. Common forms of epithelium scales.

tion, which makes their recognition practically impossible. In general the greatest importance attaches to the cells from the tubules of the kidney. The presence of albumin in more than minute traces in the urine would suggest that any smaller spherical cells present may have had their origin in the kidney rather than in the bladder or male urethra. In general it may be said that urine containing large numbers of the smaller, round tubule cells with albumin will also show casts.

Mucin Bands. Urine containing much mucus sometimes exhibits a deposit consisting of long threads or bands, curved and bent in every direction. These bands are important because they are sometimes confounded with the tube casts to be described next. They can be produced in urine highly charged with mucus by the addition of acids, and appear therefore sometimes spontaneously when the urine becomes acid. These threads are sometimes covered with a fine deposit of granular urates and then bear some resemblance to granular casts. In general, however, they are relatively longer and narrower than the true casts of the uriniferous tubules. The mucin threads can occur, and frequently do occur, in urine entirely free from albumin, while true tube casts are usually associated with albumin, although not always, as will be explained below. The length and shape of the mucin threads may generally be relied upon to distinguish them from true casts.

Casts. The structures properly termed casts are seldom found in urine which does not contain albumin. They are formed in the uriniferous tubules, and, to a certain extent, are "casts" of portions of the same. Their specific gravity differs but little from that of the urine, for which reason they remain long in suspension. It is therefore necessary to allow the urine to stand some hours at rest, over night or longer, before attempting an examination, if a centrifuge is not at hand.

True casts of the uriniferous tubules rarely appear in normal urine and their recognition is therefore a matter of the highest importance in diagnosis. Much has been written on the subject of the origin of these bodies in the kidney and several theories have been advanced to account for their formation and chemical constitution. Most of this discussion would be out of place in a work like the present dealing mainly with questions of analysis, but enough will be given to aid the student in his practical work. It must be said that few subjects are more perplexing to the beginner than that of their certain recognition, because of the fact that some varieties are so transparent as to be almost invisible, while others are closely resembled by formations of entirely different nature, not pathological. With practice, however, these difficulties can be surmounted.

Most of the bodies termed casts are formed of organized structures or the remains of such, but another and rather common form consists of crystalline matter, usually uric acid or fine granular urates.

These bunches of urates have no pathological significance and are of frequent occurrence. Urine containing them clears up by heat, and the deposits themselves are dissipated by weak alkali. While it is true that they resemble, to some degree, the so-called granular casts referred to below, there are certain well-defined points of difference. The bunches of urates lack the coherence which can be observed in the true casts, and besides, the granulation is finer and more clearly defined.

The fact that mucin bands occasionally appear covered with a precipitate of granular urates has been referred to. These aggregations are more compact than the loose bunches of urates just mentioned and much longer generally. They are also darker and therefore more easily seen than are the casts proper or the urates.

The true casts are made up of matter in which evidence of cell structure or transformation is visible. An accurate classification of these bodies cannot yet be made, and, as said, authors differ regarding the importance of several forms and their origin. But for our purpose it will be sufficient to make the following rough division, which accords in the main with what is found in the text-books of urine analysis:

1. Blood casts.
2. Epithelium casts.
3. Granular casts.
4. Fatty casts.
5. Waxy casts.
6. Hyaline casts.

What are termed blood casts consist of or contain coagulated blood, recognized by the corpuscles. Plugs of this coagulated matter are forced out from the tubules by pressure from behind, and form one of the most characteristic varieties of casts. They are generally very dark in color, and easily distinguished from other matter. A representation of blood casts is given in the following cut.

In epithelium casts the characteristic substance is the lining epithelium of the tubule. Sometimes this lining epithelium becomes detached in the form of a hollow cylinder, the walls consisting of the united cells. Again, the coagulated contents of the tubule in passing out may carry the epithelium with it as a coating. In either case a grave disorder of the kidneys is indicated, as acute nephritis, or other disease in which a profound alteration of the internal structure of the organ is involved.

What are termed granular casts, proper, appear in a variety of forms, produced probably by the disintegration of blood or epithelium casts.

There is no uniformity in the fineness of the granulation; sometimes a high amplification is necessary to disclose the structure. Occasionally blood corpuscles, epithelium, fat globules and crystals can be detected in them, and when derived from blood cast disintegration they usually have a yellowish red color, which makes their recognition comparatively easy. In outline they are generally regular, with rounded ends, one of which is somewhat pointed. Frequently, however, they appear to be broken, the ends showing irregular fracture.

Fatty casts contain oil drops produced by some variety of fatty degeneration

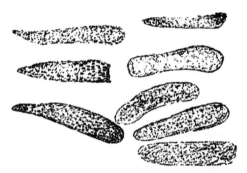

FIG. 33. Blood casts and granular casts.

of the tissues of the kidney. These oil drops may form coherent bunches, or they may be held by patches of epithelium. It also happens that epithelium or granular casts may be partially covered by oil drops. The name, fatty cast, is applied to those in which the fat globules predominate. Along with these globules the microscope sometimes shows crystals of free fatty acids, and probably also of soaps containing calcium and magnesium.

Waxy casts consist of the peculiar matter produced by amyloid degeneration of the kidney. They have a glistening wax-like or vitreous appearance, and refract light very strongly. Sometimes they reach a great length, and they frequently are found with blood corpuscles or oil drops on the surface. They have been detected in several renal disorders. Illustrations are given.

True hyaline casts are nearly transparent and hard to see unless the illumination is very carefully managed. To detect them it is often necessary to add a few drops of a dilute solution of iodine in potassium iodide to the sediment. This imparts a slight color which renders them visible.

The hyaline casts seem to be formed by the passage of homogeneous matter from the tubules, leaving the epithelium behind. A cast is rarely perfectly hyaline, as at least an occasional blood corpuscle, fat globule, or epithelium cell will usually be found attached to it. Waxy casts may be looked upon as a special form of hyaline casts. Very imperfect representations are given in the above cut.

In general, it must be said that the representation of these casts on paper is a very difficult matter. Ordinarily they are drawn and printed much too heavy and dark.

Hyaline casts do not necessarily indicate kidney disease, although this is usually the case. They have been found in urine free from albumin and under circumstances not connected with renal disorders.

The preservation of sediments containing casts is unusually difficult because of the nature of the material to be preserved. In urine of the slightest alkalinity their disintegration soon begins, so that the outlines are rendered indistinct, often making identification impossible.

For temporary preservation the addition of chloroform renders as good service as anything else. Many other sediments can be permanently mounted and kept for future comparison but with casts this can rarely be done.

Beginners are apt to overlook casts in their first examinations. It must be remembered that some of them are nearly transparent and unless brought into proper focus they may not be seen at all. At the outset students usually employ too bright a light in looking for casts. While no specific directions can be given regarding the intensity of illumination best suited for the purpose, this may be

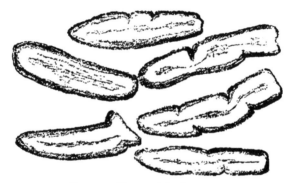

FIG. 34. Waxy and hyaline casts.

said, that the light commonly found necessary in studying ordinary histological slides is far too bright to use in the search for casts.

Practise alone, first under the direction of the instructor, will indicate what is proper here.

Spermatozoa. These minute bodies, as found in the semen of man, have a mean length of about 0.050 millimeter. Nearly one-tenth of this is in the head portion. When observed in recently discharged semen they have a characteristic spontaneous movement by which they are propelled forward rapidly. This motion is soon lost if the semen is diluted with water or similar liquid. Hence, as usually seen in urine, they are entirely motionless. They are found abundantly in the urine of men after coitus or nocturnal emissions, and also in spermatorrhea, when their presence is continuous and characteristic.

Fungi. The urine sometimes contains certain fungus growths, the recognition of which is important. These may have entered the urine after voiding, or they may have come from the bladder.

Normal urine when passed is probably free from fungi of all kinds, but in a short time certain organisms enter it from the air or from other sources and become active in producing in it characteristic changes.

The three important groups of fungi, the **schizomycetes** or **bacteria**, the **hyphomycetes** or **molds**, and the **blastomycetes** or **yeasts** are represented in the organisms sometimes found in the urine. The conditions under which they are found will be briefly explained.

Of the bacteria the following have been observed:

Micrococcus Ureæ. This is the exceedingly common form found in urine undergoing alkaline fermentation by which urea is converted into ammonium carbonate. It is usually introduced from the air and multiplies very rapidly under ordinary conditions. Nearly all old specimens of urine, unless containing some active preservative, are found infected with this small organism. The micrococci are minute spherical bodies belonging to the suborder **spherobacteria** and are found separate or in chains. They are the smallest of the organized forms occurring in urine and appear under a power of 250 diameters but little more than points.

While generally finding their way into urine after it has been voided they are

Fig. 35. Micrococci and other bacteria.

occasionally present in the bladder. It is usually held that under such circumstances they have been introduced by a dirty catheter or sound, although cases are on record where this has not been proved.

In the bladder they give rise to alkaline fermentation, so that the voided urine may show ammonium carbonate directly.

It is now recognized that the production of ammonium carbonate from urea may be brought about by several species of bacteria.

Streptococcus Pyogenes is a pathogenic form sometimes found in the urine in cases of infectious diseases.

Sarcinæ. The genus **sarcina** is frequently classed with the spherobacteria and several species have been found in urine. The cells are larger than those of micrococcus ureæ, and are arranged in groups of two or four usually. They are not pathogenic.

Bacilli. Several species of the genus **bacillus** are found in urine in disease. The most important of these are the typhoid bacillus, **bacillus typhi abdominalis**, the tubercle bacillus, **bacillus tuberculosis**, and the bacillus of glanders, **bacillus mallei**. These bacilli occur in urine only during the progress of the corresponding diseases and their detection is of the highest interest. A description of the methods to be followed for the certain demonstration of these bodies is not within the scope of this book, but must be looked for in the laboratory manuals of bacteriology. It may be said, in general, that in diseases characterized by the presence of certain bacteria in the blood, they may be found also in the urine.

Spirilla. Certain species of the genus **spirillum** have been found in urine. The best known of these is the spirillum of relapsing fever, **spirillum obermeieri**. This is only found rarely and as its habitat is the blood of relapsing fever patients it must enter the urine through a hemorrhage into the kidney. Its form is that of a long, wavy spiral, which makes its detection somewhat easy.

Although not pathogenic it is well to call attention to certain molds which may sometimes be seen in urine. The common blue-green mold, **penicillium glaucum**, is the best known of these, and is occasionally found in urine along with yeast cells. Another mold which has been found in urine is the **oidium lactis**, commonly occurring in milk and butter. It has been observed in fermenting diabetic urine. Both of these fungi enter the urine after voiding. In urine which has stood some time in a cool place the **penicillium glaucum** sometimes becomes covered with an incrustation of urates or minute crystals of uric acid.

Finally we have yeast cells in urine and sometimes in great numbers. Like other fungi they enter the urine from the air and when not very abundant have

FIG. 36. Yeast cells and common mold.

no significance. In great numbers the yeast cells suggest presence of sugar. The ordinary yeast plant, *saccharomyces cerevisiæ*, is shown, isolated and budding, in the accompanying figure. These forms are described here because their presence is often confusing to the beginner.

UNORGANIZED SEDIMENTS.

Uric Acid. Among the more common of the unorganized sediments found in urine this must be mentioned first. As was explained in the last section uric acid occurs normally in combination in all human urine.

Some time after its passage urine often undergoes what has been spoken of as the *acid fermentation* by which a precipitate of urates and even free uric acid

FIG. 37. Uric acid.

may appear. This reaction is in no case due to a ferment process in the ordinary sense of the term, but is probably brought about by a purely chemical double decomposition. Urine contains acid sodium phosphate and neutral sodium urate and it has been suggested that these react on each other according to the following equation:

$$Na_2C_5H_2N_4O_3 + NaH_2PO_4 = NaC_5H_3N_4O_3 + Na_2HPO_4.$$

The precipitate of acid urate settles out and forms a light reddish deposit. If the amount of acid phosphates present is excessive the reaction may go still further, resulting in the precipitation of free uric acid. The well-characterized crystals of uric acid are often found with the sediment of fine urates. Sometimes this liberation and precipitation of the acid takes place in the bladder, and the urine, as passed, shows the crystals or "gravel." If they are relatively large, which is sometimes the case, their passage through the urethra may cause severe pain.

As the illustrations show, uric acid occurs in a great variety of forms. The rosettes and whetstone-shaped crystals are probably the most common, while long spiculated forms are frequently seen. Pure uric acid is colorless but as deposited from urine it is always reddish yellow, because of its property of carrying down coloring-matters. The crystals are often so large that their general form can be seen by the naked eye; usually, however, they are minute.

Uric acid crystals when once deposited are not readily redissolved by heat, but they go into solution by the addition of alkali. If the urine contains extraneous matter, as specks of dust, bits of hair, cotton or wool fibers, the crystals are very apt to deposit on them.

Urates. The common fine sediments of urine are usually urates or amorphous phosphates. They can be most readily distinguished by their behavior with acids and on application of heat. Urates disappear on warming the urine containing

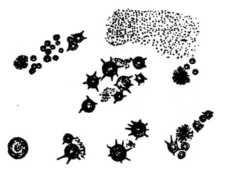

FIG. 38. Common crystalline and granular urates.

them, while a phosphate sediment is rendered more abundant. A urate sediment is little changed by acids, while the phosphates dissolve completely if the urine is made acid in reaction with hydrochloric or nitric acid. The acid urates of sodium and ammonium are the most abundant and are shown in the cut. Acid ammonium urate may exist in urine which has become alkaline from the decomposition of urea and formation of ammonium carbonate, and may therefore be seen in company with the phosphate sediments. The other urates dissolve in alkaline urine. Like uric acid the urates appear in a great variety of forms, and there is still some uncertainty about the composition of some of their crystals which have been found in urine.

Leucine and Tyrosine. These two substances are of rare occurrence in urine and appear only under pathological conditions. Urine containing them shows usually strong indications of the presence of biliary matters as they generally are found in consequence of some grave disorder of the liver in which destruction of its tissue is involved. They have been most frequently found, and associated, in acute yellow atrophy of the liver and in severe cases of phosphorus poisoning. In general they must be considered as products of disintegration and are produced in the intestine in large quantity by bacterial agency in the last stages of the digestion of proteins, as was pointed out in an early chapter.

As both bodies are slightly soluble they may not be seen directly, but only after partial concentration of the urine. In pure condition leucine crystallizes in thin plates but from urine it separates in spherical bunches made up of fine plates or needles. These bunches are sometimes so compact that it is hard to distinguish between them and other substances, particularly lime soaps and oil drops. Chemical tests must therefore be applied. If mercurous nitrate is added to a leucine solution and the mixture is warmed, metallic mercury precipitates. This test can be carried out only when the substance is abundant enough to be purified by crystallization from hot water. Pure leucine, when strongly heated with nitric acid on platinum, forms a colorless residue, which when heated with potassium hydroxide leaves an oil-like drop that does not wet the platinum.

Tyrosine is usually seen in long needles, which sometimes are bunched in the form of sheaves, and is more readily recognized than is leucine. Tyrosine heated with nitric acid on platinum turns orange-yellow, and leaves a dark residue which becomes reddish yellow by addition of caustic alkali. Solutions containing tyrosine when treated hot with mercuric nitrate and potassium nitrite, turn red and finally throw down a red precipitate.

Cystin. This is a rare sediment, although it is found constantly in the urine of certain individuals. It crystallizes in thin hexagonal plates, small ones sometimes resting upon or overlapping large ones. The crystals are regular in form but

FIG. 39. Leucine spheres, tyrosine needles and cystin plates.

variable in size and readily recognized. A rare form of uric acid crystallizes in a somewhat similar manner but the two substances differ in their behavior toward ammonia. To distinguish between them in the microscopic test place a drop of ammonia water on the slide and allow it to pass under the cover glass. Cystin dissolves but, unless heated, uric acid does not. When the ammonia evaporates cystin precipitates.

Cystin is precipitated from urine by addition of acetic acid. Mucin and uric acid may come down at the same time. The precipitate is collected on a filter, washed with water and finally dissolved in ammonia. By neutralizing the ammoniacal filtrate with acetic acid and concentrating a little, it comes down in the characteristic form suitable for microscopic recognition.

Fat Globules. These are often seen in urine, but in most cases have not been voided with it. They can come from several extraneous sources, as from a catheter, from vessels in which the urine is collected or sent for examination, from admixed sputum, etc., which facts should be borne in mind.

352	PHYSIOLOGICAL CHEMISTRY.

Fat has been found in cases of fatty degeneration of the kidney and more abundantly in chyluria where communication seems to be formed between the lymphatics and the urinary tract by the invasion of small thread worms.

Hippuric Acid. This acid is found normally in human urine in small amount. It may be found in large quantity after taking benzoic acid and may even appear in crystalline form in the sediment. It has no pathological importance, ordinarily.

Calcium Carbonate. This is sometimes observed as a coarse, granular sediment which dissolves with effervescence in acetic acid. It occasionally forms dumb-bell crystals, and is devoid of pathological importance.

Calcium Sulphate. Crystals of this substance are rarely found in urine. They form long, colorless needles, or narrow, thin plates.

Calcium Oxalate. We have here one of the commoner of the crystalline bodies observed in urine.

This may be found in neutral or alkaline urine, but more commonly in that of acid reaction. It occurs normally and sometimes is very abundant, especially after the consumption of vegetables containing oxalic acid.

Two principal forms of the crystals are found, the octahedral and dumb-bell crystals.

The octahedra have one very short axis which gives the crystals a flat appearance. When seen with the short axis perpendicular to the plane of the cover

FIG. 40.   Calcium oxalate.

glass, which is the common position, they appear as squares crossed by two bright lines. Sometimes they are seen on edge, and then present a rhomb in section with one diameter very much shorter than the other.

A form of triple phosphate bears a slight resemblance to calcium oxalate, but it is soluble in acetic acid, while the oxalate is not.

The dumb-bells are much less common than the octahedra, and are found in several modified forms, as shown in one of the figures.

The clinical significance of the oxalate is not clearly understood. It does not seem to be characteristic of any disease even when occurring in quantity. It has been found considerably increased in dyspeptic conditions, but not always, and many of the statements found concerning its significance seem to have been based on insufficient observations.

Urine may contain a large amount of oxalic acid, which does not show as a sediment, but must be found by precipitation with calcium chloride in presence of ammonium hydroxide. Acetic acid is then added in very slight excess and the mixture is allowed to stand for precipitation.

The constant or prolonged excretion of large amounts of oxalic acid is spoken of as oxaluria.

The Phosphates. It has been explained that phosphates of alkali and alkali-earth metals occur normally in the urine, and a method was given for their estimation. As sediment we know several forms of calcium and magnesium phosphates and the microscopic detection of these will be here explained. In normal fresh urine of acid reaction these phosphates are held in solution, but if the urine as passed is alkaline it is often turbid from the presence of basic phosphates held in suspension. Urine which has stood long enough to undergo the alkaline fermentation always contains phosphates in the sediment. Finally, it must be remembered that a neutral or very slightly acid urine, containing ammonium salts in abundance, may also deposit a crystalline precipitate of ammonium magnesium phosphate. The common phosphate sediments are those consisting of ammonium

FIG. 41. Triple phosphate.

magnesium phosphate (triple phosphate), normal magnesium phosphate, neutral calcium phosphate, and mixed amorphous phosphates of calcium and magnesium.

Triple Phosphate. Of the crystalline phosphate deposits this is the most abundant and at the same time the most characteristic.

The crystals are the largest found in urine, and from their shape are sometimes spoken of as coffin-lid crystals. Ordinarily they are not found in perfectly fresh urine, but after it has undergone the alkaline fermentation they are generally present in profusion.

Normal Magnesium Phosphate. Crystals having the composition, $Mg_3(PO_4)_2$, $22H_2O$, are sometimes found in urine of nearly neutral reaction. They consist of thin, transparent, rhombic plates with angles approximately $60°$ and $120°$. If urine containing this sediment becomes alkaline, triple phosphate forms.

Neutral Calcium Phosphate. This has the composition, $CaHPO_4.2H_2O$, and is found in urine of neutral or slightly acid reaction. It crystallizes frequently in rosettes formed of wedge-shaped, single crystals, uniting at their apices. The cut shows some variations in the form.

Amorphous Phosphates. Finally we have the very common, finely granular, earthy phosphates in amorphous condition. This sediment dissolves readily in weak acetic acid and is colorless. The common amorphous urate sediment is colored and does not dissolve in acetic acid. On addition of sodium carbonate or hydroxide to urine, the precipitate which forms consists mainly of this phosphate.

These several phosphates can be produced artificially and should be made for study and comparison. The normal magnesium phosphate can be made by dissolving 15 grams of crystallized common sodium phosphate in 200 cc. of water and mixing this with 3.7 grams of crystallized magnesium sulphate in 2000 cc. of water. Enough sodium bicarbonate is added to give an amphoteric reaction and then the mixture is allowed to stand a day or more for precipitation.

Crystals of triple phosphate of peculiar form are often obtained by adding ammonia to urine, and sometimes a trace of ammonia is sufficient to throw down the crystals of neutral calcium phosphate. The latter can also be obtained by

FIG. 42. Neutral calcium phosphate and amorphous phosphate.

adding to a weak solution of crystallized sodium phosphate a trace of acid and then a very little calcium chloride solution.

URINARY CALCULI.

Calculi, like the sediments just described, are formed by the precipitation of certain substances from the urine, but in compact form. Occasionally a calculus consists of a single substance, as calcium oxalate or cystin, but in the great majority of cases a mixture of bodies is present, these being deposited usually in layers around a nucleus which serves as the foundation of the concretion. Calculi are built up much as certain forms of crystals are by successive depositions on a nucleus. Uric acid is a very common nucleus on which may be deposited urates, phosphates, organic matters, etc.

Calculi are sometimes distinguished as **primary** or **secondary**. Primary calculi may be traced to an alteration of the urine of such a nature that its reaction is constantly acid. The foundation for the concretions in this case is found in the kidney and they are built up of such substances as most easily deposit from acid urine. Secondary calculi are generally formed in the bladder, and have for nuclei matters precipitated from alkaline urine, as coagulated blood or other organic substances. Sometimes fragments introduced into the bladder from without serve as the foundation for these secondary formations. Bits of catheters, remains of bougies, and other things have been found as the nuclei around which concretions have formed. The recognition of the nucleus is a matter of the first importance as this gives a clew to the determining cause active in the formation of the calculus.

In making an examination, then, of a calculus, it is first cut in two by means of a very sharp thin saw. This exposes the nucleus which may often be recognized by the eye alone. If one of the halves be polished it is often possible to discern distinctly the various layers grouped around the center.

In a large number of cases examined by Ultzmann about 80 per cent. were found to contain uric acid as the nucleus.

Chemical Examination. In the chemical examination of a calculus several methods may be employed. We may begin by applying certain preliminary tests designed to show the general nature of the stone.

Heat Test. Reduce some of the calculus to a powder and heat to bright redness on platinum foil. Two cases may arise: (a) the powder is completely consumed; (b) the powder is only partially consumed or not at all.

Case (a). If this is the result of the incineration the following substances may be suspected:

Uric Acid, which may be recognized by dissolving a little of the powder in weak alkali, precipitating by hydrochloric acid and examining the precipitate by the microscope.

Ammonium Urate. This gives the above reaction under the microscope, and is further recognized by the liberation of ammonia when heated with a little pure sodium hydroxide solution.

Cystin. Dissolve some of the powder in ammonia, filter if necessary and allow drops of the filtrate to evaporate spontaneously on a slide. Cystin is then recognized by the microscope as already explained. Cystin contains sulphur which, on burning on the platinum foil, gives rise to a disagreeable sharp odor. If a little of the powder be heated with a mixture of potassium nitrate and sodium carbonate the sulphur is oxidized to sulphate, which may be recognized by the usual tests.

Xanthine. This is a rare substance in calculi. Those consisting wholly of xanthine are brown in color and take a wax-like polish.

Organized Matter. Parts of blood cells, epithelium, precipitated mucin, pus corpuscles and similar substances may become entangled with the growing stone and even form a large part of it. On burning, these bodies are recognized by the characteristic odor of nitrogenous matter.

Case (b). When an incombustible residue is left on the platinum foil the stone may contain the following constituents:

Calcium Oxalate. Stones of this substance are very hard and break with a crystalline fracture. They are often called "mulberry calculi." When the powder is heated it decomposes, leaving carbonate, which may be recognized by its effervescence with acids.

Calcium and Magnesium Phosphates. They leave a residue in which the metals and phosphoric acid may be detected by simple tests of qualitative analysis. The ignited powder is soluble in hydrochloric acid without effervescence. When ammonia is added to this solution in quantity sufficient to give an alkaline reaction, a precipitate of triple phosphate or calcium phosphate appears, which may be recognized by the microscope.

The above tests are generally sufficient to tell all that is practically necessary about the calculus. If more detailed information is desired a systematic analysis must be made.

CHAPTER XXI.

THE GASEOUS EXCRETION. RESPIRATION.

In the last chapters the amount of nitrogen excreted with the urine was discussed at some length. With the nitrogen certain corresponding proportions of carbon, hydrogen and oxygen are excreted in the urea, uric acid and other bodies described. But the larger amounts of these elements are thrown off from the body in different form, and especially in the carbon dioxide and water vapor eliminated in respiration and perspiration. From certain classes of foods the end products formed are these two only when the oxidation is ideally complete. This is the case with the fats and carbohydrates, and supposing them wholly burned in the body the final results are represented in this way, taking typical substances for illustration:

$$C_6H_{12}O_6 + 6O_2 = 6CO_2 + 6H_2O,$$
$$C_3H_5(C_{17}H_{35}O_2)_3 + 163O = 57CO_2 + 55H_2O.$$

In the actual behavior of these compounds in the human body, however, the results are somewhat different. The oxidation is never quite as complete as here indicated, as traces of both carbohydrates and fats are left in more complex forms.

THE RESPIRATORY QUOTIENT.

In studying the completeness of oxidation of certain foods much has been learned by a consideration of the factor known as the respiratory quotient which is simply the ratio of the carbon dioxide eliminated to the oxygen absorbed, measured by volume. This quotient is therefore given by the expression CO_2/O_2. For the sugar of the above equation we require six molecules of oxygen, and the carbon dioxide produced is also six molecules. Hence $CO_2/O_2 = 1$. For all common carbohydrates the result is the same. For the fats, however, the quotient is much smaller since 57 CO_2 is the carbon dioxide volume excreted for an oxygen consumption of 81.5 O_2. In this case $CO_2/O_2 = 57/81.5 = 0.7$. For the protein bodies the factor cannot be as easily calculated, since we are not able to assign a formula to these substances, and moreover we are not familiar with all their oxidation products. But from the percentage composition, and the known facts regarding the elimination of urea, uric acid, ammonia and creatinine it is possible to calculate an approximate quotient. This is about 0.8, which factor may be used in calculations.

The use of these quotients is ordinarily based on the assumption that the oxidation is a direct one, and that corresponding to the oxygen absorbed there is almost immediately a liberation of carbon dioxide in the right proportion. But this assumption does not hold absolutely true; the breakdown of carbohydrate, for example, may yield at first, in part, products with high oxygen content, from which CO_2 separates later. In other words, there may be an apparent temporary storing up of oxygen, which would make the quotient appear low. Later a compensating excessive liberation of carbon dioxide would have the opposite result. However, in observations carried out through a period of proper length these variations would not affect the general mean.

The Carbon and Nitrogen Balance. The body is in carbon equilibrium when just as much carbon is eliminated as is consumed in the food, and a determination of this element in the various excreted products and in the food is sufficient to show whether there is gain, loss or equilibrium in body weight. All the food stuffs are organic, it will be remembered, and contain carbon as the fundamental element.

A change in weight may result from gain or loss in fat or gain or loss in protein. Nitrogen equilibrium exists when income and outgo are equal; in this case all the proteins consumed as foods are decomposed. A determination of nitrogen in the urine and feces, coupled with a knowledge of the food protein, will decide this point, since the excreted nitrogen multiplied by 6.25 gives a measure of the food protein. The most accurate method of reaching the value of the excreted nitrogen is by Kjeldahl determinations on the urine and feces, but good approximate results are secured by determination of urea alone, it being remembered that about 85 per cent of the urinary nitrogen appears in this form.

Respiration Apparatus. To determine the volume of oxygen inhaled and carbon dioxide given off, the animal or person under experiment is placed in a respiration chamber of some kind. In a form of respiration chamber sometimes used, an accurately measured volume of air with known content of moisture and carbon dioxide is forced through. The air leaving the tight chamber is analyzed and the amount of carbon dioxide, moisture and oxygen determined. This last determination may be made directly, or the loss of oxygen by the respiration of the person in the cage may be found by calculation from this basis: The sum of all the factors consumed, that is the food and the oxygen, plus the body weight, must be balanced by the weight of the body at the end of the experiment plus the various excreted matters. If A represents the body weight at the beginning of the test and A' the body weight at the end of the test, Ox the oxygen consumed, F the food consumed, Ex the total excreta by weight, then

$$A + Ox + F = A' + Ex.$$
$$Ox = A' + Ex - (A + F).$$

In some of the recent forms of respiration apparatus, especially that of Atwater and Rosa, extremely accurate results are possible in the determination of carbon dioxide and moisture produced; but with increase in size of the apparatus a direct determination of oxygen difference becomes more and more difficult.

In the Zuntz apparatus, which is often used for short experiments on the gaseous excretion only, a peculiar mouthpiece is worn which permits a collection of the carbon dioxide and vapor from the lungs, and of the total expired air. A determination of the oxygen and carbon dioxide is accurately made and this furnishes all the data necessary for the calculation. The nose is closed in this experiment; the mouthpiece is so arranged that air may be drawn in without allowing the excretory products to escape.

DEDUCTIONS FROM RESPIRATION EXPERIMENTS.

These are undertaken to answer a number of important questions. The weight of carbon dioxide excreted may reach fifteen hundred grams or more daily and it is interesting to know under what circumstances it is increased and when diminished. Very simple observations show that the body at rest produces much less of the gas than does the body at work. In the latter condition the destruction of food stuffs is called for to liberate mechanical energy. This is practically possible only through oxidation, and carbon dioxide is the first tangible result of the oxidation.

The question also comes up, what kind of organic matter is most readily or most commonly oxidized when work is done? On this question much has been written and our views have undergone various changes through the years. Liebig considered the proteins as the foods which must be burned to enable us to do mechanical work, but in a famous experiment by Fick and Wislicenus, undertaken to throw some light on this question, no great excess in the excretion of urea was found in the work of ascending the Faulhorn, and the protein oxidized was far too little to account for the work done. Other investigators reached the same conclusion, but it has been found that under certain conditions the proteins *may* be consumed to do work. Ordinarily fats and carbohydrates are used in preference, and no large amount of protein is used if the other substances are present in sufficient quantity.

The question of what kind of foodstuff is oxidized through periods of work and rest may be answered by experiment. As just intimated, examinations of the urine give us information as to the nitrogen excretion, and the extent of oxidation of fats and carbohydrates may be measured by respiration experiments. In a fasting animal at rest the respiratory quotient sinks to a value but little above 0.7, showing that the substance metabolized is mainly fat; as some proteins are also used up the quotient cannot absolutely reach 0.7. If work is done by the fasting animal, the carbohydrate bodies of the muscular juices, glycogen essentially, are called upon and their effect is added to that of the proteins in raising the respiratory quotient. On the other hand, it has been found that a well-fed animal at rest, with abundance of carbohydrates in the ration, will excrete a volume of carbon dioxide nearly as great as that of the oxygen absorbed. In this case the ratio CO_2/O_2 shows that essentially carbohydrates are burned and that fat is allowed to accumulate. When very hard work is done by the well-nourished animal the quotient sinks to an intermediate value, showing that fats are now consumed as well. This would be evident also from observations continued over a long period in which no accumulation of fat could be recorded. With moderate work there is not much change.

Some of these results are illustrated by the figures in the following table taken from observations published by Chauveau and Laulanié, in which dogs were the subjects of experiment. These figures show very clearly alteration in the respiratory quotient with work, and also by diet.

No.	Food Consumption.	Before Work.	Minutes of Work Before Observations.						Minutes of Rest Following Work.			
			30	45	60	90	120	180	45	60	120	240
1	24 hours fast.	0.790	0.943		0.905	0.900			0.789			
2	6 days fast.	0.750	0.819		0.840					0.687		0.756
3	1 day fast.	0.874			0.895		0.900	0.900		0.770	0.770	
4	2 days fast.	0.740			0.780		0.866	0.866		0.730	0.708	
5	3 days fast.	0.685			0.790		0.808	0.772		0.681	0.681	
6	After full meal.	1.033		1.017		1.044				1.052		
7	After full meal.	1.000			1.042		1.008			1.032	1.017	

Other experiments are in general good agreement with these. The effect of work in the fasting animal is seen almost immediately. In the last experiments the respiratory quotient is greater than unity. This may be due in part to slight errors in observation, but it should be remembered that there are classes of compounds in which such a result would always follow. Such compounds are not common articles of food, but often make a part of certain vegetable foods. The complete oxidation of tartaric acid, for example, would yield a quotient of 1.6.

The quotient may sometimes be high, as intimated above, if the oxygen has been at first absorbed to form compounds relatively rich in oxygen, which are later broken down rapidly, under working or other conditions. If an observation is made just at this period the excess of CO_2 liberated would present an abnormal result. For characteristic results the observation periods should be as long as possible.

Illustrative Case. Some idea of the importance of the respiratory coefficient determination may be obtained from a consideration of the following assumed case in which the conditions are made somewhat ideal for simplicity of calculation. The numerical values given are such as might be obtained from the mean of several 24-hour experiments in a large respiration chamber. The diet is assumed to be abundant and the tests begun after a condition of practical nitrogen equilibrium is reached.

Initial weight .... 75 kilograms
Final weight .... 75.05 kilograms

INCOME OBSERVED.

	Wt. in Grams.	C Per Cent.	N Per Cent.	O Per Cent.	H Per Cent.	C Total.	N Total.	O Total.	H Total.
Proteins.........	150	53.5	16.0	23.5	7.0	80.3	24.0	35.2	10.5
Fats.............	110	76.5		11.4	12.1	84.2		12.5	13.3
Carbohydrates...	440	44.2		49.6	6.2	194.5		218.2	27.3
Salts............	35								
Water...........	2,000								
Total.........	2,735					359.0	24.0	265.9	51.1

OUTGO OBSERVED.

	Weight in Grams.	C.	N.	Salts.	Vol. CO_2.
Respiration, CO_2..	932	254.2			471 l.
Respiration, H_2O..	904				
Urine, H_2O.......	1,350				
Urine, solids......	74	8.7	20.4	30	
Feces, H_2O.......	100				
Feces, solids.....	33	16.2	3.6	5	
Total	3,393	279.1	24.0	35	

The weight of the various excreted products is greatly in excess of the visible income, but the oxygen inhaled has not yet been calculated. The formula given above may be applied to find this:

$$\begin{aligned}\text{Oxygen} = &\ A' &+&\ \text{Excreta} &-&\ (A &+&\ F) \\ &\ 75{,}050 & &\ 3{,}393 & &\ 75{,}000 & &\ 2{,}735 \\ = &\ 78{,}443 & &\ & -&\ 77{,}735 \\ = &\ 708 \text{ grams} \\ = &\ 495.4 \text{ liters.}\end{aligned}$$

In the table above the volume of carbon dioxide eliminated is given as 471 liters. The respiratory quotient is therefore

$$RQ = \frac{471}{495.4} = .95.$$

This gives us the first clue as to the nature of the foods metabolized. The factor is so much larger than that corresponding to the fats that we may practically exclude these at once. In any event there is a large protein metabolism since the original nitrogen of the food is all found in the urine and the feces. In other words, we have nitrogen equilibrium, with no storing up of protein in the tissues. The respiratory quotient corresponds to the combustion of carbohydrates and proteins mixed.

If we assume for the moment that no fat is oxidized, this calculation may be made. The 254.2 gm. of carbon in the carbon dioxide of respiration calls for 678 gm. of oxygen. The difference between this and the calculated absorbed oxygen, 708 gm., amounts to 30 gm., which must be used up in oxidizing hydrogen of protein substances. This conclusion is drawn because the carbohydrates contain enough oxygen to burn their own hydrogen, and the protein nitrogen appears as urea and calls for no outside oxygen. The burning of fat hydrogen is excluded in the assumption.

The nitrogen of the feces corresponds to 22.5 gm. of original protein (6.25 × 3.6). Not all of this nitrogen is actually in the form of unchanged or residue protein; a part of it represents products of metabolism which are excreted in the feces, as explained in a previous chapter. Probably a considerable fraction may be considered in that form; but it must be counted as a loss to the body, and we have therefore as net available protein (actually used) about 127.5 gm. In the final metabolism of this the nitrogen appears in urine in several forms, but mostly as urea. The per cent of nitrogen in this is 46.7. In some of the other compounds the nitrogen is higher and in some lower. In ammonia much hydrogen (relatively) is held, and in uric acid little. In some cases there is an excess of carbon and in other cases relatively little carbon is held with the same weight of nitrogen. The various conditions balance each other pretty well, so that no great error will be made if, for our special purpose, we count all the excreted nitrogen as combined in the form of urea. We have then these relations:

	C.	N.	H.	O.
In metabolized protein	68.3	20.4	8.9	29.9
In urea	8.7	20.4	2.9	11.6
	59.6	00.0	6.0	18.3

To oxidize this remaining carbon requires 159.8 gm. of oxygen. The 18.3 gm. of protein oxygen will oxidize about 2.3 gm. of hydrogen. The remaining hydrogen from the 6 gm. will call for about 29.6 gm. of oxygen, which corresponds closely to the amount calculated above.

The ingested carbon is seen from the table to be 359 gm.; the excreted carbon is 279 gm., from which it follows that the body has gained 80 gm. If we assume this to be in the form of fat the latter must amount to about 104 gm. This represents the true gain in body weight; the weighings showed a gain of only 50 gm. The discrepancy may be accounted for by assuming an excessive excretion of urine. No such discrepancy would appear if the urine were passed from the bladder as fast as formed, but as it is collected at intervals it is not possible to obtain exactly comparable results.

In the feces there must be some carbon derived from fats; but the amount cannot be large, because for the nitrogen of the feces we must calculate at least 10 or 12 gm. to correspond. This would leave about 5 gm. of carbon from other sources, accounting for the discrepancy between consumed fat and deposited fat.

The above calculations illustrate the principles involved; in an actual practical observation the method would be the same, but the interpretation of results might not be as simple, especially with a low respiratory quotient found. In the above tables the *salts* taken with the food are assumed to include those to be formed by the oxidation of the protein, and the latter substance figured as income is assumed to consist of the organic elements only. A slight error is introduced in the calculation in this way, but that is not considered. In actual practice, of the carbohydrates some little would escape complete metabolism. The above results would correspond to a completely burned carbohydrate.

In the above table of observations the total oxygen in the consumed substances, including the water, is 2,044 gm. In the excreted products, allowing 30 gm. for the solids of the urine and feces coming from bodies other than the original salts, the oxygen appears to amount to about 2,800 gm. The difference shows an excess of 756 gm. while the calculation above gave 708 gm. of oxygen taken in. The discrepancy is due to the excess of water excreted as urine. It will be noticed also that there is a great excess of excreted water over the 2,000 gm. consumed. This amounts to over 350 gm., of which 300 gm. would come from the combustion of the carbohydrates and proteins metabolized.

SKIN RESPIRATION.

It is usually assumed that the gaseous exchange is wholly through the lungs, but this is not quite correct. Experiments with men and animals have shown an absorption of oxygen and an escape of carbon dioxide through the skin. A number of observers have put results for the latter on record which, however, are not in good agreement. For 1.6 square meters of skin surface the results found in seven observations varied from 2.2 gm. to 32 gm. in 24 hours. The last result is probably much too high. It has been noticed further that the amount of carbon dioxide escaping through the skin is increased greatly by temperature. The excretion at 30° seems to be several times as great as at 20°.

For the absorption of oxygen no exact figures are given, but the amount is very small. In some of the lower animals, however, a large part of the absorbed oxygen, as well as of the excluded carbon dioxide, *may* be by way of the skin. This has been shown especially in the frog, where after removal of the lungs a nearly normal exchange may be noted for a period of days.

The question of the excretion of other gases than carbon dioxide by the skin, and the lungs also, has been much discussed. Formerly it was held that a very appreciable quantity of organic gaseous bodies is given off through the skin and this elimination was considered necessary for the well being of the body. The unpleasant odor of the air of a crowded room was ascribed to these organic emanations. But much doubt has been thrown on this notion by various experiments, some of which are of very recent date, which seem to show that these odors come, not through the skin, but from decaying substances on the surface of the skin or from the clothing, if it is old and soiled. Experiments have been made of testing the air drawn through a small respiration chamber, enclosing the body of a man to the neck, with perfectly clean skin and clothed in fresh, clean garments. Such air is practically without odor and has no action on solutions of permanganate through which it is aspirated. It is free from ammonia. The odors of perspiration are apparently largely due to the fermentation changes of solid or semi-solid substances on the surface of the skin rather than to excreted gaseous products passing through the pores with the water. It has been found also that the whole surface of the body may be covered with varnish without harmful result if precautions are taken to prevent loss of heat.

TIME AND PLACE OF OXIDATION.

The determination of the respiratory quotient through short intervals shows considerable variations, as pointed out some pages back. The human organism has not the power of storing up oxygen in the free or combined form through a long period, as appears to be the case with some cold-blooded animals, which are able to exist for a time in an atmosphere free from oxygen. With man and warm-blooded animals in general this is not possible; with these life without oxygen may be maintained for but a few minutes at most. An exception exists in the case of those animals which pass the winter in a dormant condition (hibernating animals) and for human beings in trance. Here the absorption of oxygen and excretion of carbon dioxide are reduced to a minimum. But ordinarily man and the higher animals require some inflow of oxygen all the time.

The extent to which this oxygen is used depends on the activity of the muscles largely. In rest periods the amount of oxygen taken up by the muscles is much greater than is the carbon dioxide given off, but with the contracting or working muscle the reverse is the case. In experiments in which the changes in the blood supply of individual muscles may be followed it may be shown that for rest periods the respiratory quotient for the muscle may fall far below 0.7 or even below 0.5. From the rapidly contracting muscle, on the other hand, the evolution of carbon dioxide is relatively great. A respiratory quotient, for the muscle, of 1.5 or even 2 or more may be found. This indicates that during rest oxygen may be taken up from the blood and held or condensed in some manner by substances within the muscular tissue, but of the mechanism of this reaction unfortunately but little is known. In doing work tissue is rapidly oxidized at the expense of the stored-up oxygen, and a great excess of carbon dioxide is given off quickly. These changes follow one upon the other rather rapidly. In the oxygen-absorbing stage some intermediate products are probably built up from sugar or glycogen or other substances, which fall apart with liberation of water and carbon dioxide in the succeeding active condition of the muscle.

The problem of oxidation in the tissues is possibly somewhat like that of etherification in which alcohol yields ether indirectly through the intermediate ethyl sulphuric acid, and several suggestions have been brought forward as to the character of complexes formed in one stage of the oxidative metabolism to be decomposed in another. At the present time these suggestions are practically wholly within the realm of speculation, and not therefore suitable for presentation in this place. It is likely that all these reactions, which seem to be carried on in the tissues rather than in the fluids of the body, are incited by enzymic ferments, and to-day certain classes of *oxidases* are often assumed to be the agents active in the changes. For some of the oxidations it has been pretty well settled that an enzyme produced by the pancreas is necessary.

CHAPTER XXII.

THE ENERGY EQUATION.

We come now to a brief consideration of one of the most important questions connected with the whole animal chemistry, and this is the question of the liberation of energy from the consumption of various foods. The function of the food we eat is a multiple one. It may not only increase the body weight and maintain the various functions of the body through oxidation, but in its combustion heat is liberated to maintain also the body temperature, and energy is furnished to enable us to perform external work. It is interesting to measure the effect of the food in these several directions, which may be done with a fair degree of accuracy. The following considerations will show the basis of the calculations.

POTENTIAL ENERGY OF FOOD.

The food, consisting essentially of combustible substances, is the source of a large amount of *potential energy*. In the complete combustion of the fats, carbohydrates and proteins of the food a large amount of heat is liberated and this in turn is the equivalent of a certain amount of work. The potential energy of chemical substances may be measured in various ways, but for purposes like the present it is customary to measure this energy in terms of the units of heat liberated in the combustion of the body in question with oxygen. Certain units are in common use:

Unit of Heat, Calorie. The unit of heat or calorie may be defined as the quantity of heat required to raise the temperature of a gram of water one centigrade degree, at a mean temperature. As the heat absorption of a gram of water is not quite the same throughout the scale, the calorie is perhaps more satisfactorily defined as the one hundredth part of the quantity of heat required to raise the temperature of a gram of water from 0° to 100° C. This gives the ordinary, or *small calorie*. In dealing with large heat transfers a larger unit is preferable and one just 1,000 times as large is frequently used. In this the kilogram in place of the gram of water is warmed, and the unit is called the *large calorie*. The first may be abbreviated *cal.* and the second *Cal.*

Unit of Work and Unit of Force. The unit of *force* is called the *dyne* and may be defined as the force which, acting for 1 second on a mass of 1 gram, gives to it an acceleration of 1 centimeter per second. The force of gravity at the sea level is about 981 dynes, since this adds to a falling body an acceleration of 981 cm. per second.

The unit of *work* is the *erg*, and it may be defined as the work done in overcoming unit force through unit distance. One dyne acting through one centimeter gives us one erg of work. To lift 1 gram through 1 centimeter requires 981 ergs of work.

Mechanical Equivalent of Heat. Work may be done by the proper utilization of heat, and in turn work may be wholly converted into heat. It is possible, therefore, to express one in terms of the other. The mechanical or work equivalent of a unit of heat has been determined many times by very elaborate experiments. If a given quantity of heat could be applied wholly to the lifting of a weight it would be found, in accordance with the mean results of these experiments, that 1 calorie would be able to lift 423.5 gm. through 1 meter, or 1 gm. of substance through 423.5 meters. Conversely, if a gram of water be dropped from a height of 423.5 meters, and its energy of motion wholly converted into heat, its temperature will be found to be increased 1° C. We have then these relations:

$$1 \text{ calorie} = 42{,}350 \text{ gm. cm.}$$
$$= 41{,}500{,}000 \text{ ergs.}$$

Heats of Combustion. By means of calorimeter experiments the following heats of combustion have been determined. Results found by different workers show slight variations, but the values here are mean values and sufficient for illustration. The number of calories furnished by burning 1 gm. of substance in each case is given.

TABLE OF HEATS OF COMBUSTION.

Hydrogen	34,200	Cane sugar	4,000
Carbon	8,100	Starch	4,200
Ethyl alcohol	7,060	Casein	5,700
Glycerol	4,200	Egg albumin	5,700
Mannitol	4,000	Urea	2,500
Palmitic acid	9,300	Uric acid	2,700
Stearic acid	9,400	Leucine	6,500
Fats, average	9,400	Tyrosine	6,000
Hexoses	3,700	Creatine, anhyd	4,250

These values are for complete combustion, but as the proteins in the body are not oxidized to leave water, carbon dioxide and nitrogen, we must subtract from the given values the heats of combustion of the urea, uric acid, creatinine and other products found in the urine, in order to secure the *physiological heats of combustion*, with which we are practically concerned.

DISTRIBUTION OF FOOD ENERGY.

With these preliminary considerations we are able to look at the manner in which the energy of the consumed food is distributed. On the one side we have the substance burned, on the other the products, which may be represented diagrammatically in this way:

Potential energy of
Food.
	Potential energy of
	    Flesh gained.
	    Feces.
	    Urine.
	    Perspiration.
	Kinetic energy of
	    Work.
	    Heat.

Experimentally, the whole of the kinetic energy may be made to take the form of heat, which simplifies the observations materially. It is practically possible to determine the heat liberation in the large respiration calorimeters already referred to, and the use of such apparatus will be explained below. First, however, a general method of calculating the energy liberated as heat will be given.

CALCULATION OF KINETIC ENERGY OF FOOD.

In illustration of this we may make use of the example given in the last chapter, and employ a method which in principle is very simple. The income of energy is due to the consumption of certain weights of protein, fat and carbohydrates, the last of which we may assume is made up of 9 parts of starch and 1 part of cane sugar, all weights referring to the anhydrous condition. The effect of the oxidation of sulphur and phosphorus will be neglected here, and the protein will be assumed pure carbon, hydrogen, oxygen and nitrogen. We have then as income:

$$\begin{array}{lrr}
\text{From 150 gm. protein,} & 150 \times 5700 = & 855,000 \\
\text{110 gm. fat,} & 110 \times 9400 = & 1,034,000 \\
\text{440 gm. carbohydrate,} & 440 \times 4180 = & 1,839,200 \\
\hline
& & 3,728,200
\end{array}$$

In small calories the whole income is therefore equivalent to 3,278,200 cal.

We have next to calculate the potential energy of the food stuffs not actually consumed, which are left in the feces and the urine, and also the energy of any substance which may be put down as a gain in weight in the body. Recalling the data of the experiment in the last chapter we have

133 gm. feces with 16.2 gm. C.
1,424 gm. urine with 20.4 gm. N.

Calculating the N of the urine as urea, which in practice would not be quite accurate, we have 44 gm. of that substance. The organic matter of the feces corresponds approximately to 22.5 gm. of bodies resembling protein and 5.5 gm. of bodies resembling fats, and these data we can now employ in the calculation.

The illustration gave also a gain of 80 gm. of fat. The solid matter lost in the form of perspiration is so small that it may be ignored for the present purpose. We have then the following deductions to make:

Potential energy in 80 gm. of fat stored	752,000
133 gm. of feces.	180,000
1,424 gm. of urine	110,000
	1,042,000

This leaves as a balance to be calculated as kinetic energy

3,728,200
1,042,000
2,686,200 calories

Another method of calculation deals with the carbon and hydrogen of the food and feces only. The heat production was at one time assumed to depend on the combustion of the carbon of the fats, carbohydrates and proteins and the hydrogen of the fats and proteins. The hydrogen of the carbohydrates was not considered because it was supposed to be closely combined with the oxygen present in the same compounds in such a form as to yield no more heat on oxidation. In like manner the combustion heats of the urine and feces may be calculated from the whole carbon and hydrogen content of the organic substances. The total carbon of the food in the experiment is 395 gm., of the hydrogen in fats and proteins 23.8 gm. The carbon of the urine and feces is 24.9 gm., while the hydrogen of the urine and feces is about 5.2 gm. We have then:

Heat units from 359 gm. of food carbon	2,907,900	
Heat units from 23.8 gm. of food hydrogen	813,900	
	3,721,800	3,721,800
Heat units from 24.9 gm. of excreta carbon	201,690	
Heat units from 5.2 gm. of excreta hydrogen	177,840	
	379,530	379,530
Net calories		3,342,270

From this result the value of the energy stored as fat would have to be subtracted as before. This method of calculation gives a somewhat lower result than the other, and largely because of the uncertainty in allowing for the excreted carbon and hydrogen, but it has value as a comparison process.

Respiration Calorimeters. In experiments with men or large animals on the combustion of food and liberation of heat some kind of respiration apparatus is employed. Some modification of a type originally introduced by Pettenkofer is generally used. In this the subject is placed in a chamber with double walls through which a current of air may be forced and uniformly mixed inside. A known part of the ingoing air and of the outgoing air may be diverted for analysis so as to permit an exact determination of the amount of oxidation products liberated at any time. The Atwater and Rosa calorimeter is the most complete of all such constructions. In this the heat liberated by the subject is taken up by a current of cold water circulating through numerous coils of pipe inside the chamber and in such a way as to maintain a perfectly uniform temperature in the chamber space. The walls of the chamber are made of compartments containing two layers of air and two layers of water maintained in such relations that they prevent gain or loss of heat. The whole heat liberation is taken up by the circulating water and may be accurately measured. The respiration chamber has a capacity of about 175 cubic feet and is large enough to contain a chair and small table for the convenience of the occupant and a couch to sleep on at night. The construction is such that food may be passed in and the urine and feces removed without making any appreciable change in the temperature or content of the air inside. With such an apparatus it is possible to carry on a test of many days duration and obtain extremely accurate and important results.

In work experiments in such a calorimeter a bicycle is mounted so that work is done against friction. The final effect is increased heat liberation, measured as before.

The construction of this large calorimeter suggested the building of still larger ones of the same general type. Some of these are being used in agricultural experiment stations in metabolism experiments on large animals, from which results of great practical value may be expected.

DISTRIBUTION OF THE HEAT ENERGY.

According to the first calculation above we have from the 700 grams of food consumed a balance of 2,686,200 calories. It remains to show about how this may be dissipated. If retained in the body it would soon bring the latter to the boiling point. But the heat liberated in the combustion of the food finds several outlets, the most important of which will be now indicated. In the first place the urine and feces leave the body at a temperature much higher than that of the water consumed; the water of respiration and perspiration has to be vaporized at the expense of heat; the air inhaled is warmed to a temperature of 37°, which is in the mean 20° higher than when taken in. The specific heat of the air (at constant pressure) is about 0.25. We have then, approximately, the following relations, assuming 15 kilograms of air to be inhaled in the 24 hours:

To warm 15,000 gm. air 20°	75,000 cal.
To warm 1,557 gm. urine and feces 20°	31,140
To evaporate 904 gm. of water (904 × 580)	524,320
	630,460

THE ENERGY EQUATION.

This number of calories must be taken from the net produced calories to obtain the heat radiated or otherwise lost by the body. We have then this difference:

```
2,686,200
  630,460
─────────
2,055,740
```

That is, something over 2,000,000 calories are dissipated by radiation.

HEAT RADIATION WHEN WORK IS DONE.

All these calculations are based on the assumption that no mechanical work is being done by the person under observation, or if done it is finally all converted into heat. In experiments in the respiration calorimeter a very close agreement is found between the calculated and observed heat or energy liberations. This is illustrated by the results of one of the Atwater and Rosa experiments. The figures are the daily means from tests running through 4 days:

Total energy of food, determined	3,678 Cal.
Energy of urine and feces	264
Net energy	3,414
Energy of fat lost	488
	3,902
Energy stored as protein	38
Total energy of material actually oxidized	3,864
Heat actually measured	3,739
	125

There is, therefore, a difference of only 125 large calories in this test, which was one of the early ones with the new apparatus. In later experiments described by Atwater and his colleagues much closer results have been reported, which shows the general correctness of the method of calculation followed.

Effect of Work. To maintain the individual at work a greater expenditure of energy is necessary, and the total energy of substances metabolized must be balanced by the heat liberated and external work done. In this connection it may be well to recall some relations first pointed out by Hirn, in which a comparison is drawn between the work of man and an engine. In both cases the work is accomplished through the expenditure of the potential energy stored up in carbonaceous substances, food in the one instance, coal in the other. As the illustrations above show, the heat from the food is practically constant, whether it be evolved through oxidation in the body or in a calorimeter.

Imagine now a small steam engine burning a constant amount of coal inside a calorimeter. In one case let no work be done by the piston; the heat of the steam is not employed in expansion, but is totally absorbed by the water of the calorimeter, which takes up a certain number of calories that may be accurately noted. In a second case allow the same amount of coal to be burned under the boiler of the small engine in the same time, but let the engine do work *outside* the calorimeter, which may be accomplished, for example, by means of a small shaft passing through the walls of the calorimeter in such a way as to convey no appreciable amount of heat. It will now be found that the gain in temperature in the water of the calorimeter is less than before for the same coal consumption, and that the difference is measured by the external work alone. One calorie less in the calorimeter heat corresponds to 423.5 gram-meters of work done through the agency of the shaft.

With an animal the case is different. The doing of external work *necessitates* always the burning of more food than is the case with the fasting metabolism, when the energy requirement is for doing internal work, as will be explained below. In the engine the amount of coal burned may be constant whether work is done or not. However, with the animal this result is noticed: Increased food consumption, with increased oxidation, may not be accompanied by an increase in work; in this case there must be an increased liberation of heat. If work is done, there is still an increase in the liberation of heat, but, as with the machine, we must subtract the heat equivalent of the accomplished work. A part of the increased heat liberation is called for by increased *internal* work also. It follows, therefore, that the working animal is warmer than the passive animal, but the increase is not proportional to the food consumed or oxygen absorbed.

External Work Equivalent. Although the animal is able to convert but a limited portion of the potential energy of the food into external work, as a machine it is still much more perfect than the steam engine. This is especially true of man. In the best steam engines not more than about 12 per cent of the potential energy of the fuel can be recovered in the form of work. In animals, *through a short period,* the transformation may amount to as much as 35 per cent of the net available potential energy. In making such comparisons, however, it must be remembered that the animal can work but a limited time. In the rest periods of the animal the loss of heat, without any corresponding mechanical gain, goes on.

The law defining the *maximum* conversion of heat into work through the steam engine is well known. The extent of the limitations in the animal are not known. In the steam engine the limitation depends on the relation of the highest heat of the steam to the temperature of the condenser. If T is the absolute temperature of the live steam and t the temperature of the condenser the *maximum* transformation of heat into work cannot be greater than

$$\frac{T-t}{T}$$

THE INTERNAL WORK OF THE ANIMAL.

Even when the animal appears passive a great deal of work is going on which may be described as internal work. The nature and extent of some of this is known with a fair degree of accuracy, while for the extent of the metabolism corresponding to other kinds of work we have not much beyond conjecture. It is possible to calculate approximately the work done in maintaining the circulation of the blood, and in respiration, and some attempts have been made to estimate the work of the other muscles at rest; but to approximate the work done in masticating, digesting, transporting and transforming the food stuffs is much more difficult. The work of the heart alone has been estimated at from 20,000 to 60,000 kilogram-meters in 24 hours. Three thousand Cal. of heat liberated would correspond to 1,270,500 kilogram-meters of work; hence the heart work in forcing the blood through the vessels may amount to as much as 5 per cent of the whole metabolism.

In respiration the work done is largely the expansion of the thorax against the atmospheric pressure and the elastic tension of the rib cartilages and the lungs. The conditions for estimation are perhaps more favorable than in the other case. It has been calculated that 4 to 5 per cent of the whole metabolism is called for by this work.

In maintaining the tonus of the great mass of skeletal muscles of the body it is likely that a large metabolism is required. The muscular part of the body is not far from 10 per cent in the mean, or 40 per cent of dry substance. In the body of a man weighing 75 kilograms we have therefore about 7.5 kilograms of muscle substance. A large fraction of this falls to the so-called skeletal portion which exists always in a peculiar tense condition. In keeping up this condition without doing outside work oxidation is necessary. Glycogen is split up and water and carbon dioxide appear. The only visible effect of this metabolism is the production of heat. When the muscle does outside work, as in lifting a weight, although the heat production may be greater in the sum, it is relatively less in proportion to the oxygen consumption. A part of the energy is consumed in lifting the weight.

Heat Production Incidental. It is possible that the whole heat liberation, at times, is but a result of the various kinds of internal work done, and that no oxidation takes place for the simple production of heat. This may be the case at relatively high temperatures. At certain lower temperatures, on the other hand, it is apparent that a part of the heat liberation is called for independently of that transformed in the internal work. A certain heat production is necessarily connected with the performance of the various body functions and this down to some particular external temperature limit is sufficient for the heat demands of the body. Numerous experiments have shown that for each animal species there is an external temperature at which the general metabolism, as indicated by carbon dioxide excretion or oxygen consumption, and the resultant heat liberation, reach a minimum value. This temperature limit has been called the *critical temperature;* below it the metabolism and heat liberation increase, evidently not in consequence of a call for more work but because of the necessity for more heat.

ISODYNAMIC RATIOS.

In metabolism, fats, carbohydrates and proteins all yield kinetic energy and to a large extent each one is capable of replacing the others, a certain minimum of protein being always, of course, necessarily present. The proportions in which they may replace each other may be found by a variety of experimental methods, which yield fairly concordant results. The foods may be burned in a calorimeter and the heats of combustion noted, or their values may be compared through the amounts of oxygen required by calculation to oxidize them, or finally animal experiments in the respiration calorimeter may be resorted to to fix the relative values. The isodynamic relations are given in this table calculated from results of animal experiments.

```
     100 gm. of fat =
Lean meat, dry ................................. 243 gm.
Cane sugar .................................... 234 gm.
Glucose ....................................... 256 gm.
Starch ........................................ 232 gm.
```

For such substances the calculated and observed values, or the combustion calorimeter and the respiration calorimeter, give closely agreeing results, but it must be remembered that many compounds which show considerable value as measured by combustion are absolutely worthless as measured by nutrition. Creatinine and urea are illustrations; both are products of metabolism. On the other hand, alcohol shows about the same value in the respiration calorimeter as it shows in the combustion calorimeter, and its metabolic value would therefore appear high. However, the actual *food value* of alcohol is practically low because limited by its toxic action.

FOOD CONSUMPTION IN SEVERE MUSCULAR EXERTION.

In the last chapter reference was made to earlier discussions on the question of the kind of food which must be metabolized to enable the animal organism to do work. By many authorities *heat liberation* was looked upon as an end in itself, and hence foods were divided into the two classes: those important in the production of heat and those important in the production of outside work. The fats and carbohydrates are found in the first group and the proteins in the second. One of the earliest experimental investigations on the subject was the classic one of Fick and Wislicenus, already referred to. These two men in 1866 made the ascent of the Faulhorn in the Swiss Alps, from a known level, and determined the excretion of nitrogen, as urea, during and following the ascent. As the elevation to which they ascended was known, it was possible to make some calculations, approximately accurate, of the work done in the ascent. Two things were shown especially by the tests and calculations: there was not a great increase in the urea excreted, and secondly the protein metabolized, as indicated by the urea measured, was not at all sufficient to account for the work done. Their own conclusions were that the protein consumed would not furnish more than half or three-fourths the energy necessary to lift their bodies through the 1956 meters of ascent, to say nothing of the work done in the horizontal direction on a winding pathway, or of the internal work of the body. This experiment attracted a great deal of attention. Frankland made a new determination of the heat of combustion of protein and showed that the value assumed by Fick and Wislicenus was far too high, thus making the discrepancy still greater.

Since then many similar observations have been made which show pretty clearly that when fats or carbohydrates are abundant in the food there is no excessive destruction of protein in the performance of ordinary work. With the ingestion of a small amount of protein it is easy to cover the normal metabolism. But the case is different when the work is hard. Here, even with abundant food and plenty of protein, there appears to be some loss of nitrogen by the body. In other words, more tissue is broken down than is formed new. This is brought out clearly in the observations of Atwater on the work done by bicyclers in a six-day race some years ago, in which the food consumed and nitrogen eliminated were carefully watched. The following table gives a summary of the most important observations, with the energy in large calories:

Rider.	Average Miles per Day.	Protein Daily.			Energy Daily.		
		In Total Food. Grams.	In Available Food. Grams.	Metabolized. Grams.	In Total Food. Cal.	In Available Food. Cal.	Metabolized. Cal.
A	334.6	169	158	223	4,957	4,547	4,789
B	303.8	179	163	223	6,300	5,871	6,066
C	287.7	211	197	243	4,898	4,323	4,464

Riders A and B rode through six days. C rode three days. The energy metabolized does not include that from body fat, which may have been considerable.

DIETARIES.

The question of the proper amount of food and the character of the food for different kinds of work has been very thoroughly studied in the past few years and a large number of observations on individuals, families and communities of soldiers, prisoners and paupers have been collected. The diet in some cases is known to be sufficient, in others insufficient. From present experience it is possible to say of many dietaries that they are excessive, and probably objectionable in consequence. The following table illustrates the dietaries of a great many people living under different conditions. The figures are taken mainly from the compilations of König and Atwater:

Occupation, Etc.	Observations Av'd.	Protein, Grams.	Fat, Grams.	Carbohydrates, Grams.	Calories.
Italian laborers, Chicago	4	103	111	391	3,060
Bohemian laborers, Chicago	8	115	103	360	2,885
Russian laborers, Chicago	9	137	102	418	3,232
Laborers, crowded district, New York	19	106	117	367	3,030
Laborers, low income, Pittsburgh	2	81	97	311	2,510
Mechanics, eastern and central U. S.	14	103	150	402	3,465
French Canadians, Chicago	5	118	158	345	3,365
French Canadians, Massachusetts	5	111	193	485	4,235
American professional men	14	105	124	420	3,335
Bavarian workmen, high class	3	151	54	479	3,085
Munich prisons, work		104	38	521	2,916
Munich prisons, no work		87	22	305	1,819
Bavarian soldiers, war		145	100	500	3,575
Bavarian soldiers, garrison		120	56	500	3,063

The above results are fairly representative and show that in general over 3,000 Cal. per day must be provided in the food. In the figures the available or net calories are calculated from 1 gm. protein or carbohydrate $= 4.1$ Cal., 1 gm. fat $= 9.3$ Cal. But many extreme results are also found in the literature. For prisoners confined in cells and not working, for paupers in asylums, and even with laborers poorly paid, the foods consumed may not yield 1,500 Cal. On the other hand, workmen in the American winter lumber camps, who as a rule are well paid, workmen in the building trades on outside work in the colder weather, teamsters and car drivers who are constantly exposed to the weather, even when the work is not excessive, may consume a diet yielding 4,000 or even 5,000 Cal.

Special Diets. With such facts as the above in mind it is not difficult to understand why nutrition with a single article of food is unsatisfactory. Assume, for example, the case of a diet of potatoes of which the edible portion shows in the mean about these per cent values: protein 2.2, fat 0.1, carbohydrates 18.4. One hundred grams of potatoes would yield then the following:

Protein	2.2 gm.	9.0 Cal.
Fat	0.1	0.9
Carbohydrate	18.4	75.4
		85.3

A pound of potatoes would furnish, therefore, 386 Cal., and 6 to 8 pounds would have to be consumed to furnish energy for ordinary work. The protein in this would be considerably below the amount which has generally been held necessary, while the fat is scarcely appreciable. The storage of energy on such a diet would be practically impossible.

On the other hand, a diet of lean meat, round steak, for example, would be almost as bad. This averages about 20.9 per cent of protein and 10.6 per cent of fat, from which 100 grams would furnish:

Protein	20.9 gm.	85.7 Cal.
Fat	10.6	98.6
		184.3

A pound would furnish 835 Cal. and about 3 to 4 pounds would have to be consumed for support of the body daily. While the fat in this would be proper the protein would amount to 275 grams at least, and make the work of excretion extremely difficult.

As a third case a diet of the small white beans may be considered. We have here protein 22.5, fat 1.8, carbohydrate 59.6. In 100 grams, therefore,

Protein	22.5 gm.	92.2 Cal.
Fat	1.8	16.7
Carbohydrate	59.6	244.3
		353.2

A pound would furnish about 1600 Cal. and 2 pounds would cover the needs of the body practically. The proteins in this would be but slightly excessive, while the same would be true of carbohydrates. The trouble is with the deficiency in fat. Notice how easily this may be corrected. A pound and a half of beans cooked with one fourth pound of fat pork will yield over 3,000 Cal. and furnish a diet easily assimilated by men at moderate work. What is said of the bean is practically true of the pea; each one approaches in value a mixed meat and cereal diet.

Required Protein. A much debated question is that of the actual protein requirement, supposing the other food elements sufficiently abundant. The standards given above have not gone unchallenged. Several observers have described metabolism tests in which less than one half the 120 or more grams of protein, usually considered necessary in the daily food, appears to be sufficient for all needs and able to maintain the body in nitrogen equilibrium. Most of these experiments have been, however, too short to really prove much definitely.

But recently very elaborate and long continued investigations on groups of men have been described by Chittenden in which the evidence in favor of a low protein requirement is put in an entirely new light, and in which it is also shown that the 3,000 large calories of energy in our food is more than necessary for ordinary practical needs. In a group of five professional men Chittenden found an average nitrogen liberation corresponding to the metabolism of something over 46 gm. of protein daily and an average energy value in the whole food of about 2,300 calories through a period of six to nine months. In a group of thirteen soldiers, taking abundant exercise, there was a daily consumption of food having an average value of about 2,600 calories, and an average protein consumption of about 56 gm. through five months, fall, winter and early spring. In a group of seven student athletes a protein consumption of about 61.5 gm. daily with a total food consumption equivalent to about 2,575 calories was observed.

In all these cases the tests were continued long enough to bring the men into practical nitrogen equilibrium with good physical condition and good general health. For this reason they deserve the fullest attention and study. It is the opinion of the author of the experiments that increased food consumption, so far from being necessary, is even in most cases a detriment, since it calls for a large amount of extra internal work, in the liver and kidneys especially, in metabolizing the digestion products and in removing the waste. This is certainly a consideration of some moment. How far these findings may be applied to the case of men at hard work, in the open, in cold weather, remains to be tried. Chittenden's results are especially interesting with respect to the necessary nitrogen; but for the hard-working man probably more fat and more carbohydrate will always be found desirable.

INDEX

Abnormal colors in urine, 325
Absorption analysis, 194
   cells for spectroscope, 200
   from stomach, 143
   ratios, 201
Acetic acid, 40
   fermentation, 115
Acetoacetic acid in urine, 322
Acetone in urine, 322
Achroödextrin, 37
Acid albumin, 77
   fermentation, 159
   of urine, 349
Acid, acetic, 40
   alloxyproteic, 302
   amino-acetic, 61
      caproic, 62
      glutaric, 62
      isobutylacetic, 62
      propionic, 62
      succinic, 62
      valeric, 62
   antoxyproteic, 302
   arabic, 38
   arabonic, 18
   arachidic, 40
   aspartic, 62
   behenic, 40
   butyric, 40, 119
   capric, 40
   caproic, 40
   caprylic, 40
   carbonic, 64
   cholalic, 265
   cholanic, 265
   choleic, 265
   chondroitin sulphuric, 89
   cresyl sulphuric, 303
   dextronic, 18
   elaidic, 46
   erythritic, 18
   fellic, 265
   formic, 40
   glutaminic, 62
   glyceric, 18
   glycerophosphoric, 48

Acid, glycocholic, 264
   glycollic, 18
   hippuric, 301, 334
   hypogæic, 41
   indoxyl sulphuric, 303
   lactic, 117
      in stomach, 134
   lauric, 40
   linoleic, 41
   lithofellic, 265
   mannonic, 18
   margaric, 40
   myristic, 40
   nucleic, 86
   œnanthylic, 40
   oleic, 41
   oxalic, 18
   oxyphenyl amino propionic, 63
   oxyproteic, 301
   palmitic, 40
   parabanic, 308
   paralactic, 281
   pelargonic, 40
   pentoic, 40
   phenyl amino propionic, 63
   phenyl sulphuric, 303
   phosphoric, urinary, 304
   picric, 56
   propionic, 40
   pyrrolidine carboxylic, 62
   ricinoleic, 41
   saccharic, 18
   sarcolactic, 119
   skatoxyl sulphuric, 303
   stearic, 40
   tartaric, 18
   tartronic, 18
   taurocholic, 264
   thiolactic, 281
   trioxyglutaric, 18
   undecylic, 40
   uric, 297
   valeric, 40
Acidity of gastric juice, 137
Acids in fats, 40
Acid zone, 158

Acrose, 19
Activators, 101, 157
Addiment, 225
Adenine, 87
Adipocere, 45
Adrenalin, 273
Agar-agar, 38
Agarose, 27
Agave sugar, 27
Agglutinins, 221
Air, 12
    tests, 13
Alanine, 62
Alanylglycylglycine, 149
Albumin, 51
Albuminates, 77
Albuminoids, 53, 90
Albumin in urine, 314
Albumins proper, 66
Albumoses, 80, 81
Alcoholic fermentation, 112, 113
Alcohol in wine, 113
    production, 113
    test for, 113
Aldopentoses, 18
Alexins, 224
Alkali albumin, 77
Alkalies and protein, 61
Alkaline zone, 158
Alkaloid reagents and proteins, 56
Alloxyproteic acid, 302
Alpha naphthol test, 23, 58
Amboceptors, 226
Amino acids, 61
    as digestive products, 150
Aminocaproic acid, 62
    propionic acid, 62
    valeric acid, 62
Ammoniacal copper solution, 29
Ammonia, determination of, 330
    in urine, 296
    in water, 10
Ammonium cyanate, 296
Amniotic fluid, 233
Amorphous phosphates in urine, 353
Amount of acid in stomach, 135
    sugar in urine, 320
Amphopeptone, 83
Amygdalin, 106
Amylodextrin, 37
Amyloid degeneration, 92
    substance, 92
Amylopsin, 103

Amylase, 103
Amylose, 33
Analysis, spectrum, 197
Analyses, ash of milk, 240, 241
    ash of muscle, 283
    bile, 263
    blood, 176
    bone ash, 286
    cells of thymus, 233
    cerebrospinal liquid, 277
    colostrum, 241
    feces, 164, 166
    gall stones, 270
    hydrocele fluid, 233
    lymph, 230
    meat extract, 284
    milk of cow, 236
    mother's milk, 245
    muscle, 278
    peritoneal transudate, 233
    pleural transudate, 233
    pus cells, 234
    serum, 233
    spermatic fluid, 275
    urine, 292
Animal foods, 293
    internal work of, 373
    starch, 36
Animals and plants, 3
Anti bodies, chemical nature, 223
    development of, 220
Anti body defined, 217
Anti group, 81
Antipeptone, 83
Antitoxins, 218
Antoxyproteic acid, 302
Apparatus for freezing point, 207
Arabinose, 18, 20, 38
Arabitol, 18
Arabonic acid, 18
Arachidic acid, 40
Arginine, 61, 84
Argon in blood, 190
Aromatic products from intestinal putrefaction, 160
Artificial purification of water, 9
Ash in tissues, 14
    of milk, 240
Aspartic acid, 62
Assay of pepsin, 131
Atwater, food standards, 376
Autodigestion, 256
Autolysis, bacterial products from, 257

INDEX.

Autolysis, importance of, 257
    organic acids from, 257
    pancreas, 272
    protein in, 257
Autolytic fermentation, 255

Bacteria in feces, 163
    in urine, 348
    lactic acid, 117
Bacterial process, 158
    purification of water, 9
Bactericidal products of autolyses, 257
Bacteriolysins, 219
Bacteriolytic processes, 117
Bases in body, 16
Beckmann apparatus, 207
Beef extract, 280
    composition, 93
    pancreas, extracts from, 146
Beeswax, 49
Beet sugar, 25
Behavior of trypsin, 145
Behenic acid, 40
Benedict and Gephart, urea method, 330
    total sulphur in urine, 339
Benjamin Thompson, 6
Benzoates and hippuric acid, 301
Benzoic acid, 106
Bernard, C. L., 5
Berzelius, 4
Bicycle rider, food and work, 376
Bile, 262
    acids, 265
        in feces, 169
        optical rotation, 266
    colors in feces, 174
    composition, 263
    concretions, 270
    emulsification by, 269
    pigments, 266
Bilicyanin, 271
Bilifuscin, 271
Bilihumin, 271
Biliprasin, 271
Bilirubin, 188, 263, 267
Biliverdin, 188, 265, 267
Biology, field of, 2
Bismuth test, 23
Bitter almonds, 106
Biuret, 57
    reaction, 57
Blood, 175
    albumin, 66

Blood, analyses, 176
    and bile pigments, 185
    anti bodies in, 218
    ash of, 14
    casts, 346
    cholesterol in, 191
    conductivity, 212
    corpuscles in urine, 342
    cryoscopy, 206
    freezing point, 206
    gases of, 190
    in tissue, 191
    in urine, 326
    lecithins in, 191
    optical properties, 193
    osmotic pressure, 204
    phagocytes in, 216
    salts of, 190
    self preservation, 216
    serum tests, 219
    sugar in, 189
    tests, clinical, 201
    transfusion, 192
Boas' reagent, 134
Body, bases in, 16
    composition of, 7
Bogg's coagulometer, 180
Bone, 285
    ash of, 14, 286
    gelatin from, 90
    glue from, 90
    marrow, 287
    ossein in, 256
Brain and nerve substance, 276
Bran, pentose in, 20
Bread, composition, 94
    fermentation in, 118
British gum, 35
Brunner's glands, juice from, 156
Buchner, ferments, 99
Bunge, 248
Burchard-Liebermann test, 50
Butter, 46
    composition, 238
Butter milk, 242
Butyric acid, 40, 119
    fermentation, 117

Cadaver wax, 45
Calcium and magnesium salts in urine, 293
    in body, 8
    oxalate in sediment, 352

Calcium sulphate in sediment, 352
Calculation of food energy, 368, 369
Calculi, 354
Calves stomach, rennet in, 108
Calorie, definition, 366
Calories in foods, 93, 94
Cane sugar, 25
    group, 25
Caproic acid, 40
Caprylic acid, 40
Carbohemoglobin, 186
Carbohydrates, 17
    changes in liver, 253
    digestion, 103, 152
    group in proteins, 81
    heats of combustion, 367
    in urine, 304
Carbonates of body, 15
Carbon balance, 357
    dioxide and plant-life, 2
    in body, 8
    monoxide hemoglobin, 184
Carnine, 280
Carnitine, 280
Carnosine, 280
Cartilage, 287
    ash of, 14
    gelatin from, 90
    mucoid in, 89
Casein, 72
    in feces, 173
    preparation, 239
Caseoses, 81
Casts in urine, 345
Catalytic action, 98
Cat fat crystals, 43
Cell globulin, 69
Cells, conductivity, 214
    epithelium in urine, 343
    in general, 249
    lymph, 233
Celluloid, 39
Cellulase, 104
Cellulose, 38
    in feces, 171
Centrifuge, uses of, 341
Cereals, composition, 94
Cerebrospinal liquid, 276
Character of antibodies, 218
Charts, Tallquist's, 203
Cheese, composition, 93
Chemical nature of anti bodies, 223
Chemistry of milk, 238

Chittenden, 53
    preparation of pepsin, 130
    protein classification, 81
    protein requirement, 378
Chlorides, determination of, 337
    in water, 10
    of body, 8, 15
    tests for, 10
Chlorophyll-bearing plants, 2
Cholagogues, 268
Cholanic acid, 265
Choleic acid, 265
Cholesterol, 49, 270
    in blood, 191
    in feces, 169
    in protoplasm, 250
    optical rotation, 271
Choline, 48
Cholalic acid, 265
Chondroitin, 89
    sulphuric acid, 89, 287
Chondromucoid, 89, 287
Chromophoric group, 136
Chyle, 232
Classification of proteins, 52
Clinical blood tests, 201
    uses of hematocrit, 211
Clumping, bacteria, 222
Clupein, 75
Coagulated albumins, 76
Coagulating proteins, 69
Coagulation of blood, 177
    tests, 54
Coagulometer, 180
Coefficient, isotonic, 208
Co-enzymes, 101
Cohnheim, 156
    theory of protein metabolism, 290
Collagen, 90
    from muscle, 279
Collodion, 39
Coloring matter in urine, 323
Color of urine, 292
Colostrum, 241
Combustion, heats of, 367
Commercial bile salts, 269
    pepsin, 109, 130
Complement, 225
Complementoids, 228
Component groups in proteins, 60
Composition of body, 7
    of cells, 249
    of lymph, 230

Composition of milk, 237
Conductivity cells, 214
   of blood, 212
   of urine, 309
Congo red test, 133
Coniferin, 107
Conjugated proteins, 54
Conservation of energy, 6
Conversion of starch, 37
Corpuscles, number in blood, 211
Count Rumford, 6
Cow's milk, 236
Creatine, 279
Creatinine, 279
   determination, 333
   from urine, 300
   reducing power, 307
Crude fat in feces, 168
Cryoscopy, 206, 312
Crystallin, 69
Crystals of fats, 43
   of hemin, 181
Curtius, 64
Cystin, 92, 262
   calculi, 355
   in urine, 351
Cytase, 104
Cytosine, 87
Cytotoxins, 219
Dare's hemoglobinometer, 202
Dehydrocholeic acid, 265
Denatured proteins, 77
Derived products, proteins, 53
   protein, 54
Despretz, 5
Destruction of glycogen, 255
Detection of free acid in stomach, 133
Determination of acid in stomach, 135
   of albumin in urine, 314
   of ammonia in urine, 330
   of blood colors, 197
   of chlorides in urine, 357
   of creatinine in urine, 333
   of digestive power, pepsin, 131
   of electrical conductivity, 212
   of fats, 47
   of fats in milk, 244
   of feces fat, 169
   of freezing point, 207
   of hemoglobin, 203
   of hippuric acid, 334
   of nitrogen in feces, 172
   of nitrogen in urine, 327

Determination of osmotic pressure, 209
   of pepsin, 139
   of phosphates in urine, 335
   of proteins, 59
      of milk, 244
   of purine in urine, 333
   of specific gravity, 313
   of sugar, 27
      in milk, 244
      in urine, 320
   of sulphates in urine, 339
   of total sulphur in urine, 339
   of urea in urine, 328
   of uric acid in urine, 332
Deutero albumose, 83
Dextrin, 35
   from starch, 123
Dextronic acid, 18
Diabetes mellitus, 305
Diagram of spectroscope, 194
   Wheatstone bridge, 212
Dialyzer, 67
Diastase, 97, 103
   action of, 123
Diazo reaction, 162, 326
Diet and feces, 165
Dietaries, 376
Diets, special, 377
Digestion, 96
   of fats in stomach, 143
   of starch, 122
   pancreatic, 144
   peptic, 109, 127
   salivary, 121
   tryptic, 110
Digestive extracts, 146
Diglycylglycine, 149
Dilution test, 315
Dimethylaminoazobenzene test, 133
Diose, 18
Dioxyacetone, 18
Direct vision spectroscope, 194
Disaccharides, 17
Distearin, 41
Distribution of food energy, 368
   of heat energy, 370
   of nitrogen in urine, 295
Donne's test, 343
Dulong, 5

Edestan, 69
Edestin, 69
Edible fats, 41

Effect of work, 371
Egg albumen, 67
    and pepsin, 131
Ehrlich reaction, 162
Ehrlich's theory, 224
Elaidic acid, 47
Elaidin, 47
Elastin, 92
Electrical conductivity of blood, 212
    of urine, 309, 310, 311
Elements in body, 7, 8
Emulsin, 106
Emulsions, 42, 269
Endothermal reactions, 2
End products of digestion, 150
    of metabolism, 289
Energy balance, 289
    equation, 366
    of food, distribution, 368
Enterokinase, 157
Enzymes, 96
    as catalytic agents, 100
    of stomach, 127
Epinephrin, 273
Epithelium in urine, 343
Erepsin, 110, 156
Erg, definition, 367
Erythritic acid, 18
Erythrodextrin, 37
Erythrogranulose, 37
Erythrol, 18
Erythrose, 18
Esbach albuminometer, 139, 315
    reagent, 139
Ethereal sulphates, 162, 261
Ewald test meal, 132
Examination of stomach contents, 132
Excretion by skin, 363
    gaseous, 356
    of alkali salts, 293
    of calcium and magnesium, 293
    of nitrogen, 289
    of phosphorus, 304
    of sulphur, 302
External work equivalent, 372
Extinction coefficient, 199
Extract of meat, 283
Extracts from yeast, 114
Extractives from muscle, 277
Exudations, 232

Fat and chyle, 232
    crystals, 43

Fat from muscle, 279
    globules, 239
    in urine, 351
    in feces, 164
    in foods, 93, 94
    of milk, 238
Fats, 40
    from proteins, 44
        sugars, 44
    heats of combustion, 367
    in blood, 191
    in body, 44
    in pancreatic digestion, 154
    solubility, 44
    splitting of, 107
Fatty acids, 40
Faulhorn experiment, 375
Feces, 161, 163
    amount of, 164
    bile acids in, 169
    blood and pus in, 174
    carbohydrates in, 170
    cellulose in, 171
    composition, 164
    from various foods, 165
    lecithin in, 168, 170
    nitrogen in, 172
    proteins in, 173
    starch in, 170
    sugar in, 171
Fehling reduction, 28
    test, 22
    urine, 318
Fellic acid, 265
Fermentation, acetic, 115
    alcoholic, 113
    autolytic, 255
    butyric, 97, 117
    in intestines, 158
    lactic, 97, 117
    mucous, 119
Ferments, 96
    classification, 102
Ferric chloride test, 323
Fibrin, 70, 177
    digestion of, 147
Fibrinogen, 70
Fibronoses, 81
Fick and Wislicenus experiment, 375
Fields of study, 2
Filter paper, 38
Fischer, 64, 225
    nomenclature of purines, 298

Fish, composition, 93
Fission fungi, 112
Fleischl hemometer, 201
Flesh bases, 279
Flour, composition, 94
Fluorine in body, 8
Folin, ammonia in urine, 331
    creatinine method, 334
    sulphate method, 339
    theory of protein metabolism, 290
    urea method, 330
    uric acid method, 332
Food and work, 6
    consumption and muscular work, 375
    of plants, 3
Foods, relation to feces, 165
Food stuffs, 93
Formaldehyde condensation, 2
Formic acid, 40
Formulas for hemoglobin, 187
    spectrophotometry, 199
Fraunhofer lines, 193
Free acid in stomach, 132
Freezing point of blood, 206
    urine, 312
Fructose, 18, 24
Fruit sugar, 20
Fuel value of foods, 93, 94, 367
Functions of bile, 268
    liver cells, 251
    lymph, 231
Fungi and fermentation, 112
    in urine, 347
Furfuraldehyde, 20
Furoaniline, 20

Gadus-histone, 75
Galactose, 24, 105, 240
Gallstones, 49, 270
Gaseous excretions, 356
Gases in air, 12
    of blood, 190
Gastric juice, 126
    acidity, 137
    titration of, 138
Gelatin, 90, 285
    tests for, 91
    uses, 91
General composition of urine, 291
    relations, 1
Gland, thyroid, 274
Gliadin, 73
Globin, 74

Globulinoses, 81
Globulins, 68
    in urine, 315
Glucase, 104
Gluco-proteids, 88
Glucosamine, 63
Glucose, 21
    from starch, 21
    in blood, 189
    reducing power, 28
Glucoses, 18
Glucosides, 20
Glucoside reactions, 106
Glucoronic acid in urine, 322
Glue, 90
Glutaminic acid, 62, 84
Glutelins, 74
Gluten, 73, 94
Glutenin, 73
Glutin, 90
Glyceric acid, 18
Glyceraldehyde, 18
Glycerol, 18, 47
Glycero-phosphoric acid, 48
Glycerose, 18
Glyceryl butyrate, 46
    caproate, 46
    oleate, 46
Glycine, 61
Glycocoll, 61, 264
Glycogen, 36, 280
    destruction, 255
    formation, 253
    in flesh, 281
    in protoplasm, 250
    stored in liver, 254
Glycol, 18
Glycollic acid, 18
Glycocholic acid, 264
Gmelin, 4
Gmelin's test, 267
Goitre and iodine compounds, 274
Gower's hemoglobinometer, 203
Graham dialyzer, 67
Grain composition, 94
Granular casts, 346
Granulose, 33
Grape sugar, 21
Group, immune, 227
    zymotoxic, 228
Groups in protein, 60
Guaiacum test, 180, 327
Guanine, 87

INDEX.

Guenzberg's reagent, 133
Gum arabic, 38
   British, 35
Gums, 37
Gun cotton, 39

Hair, keratin from, 288
Hamburger, 209
Hammarsten's test, 268
Haptophorous group, 226
Hard water, 8
Heart, ash of, 14
Heat and food stuffs, 6
   energy, distribution, 370
   mechanical equivalent, 367
   of friction, 6
   production incidental, 374
   radiation, 371
   unit, definition, 366
Heats of combustion, 367
Hematin, 187
   spectrum, 196
Hematocrit, clinical uses, 211
   methods, 210
Hematogen, 73
Hematoidin, 188
Hematolin, 187, 188
Hematoporphyrin, 187
Hematuria, 326
Hemi group, 81
Hemin crystals, 181
Hemochromogen, 187, 188
Hemoglobin, 74, 181
   analysis, 182
   combinations, 182
   crystals, 183
   specific rotation, 182
Hemoglobins, 87
Hemoglobinometer, Dare's, 202
   Gower's, 203
Hemoglobinuria, 326
Hemometer, Fleischl's, 201
Hemolysins, 219
Heteroalbumose, 82
Hexitols, 18
Hexone bases, 61
   in digestion products, 149
Hexoses, 18, 20
Hippuric acid, 301
   determination, 334
   in sediment, 352
Hirn, comparison between man and machine, 371

Histidine, 61
Histones, 74
Historical sketch, 4
History of fermentation, 97
Hofmeister, 64
Hog pancreas, extracts from, 146
Hoppe-Seyler, 5
Horn, composition, 93
   keratin from, 288
   substance, 92
Human fat, 47
   milk, 245
Hyaline casts, 346
Hydrazones, 20
Hydrocele fluid, 233
Hydrochloric acid in stomach, 126
Hydrocyanic acid, 106
Hydrogen in body, 8
   peroxide test, 181
Hydrolysis of proteins, 60
   starch, 22, 122
Hydrolytic reactions, 102
Hypogæic acid, 41
Hypoxanthine, 87

Ichthulin, 73
Immune body, 226
   group, 227
Immunization, 219
Important early works, 5
   fats, 45
Indestructibility of matter, 6
Index, opsonic, 222
Indicators and stomach contents, 136
   theory of, 136
Indican, 151, 161, 324
Indol, 147, 151, 160
Indoxyl, 151, 161
Inorganic elements, 7
Inosite, 281
Insoluble ferments, 99
Intermediary body, 226
Internal work of animal, 373
Intestinal bacteria, 159
   changes, 158
   juice, 155
Inulase, 104
Inulin, 24, 35
Invertase, 25, 105
Invertin, 105
Invert sugar, 25
   reducing power, 28
Investigations, early, 5

Iodine in body, 8
  in thyroid, 274
  test, 34
Iodothyreoglobulin, 274
Iodothyrin, 275
Iron in bile, 267
  in body, 8
  masked, 86
Isinglass, 91
Isocholesterol, 49
Isodynamic ratios, 374
Isolation of pepsin, 129
Isomaltose, 27
Isotonic coefficient, 208

Joule, 6
Juice, intestinal, 155
  pancreatic, 144

Kelling's test, 135
Kephir, 26, 118
Keratin, 92, 288
Ketopentose, 18
Kidney, ash of, 14
Kinases, 101, 157
Kinds of ferments, 102
Kinetic energy of food, 368
Kjeldahl test, 327
Koeppe's hematocrit, 210
Koprosterin, 271
Kruess spectrophotometer, 198
Kuehne, 5
  protein classification, 81
Kumyss, 118
Kyrine, 84

Laborers, dietaries, 376
Laccase, 116
Lactalbumin, 67, 239
Lactic acid, 281
  bacteria, 117
  tests for, 135
  fermentation, 117
Lactase, 105
Lactose, 25, 26, 105, 240
Landwehr's animal gum, 88
Lanolin, 49
Laplace, 5
Lard, 46
Laurent polariscope, 30
Lauric acid, 40
Lavoisier, 4
Lead hydroxide test, 58

Lecithan, 48
Lecithin, 48, 68
Lecithins in blood, 191
  in cells, 250
  in feces, 168
Legumin, 74
Lehmann, 5
Leucine, 62, 92
  as urine sediment, 350
  tests for, 147
Leucocytes, 231, 233
Leucylproline, 149
Leuwenhöck, 96
Levulose, 24
Lieberkuehn's glands, juice from, 155
  jelly, 78
Liebig, 4
  theory of fermentation, 98
Lignocellulose, 39
Linoleic acid, 41
Lipase, 107, 126, 154
Lithofellic acid, 265
Liver and poisons, 258
  ash of, 14
  autolysis, 256
  chemical changes, 252
  chemistry of, 249
  ethereal sulphates, 261
  fats in, 251
  formation of urea, 259
    uric acid, 260
  glycogen in, 251
  iron in, 252
  lecithin in, 251
  mineral substances, 252
  protein in, 251
  synthetic processes, 259
  work of cells, 252
Loewe solution, 29
Loss of free acid in digestion, 142
Lymph, 230
  amount, 231
  composition, 230
  functions, 231
Lymphagogues, 231
Lysine, 60, 84

Magnesium in body, 8
  phosphate in urine, 353
Malondiamide, 57
Malt, 103
Maltase, 104
Malt extract, 123

INDEX. 389

Maltodextrin, 37
Maltose, 25, 104
    reducing power, 28
Malt sugar, 26, 123
Margarin, 45
Manufacture of starch, 33
Maple sugar, 25
Market milk, 236
Marrow, 287
Masked iron, 86
Mayer, 6
Margaric acid, 40
Meal, composition, 94
Meat, composition, 93
    extract, 283
Mechanical equivalent of heat, 367
Melibiose, 27
Melitose, 27
Meyer, blood gases, 190
Metabolism experiments, 360-362
    theories of, 290
Methemoglobin, 186
    spectrum, 197
Metaproteins, 80
Methyl orange indicator, 136
    violet test, 133
Microorganisms in fermentation, 98
Milk, 236
    albumin, 67
    ash of, 14
    composition, 95
    curdling ferment, 143
        of, 108
    fat in, 238
    flavors, 247
    human, 245
    modified, 246
    mother's, 245
    of ass, 248
    of bitch, 248
    of elephant, 248
    of goat, 248
    of mare, 248
    of sow, 248
    origin of, 237
    preservatives, 244
    salts of, 240
    sugar, 24, 26, 240
    reducing power, 28
Millon's reagent, 56
    test in digestion, 148
Mineral matters in blood, 177
    residues of organs, 14

Mineral substances in milk, 240
Modified albumin, 76
    milk, 246
Molds in urine, 348
Molisch reaction, 83
    test, 23, 58
Mannitol, 18
Mannonic acid, 18
Monosaccharides, 17
Monoses, 17
Monostearin, 41
Moore's test, 317
Mother of vinegar, 115
Mucin bands in urine, 344
    in saliva, 121
    in urine, 316
Mucins, 88
Mucoid bodies, 89
Mucors, 112
Mucous fermentation, 119
Mucus in urine, 342
Murexid test, 331
Muscle, ash of, 14
    extraction, 71
    plasma, 71
    sugar, 281
    substance, 277
Musculin, 278
Mutton tallow crystals, 43
Mycose, 27
Myogen, 70, 278
Myosin, 70, 107, 278
Myosinogen, 70
Myosinoses, 81
Myricin, 49
Myristic acid, 40

Nails, keratin from, 288
Natural fats, 40
    purification of water, 9
    waters, 8
Native albumins, 53, 65
Nature of bile, 268
Neutral sulphur, 262, 303
Nicol prism, 31
Nitrates in water, 11
Nitric oxide hemoglobin, 185
Nitrites in water, 11
Nitrocellulose, 39
Nitrogen balance, 357
    excretion of, 294
    -free extractives from muscle, 280
    in blood, 190

Nitrogen in body, 8
    in feces, 164, 172
    of urine, distribution of, 295
Normal colors in urine, 333
    feces, 163
    reduction, 307
Nucleates, 86
Nucleic acid, 85, 86, 298
Nuclein, 85, 250
Nucleo-albumin, 71
Nucleo-histone, 75, 86
Nucleo-proteids, 85
Nucleus, 249
Number of corpuscles, 211
Nutrients, 7
Nutrose, 72
Nuts, composition, 94

Occurrence of metals in body, 8
Odor of urine, 292
Oil of bitter almonds, 106
Oleic acid, 41
Olein, 45
Oleomargarin, 46
    composition, 93
Opsonins, 222
Opsonic index, 222
    treatment, 223
Optical properties of blood, 193
    rotation, 30
    sugar tests, 30
Organic acids by bacteria, 141
    from liver, 256
    in stomach, 134
    chemistry and agriculture, 4
        pathology, 4
        physiology, 4
    matter of bones, 286
Organized ferments, 99
    sediments, 341
Organs of body, ash of, 14
Origin of fats, 44
Osazones, 21
Osborne, 53, 69
Osmotic pressure, 204
    cell, 205
    tension, 209
Osones, 21
Ossein, 90, 286
Outline of topics, 6
Oxalate calculi, 355
Oxalic acid, 18
Oxaluramide, 57

Oxamide, 57
Oxidase enzymes, 115
Oxidases, 115
Oxidation reactions, 111
    tests, 10
    time and place of, 364
    value of copper solutions, 28
Oxybutyric acid in urine, 322
Oxygen absorption by blood, 184
    in blood, 190
    in body, 8
    liberation of by plants, 2
Oxyhemoglobin, 183
    spectrum, 195
Oxyproteic acid, 295, 301
Oysters, composition, 93
Ozone in air, 13

Palmitic acid, 40
Palmitin, 45
Pancreas, 272
    ash of, 14
    autolysis, 272
Pancreatic diastases, 153
    digestion, 144
        of fats, 154
    starch, 153
    ferments, 110
    juice, 144
Pancreatin and milk, 243
Paracasein, 72
Paralactic acid, 281
Parasitic plants like animals, 3
Parathyroids, importance of, 275
Pasteur, 5, 97
    theory of fermentation, 98
Pavy method, 321
    solution, 29
Pawlow, 126
Payen, 97
Peas, composition, 94
Pectase, 104
Pectin, 104
Pectinase, 104
Pelargonic acid, 40
Pentitols, 18
Pentoic acid, 40
Pentosans, 20
Pentoses, 18, 19
Pepsin, 108, 126
    amount of, 139
    peptone, 150
    preparation, 129

Pepsinogen, 108
Peptic digestion, 127
    products of, 140
Peptones, 80, 81, 83, 108, 141
    in urines, 316
Percentage variations in urine, 291
Peritoneal transudates, 233
Permanent hardness, 9
Permanganate test, 11
Peroxidases, 116
Persoz, 97
Pfeiffer phenomenon, 224
Pflueger theory of protein metabolism, 290
Phagocytes, 216
Phenylalanine, 63
Phenylalanylglycylglycine, 149
Phenyl glucosazone, 23
Phenyl hydrazine test, 21, 23
    in urine, 319
Phenol in intestinal changes, 160
Phenol-phthalein indicator, 136
Phenomenon, Pfeiffer's, 224
Phosphate, amorphous, 353
    sediments, 353
Phosphates, 14
    determination of, 335
Phosphatides, 48
Phospho-proteins, 54, 72
Phosphorus excretion, 304
    in body, 8
Physical blood tests, 204
Physiological chemistry and medicine, 5

Physiological chemistry, scope of, 2
Phytoglobulins, 69
Phytovitellins, 69
Pigments of bile, 266
Pioneer investigators, 5
Plants and animals, 3
Plasma of muscle, 70, 278
    salted, 179
Plasmon, 72
Pleural transudates, 233
Poisons and liver, 258
    from intestine, 162
Polarimeter, 31
Polariscope, 30
Polarization tests, 30
Polypeptides, 64, 85, 149
Polysaccharides, 17, 32
Pork, composition, 93
Potassium in body, 8

Potassium indoxyl sulphate, 151
Practical urine tests, 313
Precipitation by salts, 55
    limits, 55
Precipitins, 218
Preparation of bile acids, 265
Preservatives in milk, 244
Pressure, osmotic, 204
Primary albumose, 82
    phosphates, 14
Products of peptic digestion, 140
Prolamins, 74
Proline, 62
Prolonged digestion, 142
Propepsin, 108
Propionic acid, 40
Protagon, 276
Protalbumose, 82
Protamines, 75
Proteans, 69
Proteids, 53, 85
Protein classification, 52
    combination, 64
    digestion, 128
    in foods, 93, 94
    metabolism theories of, 290
    required, 378
Proteins, coagulating, 69
    determination, 59
    heats of combustion, 367
    in autolysis, 257
    in feces, 173
    in urine, 305
    of muscle, 278
    pancreatic digestion, 146
    substances, 51
    synthesis, 64
Proteolytic reactions, 107
Proteoses, 81
    in feces, 173
    in urine, 316
Prothrombin, 178
Protones, 75
Protoplasm, 250
Pseudo acids, 56
    bases, 56
    cellulose, 39
    pepsin, 156
Psychic stimulus, 126
Ptyalin, 97, 124
Purine, 298
    bodies, 297
Purines, 87

Purines, determination in urine, 333
Pus, 232
    in urine, 343
Pyrimidine bodies, 300
Pyrimidines, 87
Pyrrolidine carboxylic acid, 62

Quadriurates, 299
Quantitative composition of blood, 176
    spectrum analysis, 197
Quotient, respiratory, 356

Raffinose, 27
Ratios, isodynamic, 374
Reaction, diazo, 326
    of blood, 180
    of feces, 166
Reactions, ferments, 97
    of fats, 41
    proteins, 54
Receptors, 226
Reduced hemoglobin, 195
Reducing power of sugars, 28
    of urine, 307
Reduction tests, 22
    urine, 318
Rennet, 108
    action on milk, 242
Rennin, 72, 108, 126
Reproductive glands, 275
Required protein, 378
Resorcinol test, 24
Respiration apparatus, 357
    experiments, 358
    gases of, 13
Respiration in plants, 3
    skin, 363
Respiratory quotient, 356
    illustrations, 359
Reversible reactions with proteins, 77
Ricinoleic acid, 41
Riegel test meal, 133
Rotation, specific, 32

Saccharic acid, 18
Saccharodioses, 17
Saccharomycetes, 112
Saccharose, 25
Saccharotrioses, 17, 27
Salicin, 107
Saliva, 121
Salivary diastase, 97
    digestion, 121

Salkowski's test for cholesterol, 50
Salmin, 75
Salmo-histone, 75
Salted plasma, 179
Salt, need of, 16
Salts, and proteins, 56
    in body, 14
    in blood, 177, 190
    of casein, 72
    of milk, 240
    of muscle, 282
Saponification, 41
Sarcolactic acid, 281
Sauerkraut, acid in, 118
Scheele, 97
Schuetzenberger, 5
Schultz prism, 200
Schweitzer's reagent, 39
Scomber histone, 75
Scombrin, 75
Scope of physiological chemistry, 2
Secondary albumose, 82
    phosphates, 14
Sediments from urine, 306, 340
Self preservation of blood, 216
    purification of water, 9
    regeneration of cells, 227
Semipermeable membrane, 205
Serine, 62
Serum albumin, 66
    globulin, 66, 68
    immunity, 227
    pus, 233
    tests, 219
Side chain theory, 225
Silicon in body, 8
Silver nitrate, uses of, 338
    test, 10
Simple proteins, 54
Sizes of starch grains, 34
Skatol, 157, 160
Skimmed milk, 242
Skin, ash of, 14
Skin respiration, 363
Soap and hard water, 42
Soaps in feces, 168
Sodium in body, 8
Soft water, 8
Solids in feces, 167
    of blood, 177
    of body, 7
    of milk, 243
Solubility of fats, 44

INDEX.

Soluble ferments, 99
    starch, 33
Sorbinose, 25
Soxhlet extraction of fat, 244
    sugar values, 28
Special diets, 377
Specific gravity of urine, 313
Specific rotation, 31
    of arabinose, 20
    of bile acids, 266
    of cholesterol, 271
    of dextrins, 38
    of edestin, 69
    of egg albumin, 67
    of fibrinogen, 70
    of fructose, 24, 32
    of glucose, 24, 32
    of hemoglobin, 182
    of invert sugar, 32
    of lactic acid, 282
    of lactose, 26, 32
    of maltose, 27, 32
    of melitose, 27, 32
    of saccharose, 26, 32
    of serum albumin, 66
    of serum globulin, 69
    of xylose, 20
Spectroscope, 193
Spectrum analysis, 197
    of blood, 193
    of carbon monoxide hemoglobin, 197
    of methemoglobin, 197
    of oxyhemoglobin, 195
    of reduced hemoglobin, 195
Spermaceti, 49
Spermatozoa in urine, 347
Spermine, 276
Spleen, 234
    ash of, 14
Splitting of fats, 107
Starch digestion, 122
Starches, 33
Starch in feces, 170
    sugar, 21
Steapsin, 107, 154
Stearic acid, 40
Stearin, 45
Stercorin, 271
Stomach, acids in, 134
    actions in, 126
    contents, tests, 132
Sturin, 75
Substratum, ferment, 101

Sucrase, 105
Sudan III reagent, 48
Sugar from malt, 123
    in feces, 171
    in milk, 240
    of blood, 189
    of malt, 26
Sugars, 17
    determination, 27
    heats of combustion, 367
    in urine, 304
        determination, 320
    reducing, 22
    relations of, 18
    synthesis of, 19
    tests for in urine, 317
Sulphates, ethereal, 261
    in body, 16
    in urine, 303
Sulpho hemoglobin, 186
Sulphur compounds from proteins, 63
    distribution of in urine, 303
    excretion, 302
Sulphur in body, 8, 16
    in keratin, 92
    neutral, 262, 303
Supply of blood, 175
Suprarenin, 273
Suprarenal bodies, 273
Swedish filter paper, 38
Symbiotic processes, 118
Syntheses in liver, 259
Synthesis of ethereal sulphates, 261
    of polypeptides, 149
    of sugar, 19
    of uric acid, 260
Syntonin, 79
Syrup, glucose, 22

Table of body elements, 8
Tallow, 46
    crystals, 43
Tallquist chart, 203
Talose, 24
Tartaric acid, 18
Tartronic acid, 18
Taurin, 266
Taurocholic acid, 264
Temporary hardness, 9
Tendons, mucoids in, 89
Tension, osmotic, 209
Tertiary phosphates, 14
Test, Almen's, 327

Test, biuret, 57
  bismuth, 319
  Boas', 134
  congo red, 133
  dimethylaminoazobenzene, 133
  Donne's, 343
  double iodide, 314
  Fehling's, 22
    for urine, 318
  ferric chloride, 323
  Gmelin's, 267
  guaiacum, 180
  Guenzberg's, 133
  Hammarsten's, 268
  Heller's, 326
  hydrogen peroxide, 180
  Kelling's, 135
  lead hydroxide, 58
  Legal's, 322
  Lieben's, 322
  methyl-violet, 133
  Moore's, 317
  murexide, 331
  phenylhydrazine, 319
  picric acid, 314
  Struve's, 326
  Tanret, 314
  Trommer's, 318
  Trousseau's, 325
  Uffelmann's, 135
  xanthoproteic, 58
Test meals, 132
Tests for acetoacetic acid, 322
  acetone, 322
  air, 13
  albumins, 54 to 59
    in urine, 314
  alcohol, 113
  ammonia in water, 10
  bile colors, 267
  bile salts, 266
  blood, 180
    in urine, 326
  chlorides, 10
  cholesterol, 50
  colors in urine, 324
  creatinine, 334
  digestive products, 131
  drinking water, 9
  fats, 42, 47
    in flour, 94
    in milk, 242
  free acid in stomach, 133

Tests for gelatin, 91
  globulins in urine, 315
  glycogen, 36
  hemoglobin, 195
  indol, 152
  lactic acid, 135
  leucine, 148
  levulose, 24
  meat extract, 285
  milk constituents, 242
  mucin in urine, 316
  muscle extractives, 79
  nitrates, 11
  nitrites, 11
  organic acids in stomach, 134
  nitrogen, 54
  oxybutyric acid, 322
  pentose, 20
  pepsin, 139
  proteins, 54
    in feces, 173
  proteoses, 131
  saliva, 121
  starch, 34
  sugar, 22
    in milk, 242
    in urine, 317
  sulphur in proteins, 65
  thiocyanates in saliva, 121
  tryptophane, 147
  tyrosine, 148
  urea in urine, 328
  uric acid in urine, 331
Tests on blood, 179
  bones, 287
  calculi, 354
  pepsin, 331
Theories of fermentation, 98
  indicators, 136
  side chain, 225
Thiocyanates in saliva, 121
Thrombin, 178
Thymine, 187
Thymus cells, 233
Thyreoglobulin, 274
Thyroid gland, 274
Thyroiodine, 275
Time and place of oxidation, 364
Time of coagulation, 180
Tissue oxidation, 364
Tissues, ash in, 14
  water in, 12
Titration of gastric juice, 138

INDEX.

Titration of stomach contents, 132
Total fat in feces, 168
    hydrochloric acid in stomach, 134
    nitrogen in urine, 327
Toxins, 227
    from intestine, 162
Toxoids, 228
Toxons, 228
Transformation products, 53, 76
Transfusion of blood, 192
Transudations, 232
Treatment, opsonic, 223
Trehalose, 27
Triolein, 45
Triose, 18
Trioxyglutaric acid, 18
Tripalmitin, 45
Triple phosphate, 304
Trisaccharides, 17, 27
Tristearin, 45
Trommer test, 22
Trousseau's test, 325
True albumins, 53, 65
Trypsin, 83, 110, 145
    antipeptone, 150
Trypsinogen, 145
Tryptophane, 63, 147
Turanose, 27
Turkey, composition, 93
Tyrosine, 63, 92, 116
    in urine, 350
    group, 56
    tests for, 147
Tyrosinase, 116

Uffelmann's test, 135
Unit of force, 367
    of heat, 366
    of work, 367
Unorganized ferments, 99
    sediments, 341
Uracil, 87
Uranium solution, uses, 336
Urates, 299
    in sediment, 350
Urea, 295
    decomposition, 296
    determination, 328, 329
    fermentation, 111
    found in liver, 259
    synthesis, 296
Urease, 111
Uric acid, 297, 331

Uric acid, calculi, 355
    determination, 332
    from spleen, 234
    in liver, 260
    reducing power, 308
    sediment, 349
Urinary calculi, 354
Urine, acetoacetic acid in, 322
    actone in, 322
    albumin in, 314
    ammonia in, 330
    analysis, 313
    bacteria in, 348
    blood in, 326, 342
    calculi from, 354
    casts, 345
    chlorides in, 337
    color and odor, 292
    coloring matters in, 323
    conductivity, 309
    creatinine, 300
    cryoscopy, 312
    epithelium, 343
    fat globules in, 351
    fermentation, 111
    freezing point, 312
    fungi in, 347
    general composition, 291
    leucine and tyrosine, 351
    molds in, 348
    mucin, 316
        bands, 344
    mucus in, 342
    nitrogen compounds in, 294
    oxalate sediment, 352
    oxybutyric acid in, 322
    peptones in, 316
    phosphate in, 335
        sediment, 353
    proteins in, 305
    proteoses in, 316
    purines in, 298
    pus in, 343
    reaction, 293
    reducing power, 307
    sediments, 306, 340
    spermatozoa in, 347
    sulphates in, 339
    total nitrogen, 327
        sulphur, 339
    triple phosphate in, 353
    urates in, 350
    urea in, 295

Urine, uric acid in, 297
    sediment, 349
  xanthine bodies in, 298
Urobilin, 324
Urochrome, 324
Uroerythrin, 324
Urohematin, 324
Urophain, 324
Uses of chlorine in body, 15

Value of blood conductivity, 214
Variations in blood, 191
Vegetables, composition, 94
Vegetable proteins, 73
Vinegar, 115
Vitellin, 73
Vitreous body, mucoid in, 89
Voit, 5
  theory of protein metabolism, 290
Volhard's method, 338

Water, distillation, 9
  in body, 8
  in tissue, 12
  physiological importance, 12
  purification, 9
  tests, 9
Waxes, 49

Waxy casts, 347
Wheat flour, 73
  starch, 33
Wheatstone bridge, 213
Whey, 240
  sugar from, 26
White of egg, 67
Widal test, 222
Wood paper, 38
  sugar, 20
Wool fat, 49
Works of Liebig, 4
Wöhler, 4
Wright's coagulometer, 180

Xanthine, 87
  bodies, 280
    in urine, 298
  calculi, 355
Xanthoproteic test, 58
Xylose, 20

Yeast, 112
  action of, 95

Zymase, 99, 114
Zymotoxic group, 228

Milton Keynes UK
Ingram Content Group UK Ltd.
UKHW012135080224
437403UK00005B/64